Advanced Structured Materials

Volume 180

Common engineering materials are reaching their limits in many applications, and new developments are required to meet the increasing demands on engineering materials. The performance of materials can be improved by combining different materials to achieve better properties than with a single constituent, or by shaping the material or constituents into a specific structure. The interaction between material and structure can occur at different length scales, such as the micro, meso, or macro scale, and offers potential applications in very different fields.

This book series addresses the fundamental relationships between materials and their structure on overall properties (e.g., mechanical, thermal, chemical, electrical, or magnetic properties, etc.). Experimental data and procedures are presented, as well as methods for modeling structures and materials using numerical and analytical approaches. In addition, the series shows how these materials engineering and design processes are implemented and how new technologies can be used to optimize materials and processes.

Advanced Structured Materials is indexed in Google Scholar and Scopus.

Holm Altenbach · Svetlana M. Bauer ·
Alexander K. Belyaev · Dmitri A. Indeitsev ·
Valery P. Matveenko · Yuri V. Petrov
Editors

Advances in Solid and Fracture Mechanics

A Liber Amicorum to Celebrate the Birthday of Nikita Morozov

Springer

Editors
Holm Altenbach
Lehrstuhl für Technische Mechanik
Institut für Mechanik
Fakultät für Maschinebau
Otto-von-Guericke-Universität
Magdeburg, Germany

Alexander K. Belyaev
IPME RAS
St. Petersburg, Russia

Valery P. Matveenko
Ural Branch RAS
Institute of Continuous Media Mechanics
Perm, Russia

Svetlana M. Bauer
Faculty of Mathematics & Mechanics
St. Petersburg State University
St. Petersburg, Russia

Dmitri A. Indeitsev
IPME RAS
St. Petersburg, Russia

Yuri V. Petrov
IPME RAS
St. Petersburg, Russia

ISSN 1869-8433 ISSN 1869-8441 (electronic)
Advanced Structured Materials
ISBN 978-3-031-18392-8 ISBN 978-3-031-18393-5 (eBook)
https://doi.org/10.1007/978-3-031-18393-5

This Springer imprint is published by the registered company Springer Nature Switzerland AG
The registered company address is: Gewerbestrasse 11, 6330 Cham, Switzerland

Preface

Nikita Fedorovich Morozov, a full member of the Russian Academy of Sciences, an outstanding Russian scientist who has enriched the science with many achievements in the field of mechanics of deformable solids, a remarkable teacher, who created the country's authoritative scientific school of mechanics, celebrated his 90th birthday.

Nikita Fedorovich Morozov

Nikita Morozov was born on July 28, 1932 in Leningrad. His childhood coincided with the hard times for the country. During the Second World War, he remained in besieged Leningrad. At the age of 11 he received his first government award: the medal "For Defense of Leningrad" - for active participation in the work of fire brigades.

In 1949, he entered the Faculty of Mathematics and Mechanics of the Leningrad State University (now St. Petersburg State University - the oldest Russian university established in 1724), where he was taught by the world-wide known scientists like Y.V. Linnik, S.G. Mikhlin, V.V. Novozhilov, D.K. Faddeev and others. After getting his diploma, he continued his postgraduate course at the Leningrad University. In 1958, he defended his PhD thesis and was awarded "Candidate of physico-mathematical sciences". After successful completion of the postgraduate course in 1958, he started working as a senior engineer at the Krylov Central Research Institute, and a year later began teaching at the Leningrad Technology Institute of Pulp and Paper Industry, where he was appointed as an assistant professor and became head of the department for 13 years.

During those years, N.F. Morozov successfully combines teaching with scientific work. The main direction of his scientific interests was the study of nonlinear problems in the theory of thin plates. He proved the existence, uniqueness and solvability of the boundary and initial boundary problems of the theory of plates and shells using the rigorous mathematical approach. He solved the problem of the existence of asymmetric solutions in the symmetrically loaded circular plate and formulated the sufficient conditions of instability of symmetric solutions [1, 2, 3]. The results obtained by N.F. Morozov on the qualitative research of nonlinear problems of the thin plate theory are classical and formed a basis of his doctoral (DSc) thesis "Nonlinear problems of the theory of thin plates and shells" defended in 1967 at the Leningrad State University.

In 1971, N.F. Morozov was invited to Leningrad State University to head the laboratory of mathematical physics at the Leningrad State Research Institute of Mathematics and Mechanics named after academician V.I. Smirnov. In 1973 he was elected professor of the department of mathematical physics of the Faculty of Mathematics and Mechanics, and in 1976 he became head of the department of theory of elasticity of Leningrad State University, one of the oldest departments of the university, which he heads to date. At different times, the students, postgraduate students and employees of the department were well-known scientists. Among them there are, for example, the later President of the Academy of Sciences of USSR academician G.I. Marchuk, academicians A.S. Alekseev, E.I. Shemyakin, S.L. Sobolev and N.I. Muskhelishvili, professors V.M. Babich, K.F. Chernikh, V.A. Likhachev, and others.

The organizational work of N.F. Morozov as the head of the famous department and the combination of his bright scientific and teaching activity led to the development and growth of the scientific school of mechanics of St. Petersburg State University (SPbSU), which now rightfully occupies one of the leading positions among the Russian schools in this field. Among the graduates of the department and N.F. Morozov's students there are many well-known scientists who have made a significant contribution to mechanics and mathematical physics: Corresponding Member of the Russian Academy od Sciences Y.V. Petrov, Prof. S.A. Nazarov, Prof. M.V. Paukshto, Prof. A.E. Volkov, Prof. A.I. Razov, Prof. A.A. Utkin, and also many specialists working in the field of high technologies. Even in the most difficult times, the department under the leadership of N.F. Morozov continued its research activities, and exploited new opportunities for growth. So, thanks to his energy and organizational talent, in difficult transitional period of 1990s the department could successfully begin carrying out of both theoretical and experimental researches in a number of new essentially important directions, in particular, in the field of optimization of elastic mechanical systems, dynamic problems and problems of biomechanics.

Nikita Morozov was one of the initiators of the application of strict mathematical methods in the theory of elasticity and nonlinear mechanics. The new results he obtained in a number of scientific branches of the mathematical theory of elasticity and nonlinear problems of mechanics put N.F. Morozov among the most famous specialists in the country and abroad. N.F. Morozov and his pupils and colleagues made significant progress in the strict mathematical formulation and investigation of the strength and fracture problems, suggested by academician V.V. Novozhilov. N.F. Morozov made a significant contribution to the development of crack theory. He made a significant contribution to the development of crack theory. The application of rigorous mathematical analysis tools allowed him and his students to develop effective methods for analyzing the singular fields of elasticity theory arising in static problems with sharp stress concentrators such as cracks and sharp notches. This, in turn, led to the necessity of constructing new criteria of brittle fracture that could correctly predict the ultimate loads in cases where traditional approaches based on the classical Griffith-Irwin model do not work. One of the main achievements of N.F. Morozov and his collaborators in this field was the development of nonlocal strength criteria. In particular, it was proved that the integral fracture criterion - mean stress fracture criterion - offered earlier for special problems by H. Neuber and V. V. Novozhilov can be modified for the problems of continuum mechanics so that a very wide class of problems could be considered - from the problems for regions with cracks and sharp notches (jointly with B.N. Semenov and S.A. Nazarov) to the problems with small crack-like defects (jointly with Yu.V. Petrov). The key notion that was introduced in these studies was the notion of linear fracture size. The physical treatment and the method of determination of this parameter proposed in these works opened up the possibility for further generalizations of the theory to other fields of mechanics. This led to the formation of the SPbSU school of fracture mechanics, which is still the leading school in this field. The monograph [4] became a significant contribution to the science of fracture, which was intensively developing in many scientific centers world-wide during those years.

At the end of 1980s, N.F. Morozov together with Yu.V. Petrov turned to dynamic problems of mechanics. The then prevailing conceptions of fracture came into conflict with the latest experimental research on extreme high-speed effects in materials, which revealed a number of fundamental effects that did not fit into the traditional models. To resolve these contradictions, new ideas were required, which were developed in a number of works of the St. Petersburg School of Mechanics. The joint works [5, 6] with Yu. Petrov and A. Utkin generated a set of ideas that later served as a basis for formulating a general structural-time approach to solving dynamical problems and studying the critical extreme states of continuous media under dynamic actions. Using this approach, it was possible to formulate a number of new criteria or limit conditions for transient processes in the theory of fracture, plastic deformation, electro-physics (pulse breakdown), cavitation, the theory of phase transformations. Based on this approach and new fracture criteria, new material testing methods were proposed to meet the needs of modern industry.

Problems of free vibrations and stability of compressed transverse isotropic space, half-space and transverse isotropic compressed plate were solved jointly with P.E. Tovstik. Special attention is paid to the analysis of the forms of stability loss of both the elastic foundation itself and the plate resting on it. Using the equations of geometrically nonlinear elasticity theory, the surface stability of a transverse anisotropic elastic half-space under the action of compressive stresses is studied [7, 8]. An analysis of the loss of stability in the supercritical stage showed that the dents near the free surface of the half-space are arranged in a "staggered" manner, which corresponds to the experimental results.

N.F. Morozov is distinguished by his constant interest in new topical problems of deformable solid mechanics, in the application of solid mechanics methods to related disciplines, and in the solution of practical problems. Since 1994, he has been active in the field of applying the methods of deformable solid mechanics to the problems of nanomechanics and theoretical material science and the relationship between deformation, stability, and fracture, and the diffusion processes of phase transformations. Together with E.A. Ivanova and A.M. Krivtsov, the dependence of elastic moduli on nanocrystal size is theoretically investigated. The scale effects that arise when the continuum theory of elasticity is applied to nanoobjects were estimated and also the influence of scale effects when the method of molecular dynamics is used for modeling macroscopic objects [9, 10]. Together with I.A. Ovidko, the processes of deformation and fracture of graphene sheets with different types of structural defects are studied [11, 12].

N.F. Morozov and his co-authors modified the equilibrium equations and constitutive relations of linear theory of plates and shells with consideration of transverse shear deformations by reducing the relations of spatial elasticity theory with surface stresses to two-dimensional equations given on the middle surface of the shell and analyzing the effect of surface elastic moduli on the effective stiffness of plates and shells [13, 18, 14, 15, 16, 17]. N.F. Morozov and co-authors I.A. Ovidko, A.G. Sheynerman, and S.V. Bobylev studied the mechanisms of deformation and fracture of nanocrystalline bodies. A theoretical model describing the effect of special rotational deformation on crack growth in deformed nanocrystalline ceramics and metals was proposed and its effect on the growth of pre-existing relatively large cracks in nanocrystalline metals and ceramics was evaluated [19].

A special physical mode of plastic deformation in nanocrystalline, ultrafine-grained and polycrystalline bodies caused by grain boundary slip and nucleation of nanosized grains (resulting from grain boundary splitting and migration under stress) was proposed and theoretically described. It was shown that the special deformation mode enhanced the plasticity of nanocrystalline and ultrafine-grained solids and this enhancement effect was more pronounced compared to the effect of cooperative grain boundary slip and migration [20]. The micromechanics of plastic deformation by grain boundary migration in metal-graphene nanocomposites was also studied [21].

N.F. Morozov together with L.V. Shtukin, I.E. Berinskii, D.A. Indeitsev and D.Yu. Skubov studied the effect of nanoscale on the performance of electromechanical structures. In particular, the electromechanical model of the graphene nanoresonator was considered, taking into account the capacitor's capacity change due to the deformation of the graphene layer (one of the plates) and taking into account the nonlinear-elastic properties of the graphene sheet at a small initial stretching. Two new schemes of the graphene nanoresonators were proposed, namely the differential and parametric ones [22].

The well-known Kirsch problem and its various modifications in nanomechanical formulation was considered: a plate stretched by diametrically opposite forces with a circular or elliptical hole or inclusion made of another material, a plate weakened by various combinations of parallel cracks. The features of deformation and destruction of both defect-free and defect-containing graphene sheets were considered [23, 24, 25].

Together with A.K. Belyaev and P.E. Tovstik, N.F. Morozov conducted a series of studies on the dynamic stability of rods under longitudinal loading, known in mechanics as the Lavrent'ev-Ishlinsky problem [26, 27]. Among others, an interesting result was obtained about the loss of stability of a rod under a stepwise application of a longitudinal load smaller than the Eulerian load. In 2016, N.F. Morozov, A.K. Belyaev, and P.E. Tovstik were awarded the M.A. Lavrentyev Prize of the Russian Academy of Sciences for their series of papers "The Dynamics of a Rod in Longitudinal Compression. Development of the Idea of M.A. Lavrent'ev and A.Yu. Ishlinsky".

N.F. Morozov left his mark not only as an author. He had also acted as editor in the Springer book series "Advanced Structured Materials" [28, 29], the last book was devoted to his friend and colleague Prof. P.E. Tovstik, who passed away in December 2020.

Under the leadership of N.F. Morozov at the Institute for Problems of Mechanical Engineering of the Russian Academy of Sciences (RAS) a creative team was formed that works fruitfully on the solution of urgent problems of deformable solid mechanics and includes many talented young people. Continuing the best traditions of A.I. Lurie and V.V. Novozhilov, N.F. Morozov organized a permanent city seminar at the Institute for Problems of Mechanical Engineering RAS, where scientists from leading Russian and foreign scientific centers discuss topical problems of mechanics and physics and present their achievements. Thanks to the efforts of N.F. Morozov at the Institute for Problems of Mechanical Engineering of the Russian Academy of Sciences, the following directions were created and are actively developed: micromechanics of materials, creation and application of carbon nanostructures, methods of molecular dynamics, mechanics of nanomaterials and defect theory, and dynamics of extreme states and structural transformations.

For more than 50 years, the outstanding scientist and talented teacher has been teaching at St. Petersburg State University. The scientific school of strength mechanics that he created enjoys great authority in Russia and abroad. By the decisions of the Grants Council under the President of the Russian Federation, the Academician N.F. Morozov's Scientific School has been repeatedly recognized as a leading scientific school of Russia. The pupils of N.F. Morozov defended 8 doctoral theses and more than 50 PhD theses.

N.F. Morozov is the author of more than 450 scientific works, including 8 monographs and 3 textbooks. Since 1994 he was a corresponding member of the Russian Academy of Sciences and in 2000 he was elected as full member (academician) of the Russian Academy of Sciences. He is the chairman of the Scientific Council on mechanics of deformable solids of the Russian Academy of Sciences. In 2000, N.F. Morozov was awarded the State Prize of the Russian Federation in the field of science and technology for a cycle of papers on nonlinear problems of mechanics of deformable solids. In 2006, N.F. Morozov was awarded the "A.F. Ioffe Prize" of the Government of St. Petersburg in Physics and Mathematics for his outstanding contribution to the dynamic theory of fracture of materials. In 2017 N.F. Morozov was awarded the "Blaise Pascal Medal", established by the European Academy of Sciences in 2003, in recognition of his outstanding personal contribution to science and technology in the development of research skills in education. The decision to award was made by the Scientific Committee of the European Academy of Sciences specially created for this purpose. In 2009, N.F. Morozov was elected a member of the European Academy of Sciences.

N.F. Morozov's social and scientific activities are vast and multifaceted. He is the vice-chairman of the Russian National Committee for Theoretical and Applied Mechanics, a member of the General Assembly of IUTAM, a member of the editorial boards of leading Russian and foreign journals on mechanics. With the active participation of N.F. Morozov, the scientific journal "Advances in Mechanics" was created, of which he is the co-editor. For several decades, he has been an active member of the Editorial Board of the journal "Vestnik Sankt-Peterburgskogo Universiteta: Mathematics, Mechanics, Astronomy", which English version is published as "Vestnik St. Petersburg University, Mathematics" by Pleiades Publishing, Ltd., in cooperation with Springer Nature. He was and is an active participant of the "Advanced Problems of Mechanics" conferences and this year celebrated together with the organizers the 50th anniversary of this conference series.

N.F. Morozov's active scientific-pedagogical activity was marked with governmental awards: "Order of Honour" (1999), "Order for Merit to the Fatherland" IV degree (2003), "Order of Friendship" (2010). In 1995, N.F. Morozov was awarded a title "Honored Scientist of the Russian Federation".

Nikita Fedorovich Morozov is distinguished by his energy, enthusiasm for science, civic responsibility, ability to selflessly work and organize others, availability in communication and willingness to always come to the rescue. He meets his anniversary in the prime of his creative powers. We wish him health, further scientific successes and happiness in his personal life!

References

1. N.F. Morozov. The Uniqueness of the Symmetric Solution of the Large Deflection Problem for a Symmetrically Loaded Circular Plate. *Dokl. AS USSR*, 123(3):417-419, 1958.
2. N.F. Morozov. On the Existence of a Non-Symmetric Solution to the Problem on Large Deflections of a Round Plate Loaded by a Symmetrical Load. *Izv. Vuzov. Matem.*, (2):126-129, 1961.
3. N.F. Morozov, A Qualitative Investigation of a Round Symmetrically Compressed Plate under Large Marginal Load (Proof of Corrugation). *Dokl. AS USSR*, 147(6):1318-1319, 1962.
4. N.F. Morozov. *Mathematical Problems of Crack Theory*. Nauka, Moscow, 1984.
5. Y. Petrov and N. Morozov. On the Modeling of Fracture of Brittle Solids. *ASME J. Appl. Mech.*, 61:710-712, 1994.
6. Y.V. Petrov, N.F. Morozov and V.I. Smirnov. Structural macromechanics approach in dynamics of fracture. *Fatigue & Fracture Engng. Mater. & Struct.* 26:363-372, 2003.
7. N.F. Morozov and P.E. Tovstik. Buckling forms of a compressed plate on an elastic foundation. *Doklady Physics*, 57(9):335-339, 2012.
8. N.F. Morozov and P.E. Tovstik. Surface layer stability under force and temperature loading. *Mechanics of Solids*, 45(6):769-777, 2010.
9. A.M. Krivtsov and N.F. Morozov. On mechanical characteristics of nanocrystals. *Physics of Solid State*, 44(12):2260-2265, 2002.

Nikita F. Morozov (center) at the "Advanced Problems of Mechanics" on June 24, 2019. On the left - Anton M. Krivtsov (one of the co-organizers of the conference, on the right - professors Mikhail A. Guzev (Director of the Institute of Mathematics and Computer Science, Far Eastern Branch of the Russian Academy of Sciences, academician), Reinhold Kienzler (University of Bremen), Wolfgang H. Müller (University of Technology Berlin), and Mark Kachanov (Tufts University).

10. E.A. Ivanova, A.M. Krivtsov, N.F. Morozov and A.D. Firsova. Description of crystal packing of particles with torque interactions. *Mechanics of Solids*, 38(4):76-88, 2003.
11. A.S. Kochnev, N.F. Morozov, I.A. Ovidko and B.N. Semenov. Deformation and fracture in graphene with divacancies of the 555–777 type. *Doklady Physics*, 61(8):403-406, 2016.
12. A.S. Kochnev, N.F. Morozov, I.A. Ovidko and B.N. Semenov. Deformation and fracture processes in graphene nanoribbons with linear disclinations quadrupoles. *Doklady Physics*, 61(5):239-2422, 2016.
13. H. Altenbach, V.A. Eremeev and N.F. Morozov. On equations of the linear theory of shells with surface stresses taken into account. *Mechanics of Solids*, 45(3):331-342, 2010.
14. H. Altenbach, V.A. Eremeyev and N.F. Morozov. Linear theory of shells taking into account surface stresses. *Doklady Physics*, 54(12):531–535, 2009.
15. V.A. Eremeyev, H. Altenbach and N.F. Morozov. The influence of surface tension on the effective stiffness of nanosize plates. *Doklady Physics*, 54(2):98–100
16. H. Altenbach, V.A. Eremeyev and N.F. Morozov. Surface viscoelasticity and effective properties of thin-walled structures at the nanoscale. *International Journal of Engineering Science*, 59:83–89, 2012.
17. H. Altenbach, V.A. Eremeyev, E.A. Ivanova and N.F. Morozov. Bending of a three-layer plate with near-zero transverse shear stiffness (in Russ.), Fizicheskaya Mezomekhanika (Physical

Mesomechanics), 15(6):15–19, 2012.
18. N.F. Morozov, I.A. Ovid'ko, A.G. Sheinerman and E.C. Aifantis. Special rotational deformation as a toughening mechanism in nanocrystalline solids. *Journal of the Mechanics and Physics of Solids*, 58(8):1088-1099, 2010.
19. S.V. Bobylev. Emission of grain boundary dislocations by nanovoids in deformed polysilicon materials / S.V. Bobylev, N.F. Morozov and I.A. Ovid'ko. *Reviews on Advanced Materials Science*, 13(1):77-84, 2006.
20. S.V. Bobylev, N.F. Morozov and I.A. Ovid'ko. Cooperative grain boundary sliding and nanograin nucleation process in nanocrystalline, ultrafine-grained, and polycrystalline solids. *Physical Review B: Condensed Matter and Materials Physics*, 84(9):094103, 2011.
21. A.G. Sheinerman, N.F. Morozov and M.Yu. Gutkin. Effect of grain boundary sliding on fracture toughness of ceramic/graphene composites. *Mechanics of Materials*, 137:103126, 2019.
22. L.V. Shtukin, I.E. Berinskii, D.A. Indeitsev, N.F. Morozov and D.Yu. Skubov. Electromechanical models of nanoresonators. *Physical Mesomechanics*, 19(3):248-254, 2016.
23. S. Bauer, N. Morozov, B. Semenov. The Kirsch problem and related problems in elasticity and inelasticity (in Russ.). *Proc. Int. Sci. Symp. on Problems of Mechanics of Deformable Solids, Dedicated to the 105th Anniversary of Birth of A.A. Il'yushin.*, Moscow, pp. 27-33, 2016.
24. S.M. Bauer, S.V. Kashtanova, N.F. Morozov, B.N. Semenov. Stability of a nanoscale-thickness plate weakened by a circular hole. *Doklady Physics*,59(9):416–418, 2014.
25. S.M. Bauer, S.V. Kashtanova, N.F. Morozov, B.N. Semenov. Stability loss in an infinite plate with a circular inclusion under uniaxial tension. *Vestnik St. Petersburg University, Mathematics* 50(2):161-165, 2017.
26. N.F. Morozov, A.K. Belyaev and D.N. Iliin. Dynamic buckling of a rod under axial jump loading. *Doklady Physics*, 58(5):191-195, 2013.
27. A.K. Belyaev, N.F. Morozov, P.E. Tovstik and T.P. Tovstik. Buckling problem for a rod longitudinally compressed by a force smaller than the Euler critical force. *Mechanics of Solids*, 51(3):263-272, 2016.
28. H. Altenbach and N.F. Morozov (eds.). *Surface Effects in Solid Mechanics Models, Simulations and Applications*, Advanced Structured Materials, vol. 30, Berlin, Heidelberg: Springer, 2013.
29. H. Altenbach, H., S. Bauer, V.A. Eremeyev, G.I. Mikhasev and N.F. Morozov (eds.). *Recent Approaches in the Theory of Plates and Plate-Like Structures*, Advanced Structured Materials, vol. 151, Cham: Springer Nature, 2022.

Magdeburg, *Holm Altenbach*
St. Petersburg, *Svetlana M. Bauer, Alexander K. Belyaev, Dmitri A. Indeitsev, Yuri V. Petrov*, Boris N. Semenov
Perm, *Valery P. Matveenko*
July 2022

Contents

List of Contributors

Lenser A. Aghalovyan
Institute of Mechanics, National Academy of Sciences of Republic Armenia, 24B Baghramyan Ave., Yerevan 0019, Republic of Armenia, e-mail: lagal@sci.am

Mher L. Aghalovyan
Institute of Mechanics, National Academy of Sciences of Republic Armenia, 24B Baghramyan Ave., Yerevan 0019 & Armenian State University of Economics, Republic of Armenia, e-mail: mheraghalovyan@yahoo.com

Holm Altenbach
Otto-von-Guericke-Universität Magdeburg, Universitätsplat 2, 39106 Magdeburg, Germany, e-mail: holm.altenbach@ovgu.de

Harutyun A. Amirjanyan
Institute of Mechanics, National Academy of Sciences of Republic Armenia, 24B Baghramyan Ave., Yerevan 0019, Republic of Armenia,
e-mail: amirjanyan@gmail.com

Svetlana M. Bauer
St. Petersburg State University, 7/9 University Embankment, St. Petersburg, 199034 Russian Federation, e-mail: s.bauer@spbu.ru

Sergey O. Bondarenko
St. Petersburg State University, 7/9 University Embankment, St. Petersburg, 199034 Russian Federation, e-mail: sergey.bondarenko.051@gmail.com

Michael J. Cloud
Lawrence Technological University, Southfield, Michigan, USA,
e-mail: mjc@ieee.org

Vincenzo Cucumazzo
Loughborough University, Loughborough, UK,
e-mail: v.cucumazzo@gmail.com

Lilit L. Dashtoyan
Institute of Mechanics, National Academy of Sciences of Republic Armenia, 24B Baghramyan Ave., Yerevan 0019, Republic of Armenia,
e-mail: lilit.dashtoyan@sci.am

Isaac Elishakoff
Department of Ocean and Mechanical Engineering, Florida Atlantic University, Boca Raton, FL 33431-0991, USA, e-mail: elishako@fau.edu

Victor A. Eremeyev
University of Cagliari, Via Marengo, 2, 09123 Cagliari, Italy,
e-mail: victor.eremeev@unica.it, eremeyev.victor@gmail.com

Sergei B. Filippov
St. Petersburg State University, 7/9 University Embankment, St. Petersburg, 199034 Russian Federation, e-mail: s_b_filippov@mail.ru

Andrey Yu. Fedorov
Institute of Continuous Media Mechanics of the Ural Branch of RAS, 614018, Academician Korolev Street, 1, Perm, Russian Federation,
e-mail: fedorov@icmm.ru

Mikhail Grekov
St. Petersburg State University, 7/9 University Embankment, 199304 St. Petersburg, Russian Federation, e-mail: m.grekov@spbu.ru

Vahram N. Hakobyan
Institute of Mechanics, National Academy of Sciences of Republic Armenia, 24B Baghramyan Ave., Yerevan 0019, Republic of Armenia, e-mail: vhakobyan@sci.am

Dmitry A. Indeitsev
Institute for Problems in Mechanical Engineering, V.O., Bolshoj pr., 61, St. Petersburg, 199178, Russian Federation, e-mail: dmitry.indeitsev@gmail.com

Dmitriy V. Ivanov
Saratov State University, 83 Astrakhanskaya Street, Saratov, 410012, Russian Federation, e-mail: ivanovdv@gmail.com,

Julius Kaplunov
Keele University, ST5 5GB, e-mail: j.kaplunov@keele.ac.uk

Bhushan L. Karihaloo
School of Engineering, Cardiff University, Cardiff CF24, 3AA, UK
e-mail: KarihalooB@cardiff.ac.uk

Mikhail I. Karyakin
Southern Federal University, Russian Federation, e-mail: karyakin@sfedu.ru

Irina V. Kirillova
Saratov State University, 83 Astrakhanskaya Street, Saratov, 410012, Russian Federation, e-mail: nano-bio@sgu.ru

Tatiana O. Korepanova
Institute of Continuous Media Mechanics of the Ural Branch of RAS, 614018, Academician Korolev Street, 1, Perm, Russian Federation, e-mail: ton@icmm.ru

Sergei M. Kovalev
Arctic and Antarctic Research Institute, 28 Bering Street, St. Petersburg, 199397, Russian Federation, e-mail: skovalev@aari.ru

Leonid Yu. Kossovich
Saratov State University, 83 Astrakhanskaya Street, Saratov, 410012, Russian Federation, e-mail: nano-bio@sgu.ru

Sergey Kostyrko
St. Petersburg State University, 7/9 University Embankment, 199304 St. Petersburg, Russian Federation, e-mail: s.kostyrko@spbu.ru

Leonid P. Lebedev
Departamento de Matemáticas, National University of Colombia, Bogotá D.C., Colombia, e-mail: llebedev@unal.edu.co

Valerii P. Matveenko
Institute of Continuous Media Mechanics of the Ural Branch of RAS, 614018, Academician Korolev Street, 1, Perm, Russian Federation, e-mail: mvp@icmm.ru

Rostislav D. Nedin
Institute of Mathematics, Mechanics and Computer Sciences named after I.I. Vorovich, Southern Federal University, 8a Mil'chakova Str., 344090 Rostov-on-Don & Southern Mathematical Institute, Vladikavkaz Scientific Center of Russian Academy of Sciences, 22 Markusa Str., 362027 Vladikavkaz, Russian Federation, e-mail: rdn90@bk.ru

Grigory A. Nesterchuk
St. Petersburg State University, 7/9 University Embankment, St. Petersburg, 199034 Russian Federation, e-mail: g.nesterchuk@spbu.ru

Galina V. Pavilaynen
St. Petersburg State University, 7/9 University Embankment, St. Petersburg, 199034 Russian Federation, e-mail: g_v_pavilaynen@mail.ru

Ludmila Prikazchikova,
Keele University, ST5 5GB, e-mail: l.prikazchikova@keele.ac.uk

Oxana G. Rybakina
Krylov State Research Centre, 44 Moskovskoe shosse, St.Petersburg, Russian Federation, e-mail: rybakina37@mail.ru

Avetik V. Sahakyan
Institute of Mechanics, National Academy of Sciences of Republic Armenia, 24B Baghramyan Ave., Yerevan 0019, Republic of Armenia,
e-mail: avetik.sahakyan@sci.am

Boris N. Semenov
St. Petersburg State University, 7/9 University Embankment, St. Petersburg, 199034 Russian Federation, e-mail: b.semenov@spbu.ru

Natalja V. Sevodina
Institute of Continuous Media Mechanics of the Ural Branch of RAS, 614018, Academician Korolev Street, 1, Perm, Russian Federation, e-mail: natsev@icmm.ru

Vadim V. Silberschmidt
Mechanics of Advanced Materials Research Group, Wolfson School of Mechanical, Electrical and Manufacturing Engineering, Loughborough University, Loughborough, Leicestershire LE11 3TU, UK, e-mail: v.silberschmidt@lboro.ac.uk

Sheeru Shamsi
Keele University, ST5 5GB, e-mail: s.s.shamsi@keele.ac.uk

Igor N. Shardakov
Institute of Continuous Media Mechanics of the Ural Branch of RAS, 614018, Academician Korolev Street, 1, Perm, Russian Federation,
e-mail: shardakov@icmm.ru

Andrei L. Smirnov
St. Petersburg State University, 7/9 University Embankment, St. Petersburg, 199034 Russian Federation, e-mail: a.l.smirnov@spbu.ru

Joel Storch
Department of Mechanical Engineering, California StateUniversity, Northridge CA 91330, USA, e-mail: jas405@caa.columbia.edu

Olga A. Strogonova
Krylov State Research Centre, 44 Moskovskoe shosse, St.Petersburg, Russian Federation, e-mail:

Alexander O. Vatulyan
Institute of Mathematics, Mechanics and Computer Sciences named after I.I. Vorovich, Southern Federal University, 8a Mil'chakova Str., 344090 Rostov-on-Don & Southern Mathematical Institute, Vladikavkaz Scientific Center of Russian Academy of Sciences, 22 Markusa Str., 362027 Vladikavkaz, Russian Federation, e-mail: aovatulyan@sfedu.ru

Eva B. Voronkova
St. Petersburg State University, 7/9 University Embankment, St. Petersburg, 199034 Russian Federation,
e-mail: e.voronkova@spbu.ru

Jianxiang Wang
Department of Mechanics and Engineering Science, College of Engineering, Peking University, Beijing 100871, P.R. China,
e-mail: jxwang@pku.edu.cn

Victor O. Yurov
Institute of Mathematics, Mechanics and Computer Sciences named after I.I. Vorovich, Southern Federal University, 8a Mil'chakova Str., 344090 Rostov-on-Don & Southern Mathematical Institute, Vladikavkaz Scientific Center of Russian Academy of Sciences, 22 Markusa Str., 362027 Vladikavkaz, Russian Federation, e-mail: vyurov@sfedu.ru

Mikhail P. Yushkov
St. Petersburg State University, 7/9 University Embankment, St. Petersburg, Russian Federation, e-mail: yushkovmp@mail.ru

Tatevik V. Zakaryan
Institute of Mechanics, National Academy of Sciences of Republic Armenia, 24B Baghramyan Ave., Yerevan 0019, Republic of Armenia,
e-mail: zaqaryantatevik@mail.ru

Leonid M. Zubov
Southern Federal University, Russian Federation, e-mail: lmzubov@sfedu.ru

Chapter 1
On Forced Vibrations of Orthotropic Plates in the Presence of Internal Friction

Lenser A. Aghalovyan, Mher L. Aghalovyan, and Tatevik V. Zakaryan

Abstract The forced vibrations of orthotropic plates are considered with the taking into account the internal friction, which is proportionally to the points velocity. The plate is lying on the absolutely rigid base and on its facial surface are acting normal and tangential loads, which are harmonically changing in time. By the asymptotic method of solving singularly perturbed differential equations the solution of 3D dynamic outer problem of the elasticity theory is found. It is shown that the asymptotic solution becomes mathematically exact, when the outer actions depend on the tangential coordinates polynomially.

Key words: 3D elasticity problems, Plates, Forced vibration, Friction, Asymptotic method

1.1 Introduction

For the solution of the dynamic problems of the elasticity theory Fourier's method, the method of integral transformations, Wiener-Hopf's, Smirnova-Sobolev's method were used [24]. Until recently, there was a relatively small number of works devoted to the spatial dynamics problems of plates and shells. For solving similar static and dynamic problems, an effective turned out the asymptotic method of solving singularly perturbed differential equations. Mathematicians began to pay attention to

Lenser A. Aghalovyan · Tatevik V. Zakaryan
Institute of Mechanics, National Academy of Sciences of Republic Armenia, 24B Baghramyan Ave., Yerevan 0019, Republic of Armenia,
e-mail: lagal@sci.am,zaqaryantatevik@mail.ru

Mher L. Aghalovyan
Institute of Mechanics, National Academy of Sciences of Republic Armenia, 24B Baghramyan Ave., Yerevan 0019 & Armenian State University of Economics, Republic of Armenia,
e-mail: mheraghalovyan@yahoo.com

H. Altenbach et al. (eds.), *Advances in Solid and Fracture Mechanics*,
Advanced Structured Materials 180,
https://doi.org/10.1007/978-3-031-18393-5_1

singularly perturbed differential equations from the middle of the 20th century (K.O. Friedrichs, A.N. Tikhonov, V. Vazov, M.I. Vishik and L.A. Lyusternik, A.H. Nayfeh and others). Then it was followed by a rapid development of this direction and occurred many first-class monographs [17, 18, 22, 27, 28] et al.). However in these monographs there was not discussed the type of singularly perturbed differential equations, which were occurred in the elasticity theory for thin bodies (the small parameter is the coefficient not for the entire higher operator, but only for its part). For solving similar equations and systems an effective turned out the asymptotic method. The first works on the application of the asymptotic method for solving 3D problems for plates and shells are [10, 11, 13, 14]. It was proved that by one expansion by the small parameter, as in the regularly perturbed by the small parameter equations, the problem is impossible to solve. The solution of singularly perturbed equations and systems is consist of two qualitatively different addends: the solution of outer problem (I^{out}) and the solution of boundary layer (I_b)

$$I = I^{out} + I_b \tag{1.1}$$

Applicable to the 3D problem of the elasticity theory, by solving the outer problem, beside satisfying equilibrium equations (movements) and relations of elasticity, also are satisfying the boundary conditions on the facial surfaces of plate and shell (outer conditions). The same solution in Russian-language publications is usual to call the solution of the internal problem (I^{out}) in the sense that it is valid starting from the some distance from the side surface, i.e. inside the plate or the shell. The solution (I_b) is localized near the side surface and, as a rule, all its values decrease exponentially with receding from the side surface into the inside of the plate or the shell. These solutions can be built separately [2, 11].

The solution of the outer problem is found in the form

$$I^{out} = \varepsilon^{q_I+s} I^{(s)}, \quad s = \overline{0,N} \tag{1.2}$$

In the case of a regular perturbation $q_I = 0$. In singularly perturbed equations, the value of q_I depends on desired functions and on type of boundary conditions on facial surfaces. For example, if values of the corresponding components of the stress tensor are given on facial surfaces of the plate ($\sigma_{xz}^{\pm}, \sigma_{yz}^{\pm}, \sigma_{zz}^{\pm}$), i.e. conditions of the first boundary value problem of the elasticity theory, then we have [2, 11]

$$\begin{gathered} q_I = -2 \text{ for } \sigma_{xx}, \sigma_{xy}, \sigma_{yy}; \; q_I = -1 \text{ for } \sigma_{xz}, \sigma_{yz}; \\ q_I = 0 \text{ for } \sigma_{zz}; \; q_I = -2 \text{ for } u, v; \; q_I = -3 \text{ for } w \end{gathered} \tag{1.3}$$

In the case of the second or mixed boundary value problems, we have established [1, 2, 6]

$$\begin{gathered} q_I = -1 \; \forall \, \sigma_{ij}, \; i, j = x, y, z \\ q_I = 0 \text{ for } u, v, w \end{gathered} \tag{1.4}$$

The asymptotics (1.4) is valid also for solving 3D dynamic problems.

The asymptotic theory of isotropic plates and shells was constructed by Gol'denveizer [11, 12], and for anisotropic plates and shells by Agalovyan [2], for layered plates and shells [4]. The great contribution into the development of the asymptotic theory of plates and shells was made by A.L. Gol'denveizer's students and associates [9, 15, 16, 19, 20, 21, 25, 25, 29]. Some classes of dynamic problems of anisotropic plates were solved in [3, 5, 6, 8], and for shells in [7]. In this paper, an asymptotic solution of the 3D problem of forced vibrations of orthotropic plates in the presence of internal friction is found.

1.2 The Formulation of the Problem and Basic Equations

Let we have a plate, which occupies the area $D = \{(x,y,z) : 0 \le x \le a,\ 0 \le y \le b,\ |z| \le h, h << l, l = min(a,b)\}$. The plate is orthotropic, there is an internal friction, which is proportional to the velocity of the points [23]. The plate leans against on an absolutely rigid base. There is the Coulomb friction between the base and plate sole. On the facial surface of the plate normal and tangential loads, which change harmonically in time, are acting.

It is required to find the solution to the corresponding three-dimensional problem of the elasticity theory. For this, it is necessary to find in the region D the solution to equations of motion, taking into account the internal friction:

$$\frac{\partial \sigma_{xx}}{\partial x} + \frac{\partial \sigma_{xy}}{\partial y} + \frac{\partial \sigma_{xz}}{\partial z} - k\frac{\partial u}{\partial t} = \rho\frac{\partial^2 u}{\partial^2 t} \quad (x,y,z;u,v,w), \tag{1.5}$$

relations of the elasticity for the orthotropic body:

$$\begin{gathered}\frac{\partial u}{\partial x} = a_{11}\sigma_{xx} + a_{12}\sigma_{yy} + a_{13}\sigma_{zz}, \quad (x,y,z;u,v,w;1,2,3)\\ \frac{\partial u}{\partial y} + \frac{\partial v}{\partial x} = a_{66}\sigma_{xy}, \quad \frac{\partial w}{\partial x} + \frac{\partial u}{\partial z} = a_{55}\sigma_{xz}, \quad \frac{\partial w}{\partial y} + \frac{\partial v}{\partial z} = a_{44}\sigma_{yz}\end{gathered} \tag{1.6}$$

where σ_{ij} - components of stress tensor , u, v, w - components of displacement vector, a_{ij} - constants of elasticity, ρ - density, k - coefficient of the internal friction. Notation $(x,y,z;u,v,w)$ means that there are two more equations that are obtained from the written by cyclic permutation. The found solution must satisfy to the following boundary conditions:

$$\sigma_{\alpha z}(x,y,h,t) = -\sigma^{+}_{\alpha z}(\xi,\eta)\cos\Omega t, \quad \alpha = x,y,z \text{ when } z = h \tag{1.7}$$

$$\begin{gathered}w(x,y,-h,t) = 0, \quad \text{when } z = -h\\ \sigma_{xz}(x,y,-h,t) = f_1\sigma_{zz}(x,y,-h,t),\\ \sigma_{yz}(x,y,-h,t) = f_2\sigma_{zz}(x,y,-h,t)\end{gathered} \tag{1.8}$$

$\xi = x/l, \eta = y/l, l = \min(a,b), \Omega$ - frequency of external influence. The found solution must also satisfy to conditions on the side surface. They stipulate the appearance of the boundary layer. The corresponding solution can be built separately [2, 6, 11]. The values of the boundary layer, as a rule, exponentially decrease with removing from the side surface into the inside of the plate.

1.3 The Solution to the External Problem

The solution to the problem formulated above (1.1)-(1.4) will be sought in the form

$$\sigma_{\alpha\beta}(x,y,z,t) = \sigma_{jkI}(x,y,z)\sin\Omega t + \sigma_{jkII}(x,y,z)\cos\Omega t, \qquad (1.9)$$
$$\alpha,\beta = x,y,z,\ \ j,k = 1,2,3$$
$$u(x,y,z,t) = u_I(x,y,z)\sin\Omega t + u_{II}(x,y,z)\cos\Omega t,\ \ (u,v,w)$$

Then we pass in Eqs. (1.5) and relations (1.6) to dimensionless coordinates and dimensionless displacements:

$$\xi = x/l,\ \ \eta = y/l,\ \ \zeta = z/h,\ \ U_I = u_I/l,\ \ U_{II} = u_{II}/l,\ \ (u,v,w) \qquad (1.10)$$

As a result, we will obtain the system, singularly perturbed by the small parameter $\varepsilon = h/l$:

$$\frac{\partial\sigma_{11I}}{\partial\xi} + \frac{\partial\sigma_{12I}}{\partial\eta} + \varepsilon^{-1}\frac{\partial\sigma_{13I}}{\partial\zeta} + kl^2\Omega U_{II} = -\rho\Omega^2 l^2 U_I,\ \ (\xi,\eta;1,2;U,V)$$
$$\frac{\partial\sigma_{11II}}{\partial\xi} + \frac{\partial\sigma_{12II}}{\partial\eta} + \varepsilon^{-1}\frac{\partial\sigma_{13II}}{\partial\zeta} - kl^2\Omega U_I = -\rho\Omega^2 l^2 U_{II},\ \ (\xi,\eta;1,2;U,V)$$
$$\frac{\partial\sigma_{13I}}{\partial\xi} + \frac{\partial\sigma_{23I}}{\partial\eta} + \varepsilon^{-1}\frac{\partial\sigma_{33I}}{\partial\zeta} + kl^2\Omega W_{II} = -\rho\Omega^2 l^2 W_I,\ \ (I,II) \qquad (1.11)$$
$$\varepsilon^{-1}\frac{\partial U_I}{\partial\xi} = a_{11}\sigma_{11I} + a_{12}\sigma_{22I} + a_{13}\sigma_{33I},\quad (\xi,\eta,\zeta;U,V,W;1,2,3),\ (I,II)$$
$$\frac{\partial U_I}{\partial\eta} + \frac{\partial V_I}{\partial\xi} = a_{66}\sigma_{12I},\ \ \frac{\partial W_I}{\partial\xi} + \varepsilon^{-1}\frac{\partial U_I}{\partial\zeta} = a_{55}\sigma_{13I},\ (I,II)$$
$$\frac{\partial W_I}{\partial\eta} + \varepsilon^{-1}\frac{\partial V_I}{\partial\zeta} = a_{44}\sigma_{23I},\ \ (I,II)$$

The solution I to the singularly perturbed system according to (1.1) consists of solutions to the external problem (I^{out}) and the boundary layer (I_b).

The solution to the external problem we will seek in the form

$$\sigma^{out}_{jk\gamma} = \varepsilon^{-1+s}\sigma^{(s)}_{jkI},\ \ j,k = 1,2,3,\ \ \gamma = I,II,\ \ s = \overline{0,N} \qquad (1.12)$$
$$(U^{out}_\gamma, V^{out}_\gamma, W^{out}_\gamma) = \varepsilon^s(U^{(s)}_\gamma, V^{(s)}_\gamma, W^{(s)}_\gamma),\ \ \gamma = I,II$$

The notation $s = \overline{0,N}$ means, that by umbral (repeating) index s is taken place summed over all integer values from 0 to number of approximations N. By substituting (1.12) in (1.11), for determining unknown coefficients $(\sigma_{jkI}^{(s)}, U_\gamma^{(s)}, V_\gamma^{(s)}, W_\gamma^{(s)})$ we will obtain the system:

$$
\begin{aligned}
&\frac{\partial \sigma_{11I}^{(s-1)}}{\partial \xi} + \frac{\partial \sigma_{12I}^{(s-1)}}{\partial \eta} + \frac{\partial \sigma_{13I}^{(s)}}{\partial \zeta} + 2H\Omega U_{II}^{(s)} + \rho_1^2\Omega^2 U_I^{(s)} = 0, \ (\xi,\eta;1,2;U,V)\\
&\frac{\partial \sigma_{11II}^{(s-1)}}{\partial \xi} + \frac{\partial \sigma_{12II}^{(s-1)}}{\partial \eta} + \frac{\partial \sigma_{13II}^{(s)}}{\partial \zeta} - 2H\Omega U_I^{(s)} + \rho_1^2\Omega^2 U_{II}^{(s)} = 0, \ (\xi,\eta;1,2;U,V)\\
&\frac{\partial \sigma_{13I}^{(s-1)}}{\partial \xi} + \frac{\partial \sigma_{23I}^{(s-1)}}{\partial \eta} + \frac{\partial \sigma_{33I}^{(s)}}{\partial \zeta} + 2H\Omega W_{II}^{(s)} + \rho_1^2\Omega^2 W_I^{(s)} = 0,\\
&\frac{\partial \sigma_{13II}^{(s-1)}}{\partial \xi} + \frac{\partial \sigma_{23II}^{(s-1)}}{\partial \eta} + \frac{\partial \sigma_{33II}^{(s)}}{\partial \zeta} - 2H\Omega W_I^{(s)} + \rho_1^2\Omega^2 W_{II}^{(s)} = 0, \qquad (1.13)\\
&\frac{\partial U_I^{(s-1)}}{\partial \xi} = a_{11}\sigma_{11I}^{(s)} + a_{12}\sigma_{22I}^{(s)} + a_{13}\sigma_{33I}^{(s)}, \quad (\xi,\eta,\zeta;U,V,W;1,2,3),\ (I,II)\\
&\frac{\partial U_I^{(s-1)}}{\partial \eta} + \frac{\partial V_I^{(s-1)}}{\partial \xi} = a_{66}\sigma_{12I}^{(s)}, \quad \frac{\partial W_I^{(s-1)}}{\partial \xi} + \frac{\partial U_I^{(s)}}{\partial \zeta} = a_{55}\sigma_{13I}^{(s)},\ (I,II)\\
&\frac{\partial W_I^{(s-1)}}{\partial \eta} + \frac{\partial V_I^{(s)}}{\partial \zeta} = a_{44}\sigma_{23I}^{(s)},\ (I,II)\\
&2H = kh^2,\ \rho_1^2 = \rho h^2
\end{aligned}
$$

From system (1.13), all components of the stress tensor can be expressed through the components of the displacement vector by formulas

$$
\begin{aligned}
&\sigma_{13\gamma}^{(s)} = \frac{1}{a_{55}}\left(\frac{\partial U_\gamma^{(s)}}{\partial \zeta} + \frac{\partial W_\gamma^{(s-1)}}{\partial \xi}\right),\ \sigma_{23\gamma}^{(s)} = \frac{1}{a_{44}}\left(\frac{\partial V_\gamma^{(s)}}{\partial \zeta} + \frac{\partial W_\gamma^{(s-1)}}{\partial \eta}\right),\\
&\sigma_{12\gamma}^{(s)} = \frac{1}{a_{66}}\left(\frac{\partial U_\gamma^{(s-1)}}{\partial \eta} + \frac{\partial V_\gamma^{(s-1)}}{\partial \xi}\right),\\
&\sigma_{11\gamma}^{(s)} = \frac{1}{\Delta}\left(-A_{23}\frac{\partial W_\gamma^{(s)}}{\partial \zeta} + A_{22}\frac{\partial U_\gamma^{(s-1)}}{\partial \xi} - A_{12}\frac{\partial V_\gamma^{(s-1)}}{\partial \eta}\right), \qquad (1.14)\\
&\sigma_{22\gamma}^{(s)} = \frac{1}{\Delta}\left(-A_{13}\frac{\partial W_\gamma^{(s)}}{\partial \zeta} - A_{12}\frac{\partial U_\gamma^{(s-1)}}{\partial \xi} + A_{33}\frac{\partial V_\gamma^{(s-1)}}{\partial \eta}\right),\\
&\sigma_{33\gamma}^{(s)} = \frac{1}{\Delta}\left(A_{11}\frac{\partial W_\gamma^{(s)}}{\partial \zeta} - A_{23}\frac{\partial U_\gamma^{(s-1)}}{\partial \xi} - A_{13}\frac{\partial V_\gamma^{(s-1)}}{\partial \eta}\right),\ \gamma = I, II
\end{aligned}
$$

where

$$A_{11}=a_{11}a_{22}-(a_{12})^2,\ A_{22}=a_{22}a_{33}-(a_{23})^2,\ A_{33}=a_{11}a_{33}-(a_{13})^2,$$
$$A_{12}=a_{12}a_{33}-a_{23}a_{13},\ A_{13}=a_{11}a_{23}-a_{13}a_{12},\ A_{23}=a_{13}a_{22}-a_{12}a_{23},\quad (1.15)$$
$$\Delta=a_{11}A_{22}-a_{12}A_{12}-a_{13}A_{23}$$

For determining components of the displacement vector, we will obtain equations:

$$\frac{\partial^4 U_I^{(s)}}{\partial\zeta^4}+2\rho_1^2\Omega^2 a_{55}\frac{\partial^2 U_I^{(s)}}{\partial\zeta^2}+(4H^2+\rho_1^4\Omega^2)\Omega^2 a_{55}^2 U_I^{(s)}=R_u^{(s)}\ (U,V;a_{55},a_{44}),$$
$$R_u^{(s)}=a_{aa}[2H\Omega a_{55}\left(\frac{\partial\sigma_{11II}^{(s-1)}}{\partial\xi}+\frac{\partial\sigma_{12II}^{(s-1)}}{\partial\eta}\right)-\frac{\partial^2 U_{I*}^{(s)}}{\partial\zeta^2}+2H\Omega\frac{\partial^2 W_{II}^{(s-1)}}{\partial\xi\partial\zeta}$$
$$-\rho_1^2\Omega^2 a_{55}U_{I*}^{(s)}],$$
$$U_{I*}^{(s)}=\frac{\partial\sigma_{11II}^{(s-1)}}{\partial\xi}+\frac{\partial\sigma_{12II}^{(s-1)}}{\partial\eta}+\frac{1}{a_{55}}\frac{\partial^2 W_I^{(s-1)}}{\partial\xi\partial\zeta}\quad(U,V;a_{55},a_{44};\xi,\eta;1,2)$$
$$(1.16)$$

$$\frac{\partial^4 W_I^{(s)}}{\partial\zeta^4}+\frac{2\rho_1^2\Omega^2\Delta}{A_{11}}\frac{\partial^2 W_I^{(s)}}{\partial\zeta^2}+\frac{(4H^2+\rho_1^4\Omega^2)\Omega^2\Delta^2}{A_{11}^2}W_I^{(s)}=R_w^{(s)}$$
$$R_w^{(s)}=\frac{2H\Omega\Delta^2}{A_{11}^2}\left[\frac{\partial\sigma_{13II}^{(s-1)}}{\partial\xi}+\frac{\partial\sigma_{23II}^{(s-1)}}{\partial\eta}-\frac{1}{\Delta}\left(A_{23}\frac{\partial^2 U_{II}^{(s-1)}}{\partial\xi\partial\zeta}+A_{13}\frac{\partial^2 V_{II}^{(s-1)}}{\partial\eta\partial\zeta}\right)\right]-$$
$$\frac{\Delta}{A_{11}}\frac{\partial^2 W_{I*}^{(s)}}{\partial\zeta^2}-\frac{\rho_1^2\Omega^2\Delta^2}{A_{11}^2}W_{I*}^{(s)},r$$
$$W_{I*}^{(s)}=\frac{\partial\sigma_{13II}^{(s-1)}}{\partial\xi}+\frac{\partial\sigma_{23II}^{(s-1)}}{\partial\eta}+\frac{1}{\Delta}\left(A_{23}\frac{\partial^2 U_I^{(s-1)}}{\partial\xi\partial\zeta}+A_{13}\frac{\partial^2 V_I^{(s-1)}}{\partial\eta\partial\zeta}\right),$$
$$(1.17)$$

The other components of the vector of displacement are determined by formulas:

$$U_{II}^{(s)}=-\frac{1}{2H\Omega}\left[\frac{1}{a_{55}}\frac{\partial^2 U_I^{(s)}}{\partial\zeta^2}+\rho_1^2\Omega^2 U_I^{(s)}+U_{I*}^{(s)}\right],$$
$$V_{II}^{(s)}=-\frac{1}{2H\Omega}\left[\frac{1}{a_{44}}\frac{\partial^2 V_I^{(s)}}{\partial\zeta^2}+\rho_1^2\Omega^2 V_I^{(s)}+V_{I*}^{(s)}\right],\quad(1.18)$$
$$W_{II}^{(s)}=-\frac{1}{2H\Omega}\left[\frac{A_{11}}{\Delta}\frac{\partial^2 W_I^{(s)}}{\partial\zeta^2}+\rho_1^2\Omega^2 W_I^{(s)}+W_{I*}^{(s)}\right].$$

The solution to equation (1.16) is

$$U_I^{(s)}=D_1^{(s)}(\xi,\eta)\varphi_{1u}(\zeta)+D_2^{(s)}(\xi,\eta)\varphi_{2u}(\zeta)+\quad(1.19)$$
$$D_3^{(s)}(\xi,\eta)\varphi_{3u}(\zeta)+D_4^{(s)}(\xi,\eta)\varphi_{4u}(\zeta)+\overline{U}_I^{(s)}$$

where $\overline{U}_I^{(s)}$ – particular solution to the Eq. (1.16), and

$$\varphi_{1u} = \sinh c_u\zeta \sin d_u\zeta,\ \varphi_{2u} = \sinh c_u\zeta \cos d_u\zeta,$$
$$\varphi_{3u} = \cosh c_u\zeta \sin d_u\zeta,\ \varphi_{4u} = \cosh c_u\zeta \cos d_u\zeta, \tag{1.20}$$
$$c_u = \sqrt{\frac{\Omega a_{55}}{2}}\sqrt{-\rho^2\Omega + \sqrt{\rho^4\Omega^2 + 4H^4}},$$
$$d_u = \sqrt{\frac{\Omega a_{55}}{2}}\sqrt{\rho^2\Omega + \sqrt{\rho^4\Omega^2 + 4H^4}}$$

Then according to formulas (1.14), (1.18)

$$U_{II}^{(s)} = -D_1^{(s)}(\xi,\eta)\varphi_{4u}(\zeta) + D_2^{(s)}(\xi,\eta)\varphi_{3u}(\zeta) -$$
$$D_3^{(s)}(\xi,\eta)\varphi_{2u}(\zeta) + D_4^{(s)}(\xi,\eta)\varphi_{1u}(\zeta) + \overline{U}_{II}^{(s)}$$
$$\sigma_{13I}^{(s)} = \frac{1}{a_{55}}[(c_u\varphi_{3u} + d_u\varphi_{2u})D_1^{(s)} + (c_u\varphi_{4u} - d_u\varphi_{1u})D_2^{(s)} +$$
$$(c_u\varphi_{1u} + d_u\varphi_{4u})D_3^{(s)} + (c_u\varphi_{2u} - d_u\varphi_{3u})D_4^{(s)}] + \overline{\sigma}_{13I}^{(s)}(\xi,\eta,\zeta) \tag{1.21}$$
$$\sigma_{13II}^{(s)} = \frac{1}{a_{55}}[-(c_u\varphi_{2u} - d_u\varphi_{3u})D_1^{(s)} + (c_u\varphi_{1u} + d_u\varphi_{4u})D_2^{(s)} -$$
$$(c_u\varphi_{4u} - d_u\varphi_{1u})D_3^{(s)} + (c_u\varphi_{3u} + d_u\varphi_{2u})D_4^{(s)}] + \overline{\sigma}_{13II}^{(s)}(\xi,\eta,\zeta)$$
$$\overline{U}_{II}^{(s)} = -\frac{1}{2H\Omega}[\frac{1}{a_{55}}\frac{\partial^2\overline{U}_I^{(s)}}{\partial\zeta^2} + \rho_1^2\Omega^2\overline{U}_I^{(s)} + U_{I*}^{(s)}],$$
$$\overline{\sigma}_{13n}^{(s)} = \frac{1}{a_{55}}(\frac{\partial\overline{U}_n^{(s)}}{\partial\zeta} + \frac{\partial W_n^{(s-1)}}{\partial\xi}),\ n = I, II$$

The solution to Eq. (1.16) for the displacement V is

$$V_I^{(s)} = F_1^{(s)}(\xi,\eta)\varphi_{1v}(\zeta) + F_2^{(s)}(\xi,\eta)\varphi_{2v}(\zeta) + \tag{1.22}$$
$$F_3^{(s)}(\xi,\eta)\varphi_{3v}(\zeta) + F_4^{(s)}(\xi,\eta)\varphi_{4v}(\zeta) + \overline{V}_I^{(s)}$$

where functions φ_{jv} are determined by formulas (1.20) with the difference that, it needs to replace U, a_{55} with V, a_{44}.

For Eq. (1.17), we similarly have

$$W_I^{(s)} = B_1^{(s)}(\xi,\eta)\varphi_{1w}(\zeta) + B_2^{(s)}(\xi,\eta)\varphi_{2w}(\zeta) + \tag{1.23}$$
$$B_3^{(s)}(\xi,\eta)\varphi_{3w}(\zeta) + B_4^{(s)}(\xi,\eta)\varphi_{4w}(\zeta) + \overline{W}_I^{(s)}$$

where $\varphi_{jw}(\zeta), c_w$ are calculated by formulas (1.20) by replacing $\varphi_{ju}(\zeta), c_u, a_{55}$ with $\varphi_{jw}(\zeta), c_w, \Delta/A_{11}$.

By satisfying to the boundary conditions (1.7), (1.8) we will obtain the algebraic system

$$\sigma_{j3I}^{(s)}(\xi,\eta,1)=0\ \sigma_{j3II}^{(s)}(\xi,\eta,1)=-\sigma_{\alpha z}^{+(s)}(\xi,\eta),\ j=1,2,3,\ \alpha=x,y,z$$
$$\sigma_{jz}^{+(0)}=\varepsilon\sigma_{jz}^{+},\ \sigma_{jz}^{+(s)}=0,\ s\neq 0 \quad (1.24)$$
$$w_I(\xi,\eta,-1)=0,\ \sigma_{j3I}^{(s)}(\xi,\eta,-1)=f_j\sigma_{33I}^{(s)}(\xi,\eta,-1),\ (I,II),\ j=1,2$$

Which is decomposing into three systems. The first system corresponds to conditions.

$$\sigma_{33I}^{(s)}(\xi,\eta,1)=0,\ \sigma_{33II}^{(s)}(\xi,\eta,1)=-\sigma_{zz}^{+(s)}(\xi,\eta), \qquad (1.25)$$
$$w_I(\xi,\eta,-1)=0,\ w_{II}(\xi,\eta,-1)=0$$

and looks like

$$\begin{aligned}
C_{1w}B_1^{(s)}+C_{2w}B_2^{(s)}+C_{3w}B_3^{(s)}+C_{4w}B_4^{(s)}&=p_{1w}^{(s)}\\
-C_{4w}B_1^{(s)}+C_{3w}B_2^{(s)}-C_{2w}B_3^{(s)}+C_{1w}B_4^{(s)}&=p_{2w}^{(s)}\\
\psi_{1w}B_1^{(s)}-\psi_{2w}B_2^{(s)}-\psi_{3w}B_3^{(s)}+\psi_{4w}B_4^{(s)}&=p_{3w}^{(s)}\\
-\psi_{4w}B_1^{(s)}+\psi_{3w}B_2^{(s)}-\psi_{2w}B_3^{(s)}+\psi_{1w}B_4^{(s)}&=p_{4w}^{(s)}
\end{aligned} \qquad (1.26)$$

where

$$C_{1w}=c_w\psi_{3w}+d_w\psi_{2w},\ C_{2w}=c_w\psi_{4w}-d_w\psi_{1w}$$
$$C_{3w}=c_w\psi_{1w}+d_w\psi_{4w},\ C_{4w}=c_w\psi_{2w}-d_w\psi_{3w},\ \psi_{iw}=\varphi_{iw}(\zeta=1) \qquad (1.27)$$
$$p_{1w}^{(s)}=-\frac{A_{11}}{\Delta}\overline{\sigma}_{33I}^{(s)}(\xi,\eta,1),\ p_{2w}^{(s)}=-\frac{A_{11}}{\Delta}(\sigma_{zz}^{+(s)}+\overline{\sigma}_{33II}^{(s)}(\xi,\eta,1),$$
$$p_{3w}^{(s)}=-\overline{w}_I^{(s)},\ p_{4w}^{(s)}=-\overline{w}_{II}^{(s)}$$

By Cramer's formula from system (1.26), we determine $B_i^{(s)}$

$$B_i^{(s)}=\frac{\Delta_{iw}^{(s)}}{\Delta_w}$$

$$\Delta_w=\begin{vmatrix} C_{1w} & C_{2w} & C_{3w} & C_{4w}\\ -C_{4w} & C_{3w} & -C_{2w} & C_{1w}\\ \psi_{1w} & -\psi_{2w} & -\psi_{3w} & \psi_{4w}\\ -\psi_{4w} & \psi_{3w} & -\psi_{2w} & \psi_{1w}\end{vmatrix},\ P_w^{(s)}=\begin{vmatrix} p_{1w}^{(s)}\\ p_{2w}^{(s)}\\ p_{3w}^{(s)}\\ p_{4w}^{(s)}\end{vmatrix} \qquad (1.28)$$

$\Delta_{iw}^{(s)}$ cofactor corresponding to the ith column, i.e. it is obtained from Δ_w by replacing ith column with a column of free terms.

According to conditions

$$\sigma_{13I}^{(s)}(\xi,\eta,1)=0\ \sigma_{13II}^{(s)}(\xi,\eta,1)=-\sigma_{xz}^{+(s)}(\xi,\eta), \qquad (1.29)$$
$$\sigma_{13I}^{(s)}(\xi,\eta,-1)=f_1\sigma_{33I}^{(s)}(\xi,\eta,-1),\ (I,II)$$

we have

$$
\begin{aligned}
C_{1u}D_1^{(s)}+C_{2u}D_2^{(s)}+C_{3u}D_3^{(s)}+C_{4u}D_4^{(s)}&=p_{1u}^{(s)}\\
-C_{4u}D_1^{(s)}+C_{3u}D_2^{(s)}-C_{2u}D_3^{(s)}+C_{1u}D_4^{(s)}&=p_{2u}^{(s)}\\
-C_{1u}D_1^{(s)}+C_{2u}D_2^{(s)}+C_{3u}D_3^{(s)}-C_{4u}D_4^{(s)}&=p_{3u}^{(s)}\\
C_{4u}D_1^{(s)}+C_{3u}D_2^{(s)}-C_{2u}D_3^{(s)}-C_{1u}D_4^{(s)}&=p_{4u}^{(s)}
\end{aligned}
\tag{1.30}
$$

where

$$
\begin{aligned}
p_{1u}^{(s)}&=-a_{55}\overline{\sigma}_{13I}^{(s)}(\xi,\eta,1),\ p_{2u}^{(s)}=-a_{55}(\sigma_{xz}^{+(s)}+\overline{\sigma}_{13II}^{(s)}(\xi,\eta,1)\\
p_{3u}^{(s)}&=a_{aa}[f_1\frac{\Delta}{A_{11}}(-C_{1w}B_1^{(s)}+C_{2w}B_2^{(s)}+C_{3w}B_3^{(s)}-C_{4w}B_4^{(s)})+\\
&\qquad f_1\overline{\sigma}_{33I}^{(s)}(\xi,\eta,-1)-\overline{\sigma}_{13I}^{(s)}(\xi,\eta,-1)],\\
p_{4u}^{(s)}&=a_{aa}[f_1\frac{\Delta}{A_{11}}(C_{4w}B_1^{(s)}+C_{3w}B_2^{(s)}-C_{2w}B_3^{(s)}-C_{1w}B_4^{(s)})+\\
&\qquad f_1\overline{\sigma}_{33II}^{(s)}(\xi,\eta,-1)-\overline{\sigma}_{13II}^{(s)}(\xi,\eta,-1)],
\end{aligned}
\tag{1.31}
$$

consequently

$$
D_i^{(s)}=\frac{\Delta_{iu}^{(s)}}{\Delta_u}
$$

$$
\Delta_u=\begin{vmatrix}C_{1u}&C_{2u}&C_{3u}&C_{4u}\\-C_{4u}&C_{3u}&-C_{2u}&C_{1u}\\-C_{1u}&C_{2u}&C_{3u}&-C_{4u}\\C_{4u}&C_{3u}&-C_{2u}&-C_{1u}\end{vmatrix},\quad P_u^{(s)}=\begin{vmatrix}p_{1u}^{(s)}\\p_{2u}^{(s)}\\p_{3u}^{(s)}\\p_{4u}^{(s)}\end{vmatrix}
\tag{1.32}
$$

Data corresponding to conditions

$$
\begin{aligned}
\sigma_{23I}^{(s)}(\xi,\eta,1)=0,\quad \sigma_{23II}^{(s)}(\xi,\eta,1)&=-\sigma_{yz}^{+(s)}(\xi,\eta),\\
\sigma_{23I}^{(s)}(\xi,\eta,-1)=f_2\sigma_{33I}^{(s)}(\xi,\eta,-1),&\ (I,II)
\end{aligned}
\tag{1.33}
$$

can be obtained from (1.30) - (1.32) by the cyclic permutation

$$
(u,v;a_{55},a_{44};f_1,f_2;\sigma_{13}^{(s)},\sigma_{23}^{(s)}).
$$

1.4 On Mathematically Precise Solutions

If functions σ_{jz}^{+} included in boundary conditions are polynomials from ξ,η, the iteration breaks at the certain approximation, as a result we will obtain the mathematically precise solution. For illustrating the above mentioned, let consider the case when $\sigma^+_{jz} = a_{1j}\xi + a_{2j}\eta + a_{3j},\ j = x, y, z$. When $s = 0$ for determining $W^{(0)}$ it is necessary to substitute into formulas (23), (28)

$$p^{(0)}_{1w} = p^{(0)}_{3w} = p^{(0)}_{4w} = 0,\ p^{(0)}_{2w} = -\frac{A_{11}}{\Delta}\sigma^+_{zz},\ \overline{W}^{(0)}_I = 0 \tag{1.34}$$

as a result we will obtain

$$\begin{aligned} W^{(0)}_I &= B^{(0)}_1(\xi,\eta)\varphi_{1w}(\zeta) + B^{(0)}_2(\xi,\eta)\varphi_{2w}(\zeta) \\ &+ B^{(0)}_3(\xi,\eta)\varphi_{3w}(\zeta) + B^{(0)}_4(\xi,\eta)\varphi_{4w}(\zeta) \end{aligned} \tag{1.35}$$

with

$$B^{(0)}_1 = \frac{-P^{(0)}_{2w}M^2_1}{\Delta_w},\ B^{(0)}_2 = \frac{-P^{(0)}_{2w}M^2_2}{\Delta_w},\ B^{(0)}_3 = \frac{-P^{(0)}_{2w}M^2_3}{\Delta_w},\ B^{(0)}_4 = \frac{-P^{(0)}_{2w}M^2_4}{\Delta_w},$$

where M^J_i - is the minor of the determinant element Δ_w at the intersection of the ith column and the jth row.

$$W^{(0)}_{II} = -\frac{A_{11}}{2H\Omega\Delta}\frac{\partial^2 W^{(0)}_I}{\partial\zeta^2},$$

$$\sigma^{(0)}_{11\gamma} = -\frac{-A_{23}}{\Delta}\frac{\partial W^{(0)}_\gamma}{\partial\zeta},\ \sigma^{(0)}_{22\gamma} = \frac{-A_{13}}{\Delta}\frac{\partial W^{(0)}_\gamma}{\partial\zeta},\ \sigma^{(0)}_{33\gamma} = \frac{A_{11}}{\Delta}\frac{\partial W^{(0)}_\gamma}{\partial\zeta},\ \gamma = I, II$$

$U^{(0)}_j, \sigma^{(0)}_{13j}$ are calculated by formulas (1.19), (1.21), (1.32) when

$$\begin{aligned} p^{(0)}_{1u} &= 0,\ p^{(0)}_{2u} = -a_{55}\sigma^+_{xz}, \\ P^{(0)}_{3u} &= \frac{\Delta a_{55} f_1}{A_{11}}(-C_{1w}B^{(0)}_1 + C_{2w}B^{(0)}_2 + C_{3w}B^{(0)}_3 - C_{4w}B^{(0)}_4) \\ P^{(0)}_{4u} &= \frac{\Delta a_{55} f_1}{A_{11}}(C_{4w}B^{(0)}_1 + C_{3w}B^{(0)}_2 - C_{2w}B^{(0)}_3 - C_{1w}B^{(0)}_4) \\ \overline{\sigma}^{(0)}_{13I}\overline{\sigma}^{(0)}_{13I}\overline{\sigma}^{(0)}_{13I} &= \overline{\sigma}^{(0)}_{13I} = \overline{\sigma}^{(0)}_{13I} = \overline{\sigma}^{(0)}_{13II} = \overline{\sigma}^{(0)}_{23I} = \overline{\sigma}^{(0)}_{33II} = \overline{\sigma}^{(0)}_{33II} = 0, \\ U^{(0)}_I &= D^{(0)}_1(\xi,\eta)\varphi_{1u}(\zeta) + D^{(0)}_2(\xi,\eta)\varphi_{2u}(\zeta) + \\ &D^{(0)}_3(\xi,\eta)\varphi_{3u}(\zeta) + D^{(0)}_4(\xi,\eta)\varphi_{4u}(\zeta) \\ D^{(0)}_1 &= 1/\Delta_w(-P^{(0)}_{2u}M^2_1 + P^{(0)}_{3u}M^3_1 - P^{(0)}_{4u}M^4_1), \\ D^{(0)}_2 &= 1/\Delta_w(P^{(0)}_{2u}M^2_2 - P^{(0)}_{3u}M^3_2 + P^{(0)}_{4u}M^4_2), \\ D^{(0)}_3 &= 1/\Delta_w(-P^{(0)}_{2u}M^2_3 + P^{(0)}_{3u}M^3_3 - P^{(0)}_{4u}M^4_3), \\ D^{(0)}_4 &= 1/\Delta_w(P^{(0)}_{2u}M^2_4 - P^{(0)}_{3u}M^3_4 + P^{(0)}_{4u}M^4_4), \\ U^{(0)}_{II} &= -\frac{1}{2H\Omega a_{55}}\frac{\partial^2 U^{(0)}_I}{\partial\zeta^2},\ \sigma^{(0)}_{13\gamma} = \frac{1}{a_{55}}\frac{\partial U^{(0)}_\gamma}{\partial\zeta},\ \gamma = I, II \end{aligned} \tag{1.36}$$

$V_{\gamma}^{(0)}, \sigma_{13\gamma}^{(0)}$, $\gamma = I, II$are calculated by Eqs. (1.14), (1.18), (1.22), (1.31), (1.32) by cyclic replacement $(u, v; a_{55}, a_{44}; f_1, f_2)$. When $s = 1$ for $W^{(1)}$ $p_{iw}^{(1)} \neq 0$, $i = 1,2,3,4$ and calculated by formulas (1.27). Are different from zero also $p_{iu}^{(1)}$, $p_{iv}^{(1)}$ which are calculated by Eqs. (1.31). When $s = 2$ the iteration breaks i.e. the corresponding magnitudes are equal to zero. As a result we will obtain the mathematically precise solution to the outer problem.

$$u = l(U^{(0)} + \varepsilon U^{(1)}),\ (u, v, w) \tag{1.37}$$
$$\sigma_{\alpha,\beta} = \varepsilon^{-1} \sigma_{\alpha,\beta}^{(0)} + \sigma_{\alpha,\beta}^{(1)},\ \alpha, \beta = x, y, z$$

1.5 Conclusions

It is known that if the plate is isolated, then oscillations informed to it, are damping with time anyway. This is explained by the presence of friction among the particles of the plate (the internal friction). In the paper forced 3D vibrations of orthotropic plates were considered in the presence of internal friction, proportional to the velocity of points. The plate is lying on an absolutely rigid base, the facial surface of the plate is loaded with normal and tangential loads, which are harmonically changing in time. It is obtained the asymptotic solution of the corresponding 3D problem of the elasticity theory. Are indicated cases, when the found solution becomes mathematically precise.

References

1. L.A. Aghalovyan. On the structure of solution of a class of plane problems if anisotropic elastic solids (in Russ.). *Mezhvuz. Sb., EGU, Mechanics* 2, 7-12, 1982.
2. L.A. Aghalovyan. *Asymptotic Theory of Anisotropic Plates and Shells (in Russ.)*. Moscow, Nauka, 1997.
3. L.A. Aghalovyan. On an asymptotic method for solving dynamic mixed problems of anisotropic strips and plates (in Russ.). *Izv. Vyssh. Uchebn. Zaved. Sev.-Kavkazsk. Reg. Estestv. Nauki*, 3:8-11, 2000.
4. L.A. Aghalovyan, R.S. Gevorgyan. *Nonclassical Boundary-Value Problems of Anisotropic Layered Beams, Plates and Shells*. Gitutyun, NAS RA, 2005.
5. L.A. Aghalovyan, M.L. Aghalovyan. Asymptotics of free vibrations of anisotropic plates fastened with an absolutely rigid base. In: *Modern Problems of Deformable Bodies Mechanics*. Yerevan, Gitutyun NAS RA, 1:8-19, 2005.
6. L.A. Aghalovyan. *Asymptotic Theory of Anisotropic Plates and Shells*. Singapore, World Scientific Publishing, 2015.
7. L.A. Aghalovyan, L.G. Ghulghazaryan. About Defining Solutions of a Class of Dynam0ic Space Problems of Mathematical Theory of Elasticity for Orthotropic Shells. *Scientific News Armenian State Pedagogical University After Kh. Aboyan*. 2(17):29-42, 2012.

8. M.L. Aghalovyan, T.V. Zakaryan. Asymptotic solution of the first 3D dynamic elasticity theory problem on forced vibrations of a three-layer plate with an asymmetric structure. *Mechanics of Composite Materials*, 55(1):3-18, 2017.
9. S.M. Bauer, A.L. Smirnov, P.E. Tovstik, S.B. Filippov. *Asymptotic Methods in Examples and Problems (in Russ.)*. Ed. SPbSU, 1997.
10. K.O. Friedrichs, R.F. Dressler. A boundary-layer theory for elastic plates. *Comm. Pure and Appl. Math.* 14(1):1-33, 1961.
11. A.L. Goldenveizer. Derivation of an approximate theory of bending of a plate by the method of asymptotic integration of the equations of the theory of elasticity. *J. Appl. Math. Mech.* 26(4):1000-1025, 1962.
12. A.L. Goldenveizer. *Theory of Thin Ellastic Shells (in Russ.)*. Moscow, Nauka, 1976.
13. A.E. Green. On the linear theory of thin elastic shells. *Proc. Roy. Soc. Ser. A.*, 266(1325):143-160, 1962.
14. A.E. Green. Boundary layer equations in the linear theory of thin elastic shells. *Proc. Roy. Soc. Ser. A.*, 269(1339):481-491, 1962.
15. J.D. Kaplunov, I.V. Kirilova, L.Yu. Kossovich. Asymptotic Integration of dynamic equations of elasticity theory for the case of thin shells (in Russ.). *J. Appl. Math. Mech.* 57(1):83-91, 1993.
16. J.D. Kaplunov, L.Yu. Mossovich, E.V. Nolde. *Dynamics Thin Walled Elastic Bodies*. Academic Press, San-Diego, 1998.
17. S.A. Lomov. *Introduction to General Theory of Singular Perturbations (in Russ.)*. Moscow. Nauka, 1981.
18. V.P. Maslov. *The Perturbation Theory and Asymptotic Methods (in Russ.)*. Moscow, Publishing House of Moscow State University, 1965.
19. G.I. Mikhasev, P.E. Tovstik. *Localized Oscillations and Waves in Thin Shells. Asymptotic Methods (in Russ.)*. Moscow, Nauka, 2009.
20. N.F. Morozov, P.E. Tovstik. Bending of a two-layer beam with non-rigid contact between the layers. *Journal of Applied Mathematics and Mechanics* 75(1):77-84, 2011.
21. N.F. Morozov, P.E. Tovstik, T.P. Tovstik. A continuum model of a multi-layer nanosheet. *Doklady Physics*, 61(11):567-570, 2016.
22. A.H. Nayfeh. *Perturbation Methods*, John Wiley and Sons, 1973.
23. Ya.G. Panovko. *Internal Friction at Vibrations of Elastic Systems (in Russ.)*. Moscow, Fizmatgiz, 1960.
24. V.B. Poruchikov. *Methods of Dynamic Theory of Elasticity (in Russ.)*. Moscow, Nauka, 1986.
25. P.E. Tovstik. *Stability of Thin Shells: Asymptotic Methods (in Russ.*. Moscow, Nauka, 1995.
26. P.E. Tovstik, A.L. Smirnov. *Asymptotic Methods in the Buckling Theory of Elastic Shells*. World Scientific Publishin Co Ltd., Singapore, New Jersey, London, Hong Kong, 2001.
27. A.B. Vasilieva, V.F. Butuzov. *Asymptotic Expansions of Solutions of Singularly Perturbed Equations (in Russ.*. Moscow, Nauka1973.
28. V. Vazov. *Asymptotic Expansions of Solutions of Ordinary Differential Equations (in Russ.)*. Moscow, Mir 1968.
29. M.V. Vilde, J.D. Kaplunov, L.Yu. Kossovich. *Edge and Interface Resonance Phenomena in elastic Bodies (in Russ.)*. Moscow, Fizmatlit, 2010

Chapter 2
Asymmetric Buckling of Orthortropic Plates Under Normal Pressure

Svetlana M. Bauer, Dmitry A. Indeitsev, Boris N. Semenov, and Eva B. Voronkova

Abstract The work is concerned with asymmetric buckling of clamped orthotropic circular plates under normal pressure. The effect of degree of orthotropy on the buckling load and buckling modes is examined. The problem is solved analytically and using the finite element method in the ANSYS package. It is shown that the critical load increases as the ratio of the circumferential to radial elastic modulus increases. The number of waves in the circumferential direction also increases.

Key words: Circular plate, Orthotropy, Buckling, Finite element modelling

2.1 Introduction

The question of the existence of asymmetric solutions for a symmetrically loaded circular isotropic plate first was considered by Panov and Feodos'ev in 1948 [1]. Authors used the Galerkin method to solve the problem, and estimated the axisymmetric deflection prior to buckling by a function with one unknown parameter. Later, Cheo and Reiss emphasized that approximation functions with one unknown parameter was "too inaccurate" to adequately describe the large axisymmetric deformation of plate before wrinkling [2]. It was supposed that Panov and Feodos'ev had found unstable unsymmetric state in [1]. The value of the critical load obtained by Cheo and Reiss [2] for a plate uniformly loaded with normal pressure is much larger than

Svetlana M. Bauer · Boris N. Semenov · Eva B. Voronkova
St. Petersburg State University, 7/9 University Embankment, St. Petersburg, 199034 Russian Federation,
e-mail: s.bauer@spbu.ru,b.semenov@spbu.ru,e.voronkova@spbu.ru

Dmitry A. Indeitsev
Institute for Problems in Mechanical Engineering, V.O., Bolshoj pr., 61, St. Petersburg, 199178, Russian Federation,
e-mail: dmitry.indeitsev@gmail.com

H. Altenbach et al. (eds.), *Advances in Solid and Fracture Mechanics*,
Advanced Structured Materials 180,
https://doi.org/10.1007/978-3-031-18393-5_2

the load determined by Panov and Feodos'ev [1]. The form of buckling is also different: 14 waves are formed along the edge of the plate, and not 8, as in [1]. It should be noted that back in 1958–1962 N.F. Morozov proved rigorously the existence and the uniqueness of an asymmetric solution in this case [3, 4, 5].

The problems of the loss of the axisymmetric form of equilibrium of circular plates under the normal pressure and shallow spherical shells under the action of internal pressure were solved by asymptotic methods by Coman [6, 7]. In [8, 9, 10], similar problems were considered for plates and shells with spatial inhomogeneous in the radial direction. It has been found that when the rigidity of a plate or a spherical shallow shells decreases towards the edge, the loss of stability occurs at a lower load and with the formation of a larger number of waves. The asymmetrical buckling of the circular steel plates subjected to uniformly distributed uniform normal or concentrated load was investigated by the finite element method in NASTRAN and ANSYS software packages in [11, 12, 13].

In this paper we consider the problem of loss of stability of the axisymmetric equilibrium of orthotropic circular plate under the action of normal pressure. The problem is solved analytically and using the finite element method in the ANSYS package.

2.2 Problem Formulation

Let us consider a thin plate of outer radius R and uniform thickness $h > 0\,(h/R \ll 1)$, subjected to uniformly distributed normal pressure p. We introduce the orthogonal coordinate system (r,θ,z), in which r and θ are polar coordinates in the plate plane and z is the distance along the normal to the mid-surface.

The stress–strain relations for the orthotropic plate under the assumption of the classical theory are given by

$$e_{rr} = \frac{\sigma_r}{E_r} - \frac{\nu_r}{E_r}\sigma_\theta, \quad e_{\theta\theta} = -\frac{\nu_\theta}{E_\theta}\sigma_r + \frac{\sigma_\theta}{E_\theta}, \quad e_{r\theta} = \frac{\sigma_{r\theta}}{G_{r\theta}}, \tag{2.1}$$

with Young's moduli in the r- and θ-directions E_r, E_θ, respectively, the Poisson ratios ν_r, ν_θ, the shear modulus in r-θ direction $G_{r\theta}$. The equality $E_r\nu_\theta = E_\theta\nu_r$ must be satisfied due to the symmetry.

The fundamental equations can be written in the form [14]

$$\begin{aligned} D_r L_w(w) &= P + L(w,F), \\ L_F(F)/(hE_\theta) &= -L(w,w)/2, \end{aligned} \tag{2.2}$$

where $w(r,\theta)$ is the displacement in the z direction, $F(r,\theta)$ is the Airy stress function, $D_r = E_r h^3/12(1-\nu_r\nu_\theta)$ is the bending stiffness. The Laplacian and differential operators L_w, L_F, L are defined as

$$\Delta = (\ )'' + \frac{1}{r}(\ )' + \frac{1}{r^2}(\ddot{\ }),$$

$$L_w(y) = y'''' + 2\frac{y'''}{r} + 2\frac{\lambda_{r\theta}}{r^2}\left(\ddot{y}'' - \frac{\ddot{y}'}{r} + \frac{\ddot{y}}{r^2}\right) - \frac{\lambda}{r^2}\left[y'' - \frac{y'}{r} - \left(2\frac{y}{r^2} + \frac{\ddot{y}}{r^2}\right)^{\cdot\cdot}\right],$$

$$L_F(y) = y'''' + 2\frac{y'''}{r} + \frac{\kappa}{r^2}\left(\ddot{y}'' - \frac{\ddot{y}'}{r} + \frac{\ddot{y}}{r^2}\right) - \frac{\lambda}{r^2}\left[y'' - \frac{y'}{r} - \left(2\frac{y}{r^2} + \frac{\ddot{y}}{r^2}\right)^{\cdot\cdot}\right],$$

$$L(x, y) = x''\left(\frac{y'}{r} + \frac{\ddot{y}}{r^2}\right) + y''\left(\frac{x'}{r} + \frac{\ddot{x}}{r^2}\right) - 2\left(\frac{\dot{x}}{r}\right)'\left(\frac{\dot{y}}{r}\right)',$$

with the short-hand notations $(\ )' \equiv \partial(\ )/\partial r$, $(\dot{\ }) \equiv \partial(\ )/\partial\theta$ and

$$\lambda = \frac{D_\theta}{D_r} = \frac{E_\theta}{E_r}, \quad \lambda_{r\theta} = \frac{D_{r\theta}}{D_r}, \quad \kappa = \frac{E_\theta}{G_{r\theta}} - 2\nu_\theta = 2\frac{\lambda - \lambda_{r\theta}\nu_\theta}{\lambda_{r\theta} - \nu_\theta},$$

$$D_i = \frac{h^3 E_i}{12(1 - \nu_r\nu_\theta)}, (i = r, \theta), \quad D_k = \frac{h^3}{12}G_{r\theta}, \quad D_{r\theta} = D_r\nu_\theta + 2D_k.$$

For an isotropic plate we have

$$E_r = E_\theta, \nu_r = \nu_\theta, G_{r\theta} = E_r/2(1 + \nu_\theta), \lambda = \lambda_{r\theta} = 1, \kappa = 2,$$

and the operators L_w, L_F are

$$L_w(y) = L_F(y) = \Delta\Delta y.$$

We rescale Eqs. (2.2) by introducing dimensionless quantities as

$$r^* = \frac{r}{R}, \; w^* = \beta\frac{w}{h}, \; P^* = \beta^3\frac{PR^4}{E_r h^4}, \; F^* = \beta^2\frac{F}{E_r h^3}, \; \beta^2 = 12(1 - \nu_r\nu_\theta). \quad (2.3)$$

Dropping asterisks for notational simplicity, we obtain the dimensionless forms of Eqs. (2.2)

$$\begin{aligned} L_w(w) &= P + L(w, F), \\ L_F(F)/\lambda &= -L(w, w)/2. \end{aligned} \quad (2.4)$$

We assume that the outer edge of the plate is clamped but can move freely in the radial direction without rotation. To avoid singularity in the center of the plate we consider an annular plate with a small inner radius $r = \delta = R_{\text{in}}/R$. The inner edge of the plate is assumed to be supported by a roller which can slide along a vertical wall. For numerical simulations we adjust the δ parameter so that, in the case of an isotropic plate, the magnitudes of the buckling load of an annular and solid plates are closed, and the annular isotropic plate buckles into the asymmetric state with the same number of waves in the circumferential direction as a solid isotropic plate.

The constrains on the inner and outer edges of the plate can be written as

$$
\begin{aligned}
&w = w' = N_{rr} = N_{r\theta} = 0 \quad \text{for } r = 1, \\
&u = w' = N_{r\theta} = Q_r^* = 0 \quad \text{for } r = \delta.
\end{aligned} \tag{2.5}
$$

Here u denotes the horizontal radial components of displacement, $N_{rr}, N_{r\theta}$ are stress resultants, Q_r^* is the generalized shear force, that can be expressed via the shear force Q_r, the bending M_r, M_θ and twisting $M_{r\theta}$ moments as is stated below:

$$
Q_r^* = Q_r + \dot{M}_{r\theta}/r = M_r' + (M_r - M_\theta)/r + 2\dot{M}_{r\theta}/r.
$$

In the terms of the displacement component w and stress function F, the boundary conditions (2.5) are equivalent to

$$
\begin{aligned}
&\text{for } r = 1 \quad w = w' = F'/r + \ddot{F}/r^2 = -(\dot{F}/r)' = 0, \\
&\text{for } r = \delta \quad w' = 0, \quad \dot{F}/r^2 - \dot{F}'/r = 0, \\
&\qquad w''' + \frac{w''}{r} - \frac{\lambda}{r^2}\left(w'' + \frac{\ddot{w}}{r}\right) - \frac{2\lambda_{r\theta} - \nu_\theta}{r^2}\left(\frac{\ddot{w}}{r} - \ddot{w}'\right) = 0, \\
&\qquad F''' + 2\frac{F''}{r} - \frac{\lambda}{r^2}\left(F' + \frac{\ddot{F}}{r}\right) - \frac{\nu_\theta}{r^2}(F + \ddot{F})' = 0.
\end{aligned} \tag{2.6}
$$

For the axisymmetrical problem Eqs. (2.2) are reduced to

$$
\begin{aligned}
&\Theta_0'' + \frac{\Theta_0'}{r} - \lambda\frac{\Theta_0}{r^2} = \frac{Pr}{2}\left(1 - \frac{\delta^2}{r^2}\right) + \frac{\Theta_0\Phi_0}{r}, \\
&\Phi_0'' + \frac{\Phi_0'}{r} - \lambda\frac{\Phi_0}{r^2} = -\frac{\Theta_0^2}{2r},
\end{aligned} \tag{2.7}
$$

where $\Theta_0(r) = w'(r)$, $\Phi_0(r) = F'(r)$. The system (2.7) needs to be solved together with the boundary conditions

$$
\Theta_0(1) = \Phi_0(1) = 0, \quad \Phi'(\delta) - \nu_\theta\Phi(\delta) = \Theta(\delta) = 0. \tag{2.8}
$$

2.3 Equations for Buckling

Asymmetrical solutions of problem (2.2) with appropriate boundary conditions branch from a solution of axisymmetric states [2]. To detect the occurrence of wrinkling we seek for a solution of Eqs. (2.2) in the form

$$
w(r,\theta) = w_s(r) + \varepsilon w_{ns}\cos(n\theta), \quad F(r,\theta) = F_s(r) + \varepsilon F_{ns}\cos(n\theta), \tag{2.9}
$$

where $w_s(r)$, $F_s(r)$ describe prebuckling axisymmetric state, ε is infinitesimal parameter, n is a mode number and $w_n(r)$, $F_n(r)$ are the non-symmetrical components.

After substitution of (2.9) in (2.2), using Eqs. (2.7) and linearization with respect to ε we obtain

$$
\begin{aligned}
L_w^n(w_n) &= \frac{w_n''}{r}\Phi_0 + \frac{F_n''}{r}\Theta_0 + \Theta_0'\left(\frac{F_n'}{r} - \frac{n^2}{r^2}F_n\right) - \Phi_0'\left(\frac{w_n'}{r} - \frac{n^2}{r^2}w_n\right),\\
L_F^n(F_n) &= -\lambda^2\left(\frac{w_n''}{r}\Theta_0 + \Theta_0'\left(\frac{w_n'}{r} - \frac{n^2}{r^2}w_n\right)\right),
\end{aligned}
\tag{2.10}
$$

where the operators L_w^n, L_F^n are

$$
\begin{aligned}
L_w^n(y) &= y'''' + 2\frac{y'''}{r} - 2\lambda_{r\theta}\frac{n^2}{r^2}\left(y'' - \frac{y'}{r} + \frac{y}{r^2}\right) - \frac{\lambda}{r^2}\left(y'' - \frac{y'}{r} - \frac{n^4 - 2n^2}{r^2}y\right),\\
L_F^n(y) &= y'''' + 2\frac{y'''}{r} - \kappa\frac{n^2}{r^2}\left(y'' - \frac{y'}{r} + \frac{y}{r^2}\right) - \frac{\lambda}{r^2}\left(y'' - \frac{y'}{r} - \frac{n^4 - 2n^2}{r^2}y\right).
\end{aligned}
$$

Boundary conditions (2.6) are reduced to

$$
\begin{aligned}
&\text{for } r = 0 \quad w_n = w_n' = F_n = F_n' = 0\\
&\text{for } r = \delta \quad w_n' = 0, \quad F_n' - rF_n = 0,\\
&\qquad w_n''' + \frac{w_n''}{r} - \frac{\lambda}{r^2}\left(w_n' - n^2\frac{w_n}{r}\right) + (2\lambda_{r\theta} - \nu_\theta)\frac{n^2}{r^2}\left(\frac{w_n}{r} - w_n'\right) = 0,\\
&\qquad F_n''' + 2\frac{F_n''}{r} - \frac{\lambda}{r^2}\left(F_n' - n^2\frac{F_n}{r}\right) + \nu_\theta(n^2 - 1)\frac{F_n'}{r^2} + \lambda A w_n = 0,
\end{aligned}
\tag{2.11}
$$

Buckling equations (2.10) with boundary conditions (2.11) constitute an eigenvalue problem, in which the parameter P is implicit and appears in the equations through the functions Θ_0 and Φ_0. The axisymmetric problem (2.7)-(2.8) was solved by standard MATLAB functions. The value of P, for which (2.10) with (2.11) have nontrivial solution, was found by using the finite difference method [2]. We regard the smallest of these eigenvalues as the buckling load.

2.4 Numerical Results

As described in [1, 2], a ring of large circumferential compressive stress, developed near the edge of the shell or the plate, indicates the possibility of buckling about the axisymmetric state into an unsymmetric equilibrium state. Figure 2.1 depicts the dimensionless normal displacement (left) and circumferential stress (right) prior to buckling for different values the elastic moduli ratio $\lambda = E_\theta/E_r$. The intensity of the compressive stress near the plate's edge increases and the width of the compressive ring decreases with the growth of the parameter λ (see Fig. 2.1, right).

Figure 2.2 (left and right) illustrates dependence of the normalized critical load P_{n} on the critical mode number n for different values of the orthotropy ratio λ

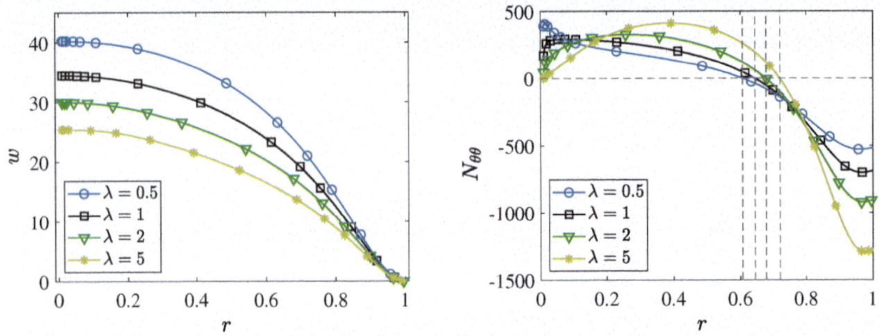

Fig. 2.1 Dimensionless axisymmetrical normal displacement w (left) and circumferential stress resultant $N_{\theta\theta}$ (right) for different values of $\lambda = E_\theta/E_r$; here $P = 30000$, $\delta = 0.01$, $\nu_\theta = 0.4$, $\lambda_{r\theta} = \lambda$. The value of $\lambda = 1$ corresponds to an isotropic plate.

(Fig. 2.2 (left) for $\lambda \in [0.5, 2]$, Fig. 2.2 (right) for $\lambda \in \{1, 5, 10\}$). The lines with

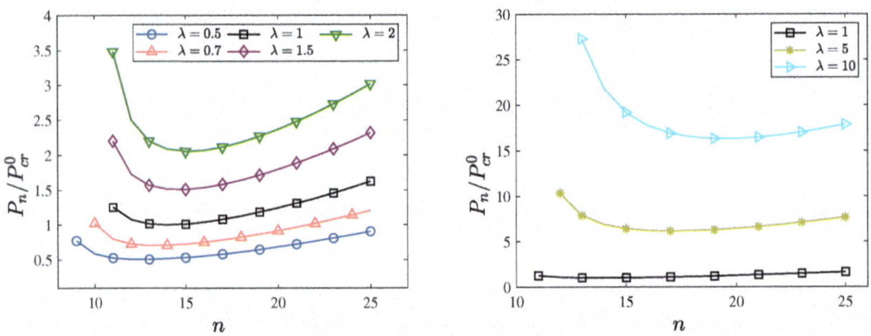

Fig. 2.2 Dependence of the normalized critical load P_n on the mode number n for different values of the orthotropy degree λ (left: $\lambda \in [0.5, 2]$, right: $\lambda \in \{1, 5, 10\}$). The lines with square markers correspond to the isotropic plate. Here $\nu_\theta = 0.4$, $\lambda_{r\theta} = \lambda$; $P^0_{\rm cr}$ denotes the buckling pressure for an isotropic circular plate.

square markers correspond to the isotropic plate. The illustrations of asymmetric wrinkling experienced by an isotropic and orthotropic plates are presented in Fig. 2.3.

The values of the buckling pressure $P_{\rm cr}$ increase by a factor of 2 when the degree of orthotropic λ increases to 2 and by a factor of 16 when λ increases to 10. When the radial modulus E_r becomes larger than the circumferential one E_θ i.e. $\lambda < 1$, the plate buckle to asymmetrical state under twice smaller load, comparing to the isotropic plate (see Fig. 2.4 and Table 2.1). The buckling mode n decreases has the same behavior as the buckling pressure $P_{\rm cr}$: it increases when λ increases.

We employed the finite element package ANSYS to study the stability of the symmetric equilibrium state of a circular plate with a radius of 50 cm and a thickness of 0.5 cm. The plate was partitioned into 5994 quadrangular elements of the

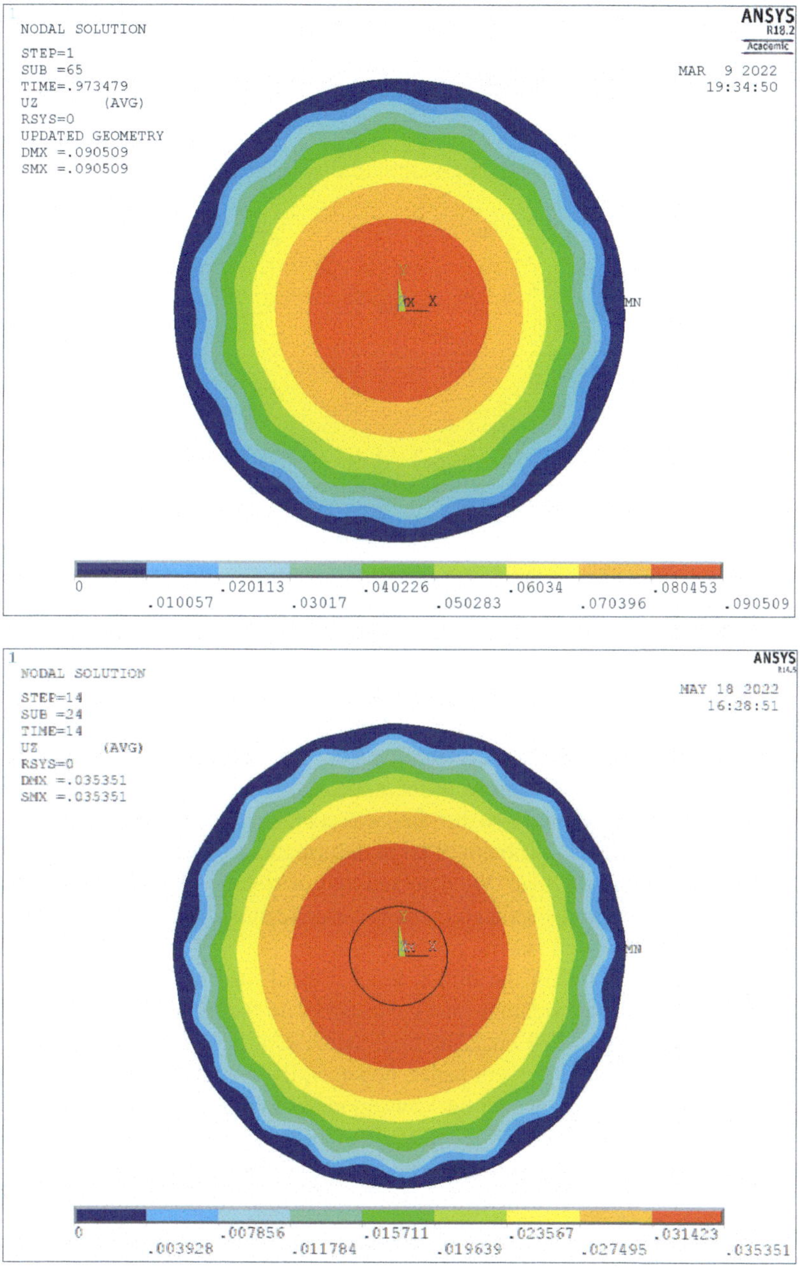

Fig. 2.3 Wrinking of an isotropic plate with 14 waves (top) and orthotropic plate (bottom) obtained by during finite element simulations

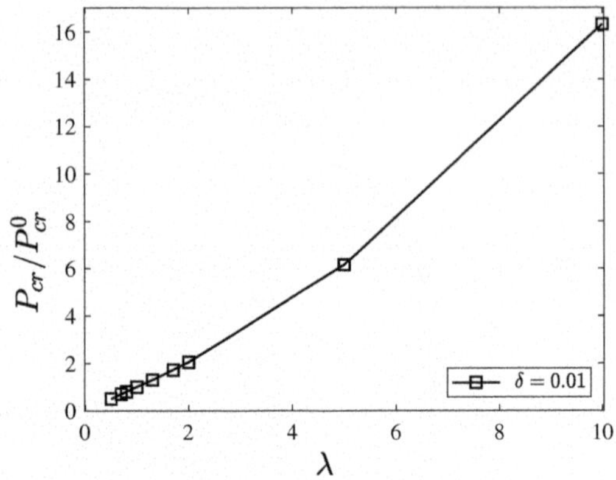

Fig. 2.4 Normalized buckling load for different values of the orthotropy degree λ. Here $\nu_\theta = 0.4$, $\lambda_{r\theta} = \lambda$; P^0_{cr} denotes the buckling pressure for an isotropic circular plate.

Table 2.1 The buckling load and buckling mode for different values of the orthotropy degree

$\lambda = E_\theta/E_r$	$\lambda = 0.5$	$\lambda = 1$	$\lambda = 2$	$\lambda = 5$	$\lambda = 10$
$P_{\mathrm{cr}}/P^0_{\mathrm{cr}}$	0.51	1	2.06	6.15	16.31
Mode number, n	13	14	15	17	20

The buckling load for an isotropic plate is $P^0_{\mathrm{cr}} = 64956$.

SHELL281 type. To study the nonlinear stability of the plate we randomly apply minor imperfections to a set of nodes, which means that the nodes can go out of the plate plane. These random deviations of the nodes from the $z = 0$ plane do not exceed 2% of the plate thickness. Then the plate was subjected to uniformly distributed normal load, and a nonlinear problem of large deformations of the plate was solved with a linear increase in the load. We detected loss of stability and determined the buckling load when a rapid localized increase in deflections was observed with a slight increase in a pressure load.

For a steel plate we obtain the dimensional critical load of 3.6 MPa with 14 waves the plate edge. The buckling load corresponds to the dimensionless load of 63370 (see Fig. 2.3 (top)). The buckling load value, determined in [11], coincides with the one obtained in the present paper, however, the buckling mode number is less and equal to 13. Difference in buckling mode number, obtained by [11] and numerically and asymptotically in [6, 8, 9, 10] can be explained by the fact that the critical load values corresponding to the wave numbers $n = 13$ and $n = 15$ differ from the load for $n = 14$ by 0.7% and 1.8%, respectively. As it was mentioned in [11] the finite element discretization of the plate could impact on the magnitude of the buckling load. For an isotropic plate with $E = 10$ MPa, $\nu = 0.4$ we get the dimensional wrinkling load of 200 Pa, which corresponds to the dimensionless load of 64006. For an orthotropic plate with $E_r = 10$ MPa, $E_\theta = 20$ MPa, $\nu_\theta = 0.1$ we get the dimensional wrinkling load of 490 Pa with formation of 17 waves near the plate edge, which corresponds to the dimensionless load of 191591 (see Fig. 2.3 (bottom)).

2.5 Conclusion

The wrinkling of the orthotropic circular plate under normal pressure has been studied in this work. It is shown that the degree anisotropy has a strong impact on the magnitude of buckling load and corresponding mode number. Analytical results and finite elements simulations show that the critical load increases as the ratio of the circumferential to radial elastic modulus increases. The number of waves in the circumferential direction also increases. The buckling load corresponds to sufficiently large deflections in the center of the plate and it means that this effect take place for soft materials.

References

1. D.Y. Panov, and V.I. Feodos'ev. On the equilibrium and loss of stability of shallow shells in the case of large displacement (in Russ.). *Prikladnaya Matematika Mekhanika*, 12:389-406, 1948.
2. L.S. Cheo and E.L. Reiss. Unsymmetric wrinkling of circular plates. *Quarterly of Applied Mathematics*, 31(1):75-91, 1973.
3. N.F. Morozov. The uniqueness of the symmetrical solution in the problem of the large deflection of a symmetrically loaded circular plate (in Russ.). *Dokl. Akad. Nauk SSSR*, 123(3):417–419, 1958.
4. N.F. Morozov. On the existence of a non-symmetric solution in the problem of large deflections of a circular plate with a symmetric load (in Russ.). *Izvestiya Vysshikh Uchebnykh Zavedenii. Matematika* , 2:126–129, 1961.
5. N.F. Morozov. A qualitative investigation of a circular, symmetrically compressed plate under a large edge load (proof of the appearance of crimping) (in Russ.). *Dokl. Akad. Nauk SSSR*, 147(6):1318–1319, 1962.
6. C.D. Coman and A.P. Bassom. Asymptotic limits and wrinkling patterns in a pressurised shallow spherical cap. *International Journal of Non-Linear Mechanics*, 81:8-18, 2016.
7. C.D. Coman. On the asymptotic reduction of a bifurcation equation for edge-buckling instabilities. *Acta Mech.*, 229, 1099-1109, 2018.
8. S.M. Bauer, E.B. Voronkova and N.F. Morozov. On unsymmetrical equillibrium states of circular plates under normal pressure (in Russ.). In: *Chigarev, A.B. (ed.) Theoretical and Applied Mechanics: International Scientific and Technical Collection*. pp. 31-35, Minsk, 2012.
9. S.M. Bauer and E.B. Voronkova. Unsymmetrical wrinkling of nonuniform annular plates and spherical caps under internal pressure. In: *Altenbach, H., Chroscielewski, J., Eremeyev V., Wisniewski, K. (eds.) Recent Developments in the Theory of Shells*, pp. 79-89. Springer, Cham, 2019.
10. S.M. Bauer and E.B. Voronkova: On buckling behavior of inhomogeneous shallow spherical caps with elastically restrained edge. In: *Altenbach, H., Chinchaladze, N., Kienzler, R., Müller, W. (eds.) Analysis of Shells, Plates, and Beams*, pp. 65-74. Springer, Cham, 2020.
11. G.A. Manuylov, S.B. Kosytsyn and M.M. Begichev. About initial imperfection sensitivity of some thin-walled strustures (in Russ.). *International Journal for Computational Civil and Structural Engineering*, 10(1):111-117, 2014.

12. G.A. Manuylov, S.B. Kosytsyn, and M.M. Begichev. Critical and postcritical equilibria in stability problems of elastic systems (in Russ.). *Structural Mechanics of Engineering Constructions and Buildings*, 5:47–54, 2015.
13. G.A. Manuylov, S.B. Kosytsyn and M.M. Begichev. On computational differences of critical points on equilibrium curve (in Russ.). *International Journal for Computational Civil and Structural Engineering*, 13(2):125-135, 2017.
14. S.A. Ambartsumyan. *The General Theory of Anisotropic Shells* (in Russ.). Moscow, 1974.

Chapter 3
On Ladyzhenskaya's Inequality and its Applications

Michael J. Cloud, Victor A. Eremeyev, and Leonid P. Lebedev

Abstract The term Ladyzhenskaya's inequality refers to any of a class of related functional inequalities that have played important roles in the theory of the Navier–Stokes equations. Here we present their detailed derivation and discuss potential applications beyond Navier–Stokes hydrodynamics.

Key words: Inequalities, Fluid dynamics, Nonlinear problems, Sobolev spaces

3.1 Introduction

Inequalities play important roles in mathematical physics. One might even say they are more important than equations. In the theory of Sobolev spaces, for example, the famous Sobolev embedding theorems are formulated in terms of inequalities [1, 15]. These theorems are widely used to study well-posedness of the boundary-value problems of mathematical physics; see, e.g., [2, 10, 12, 13]. In particular, one can prove certain properties of nonlinear operators which can establish the existence of solutions to nonlinear problems. Typical examples of nonlinear systems are Navier–Stokes hydrodynamics [11] and the nonlinear theory of plates and shells [3, 27]. Worthy of mention are the early papers by Morozov [17, 18, 19, 21], devoted to the mathematical studies of nonlinear plates.

Michael J. Cloud
Lawrence Technological University, Southfield, Michigan, USA,
e-mail: mjc@ieee.org

Victor A. Eremeyev
University of Cagliari, Via Marengo, 2, 09123 Cagliari, Italy,
e-mail: victor.eremeev@unica.it,eremeyev.victor@gmail.com

Leonid P. Lebedev
Departamento de Matemáticas, National University of Colombia, Bogotá D.C., Colombia,
e-mail: llebedev@unal.edu.co

H. Altenbach et al. (eds.), *Advances in Solid and Fracture Mechanics*,
Advanced Structured Materials 180,
https://doi.org/10.1007/978-3-031-18393-5_3

The chapter is organized as follows. After mathematical preliminaries in Sect. 3.2, we present the derivations of Ladyzhenskaya's inequality in two- and three-dimensional space in Sects. 3.3 and 3.4, respectively. This inequality relates the norm of a function u in L_p, with a certain p, to the norm of its gradient ∇u. Finally, in Sect. 3.5 we briefly discuss some potential applications for inequalities of this type.

3.2 Preliminaries

3.2.1 Some Standard Inequalities

A few classic results are needed to follow the development of Ladyzhenskaya's inequalities. They are merely listed here; further background is available, e.g., in [4]. First is the inequality for integrals:

$$\left| \int_a^b f(x)\,\mathrm{d}x \right| \le \int_a^b |f(x)|\,\mathrm{d}x .$$

Next is the Schwarz inequality for integrals:

$$\left| \int_a^b f(x)g(x)\,\mathrm{d}x \right| \le \left(\int_a^b f^2(x)\,\mathrm{d}x \right)^{1/2} \left(\int_a^b g^2(x)\,\mathrm{d}x \right)^{1/2} .$$

These two inequalities hold not only in the one-dimensional forms as stated, but also with multiple integrals (over domains in $\mathbb{R}^2$ and $\mathbb{R}^3$, for instance). They also hold for improper integrals provided the terms in the right members all exist as finite values. In its discrete version, the inequality between the arithmetic and geometric means is

$$\left(\prod_{k=1}^{n} a_k \right)^{1/n} \le \frac{1}{n} \sum_{k=1}^{n} a_k \qquad (a_k \ge 0) .$$

Finally, simple observations of the types

$$\int_a^b |f(x)g(x)|\,\mathrm{d}x \le \max_x |f(x)| \int_a^b |g(x)|\,\mathrm{d}x,$$

$$\max_x \int_a^b \int_c^d f(x,y,z)\,\mathrm{d}y\,\mathrm{d}z \le \int_a^b \int_c^d \max_x f(x,y,z)\,\mathrm{d}y\,\mathrm{d}z$$

will be used. The second of these can be obtained by integrating both sides of the obvious inequality $f(x,y,z) \le \max_x f(x,y,z)$ over y and z to get a relation

$$\int_a^b\int_c^d f(x,y,z)\,\mathrm{d}y\,\mathrm{d}z \le \int_a^b\int_c^d \max_x f(x,y,z)\,\mathrm{d}y\,\mathrm{d}z$$

that holds for all x; since the right-hand side is a constant, the result follows from taking the maximum of both sides over x.

3.2.2 Functions of Compact Support

Let $u = u(x_1,\ldots,x_n)$ be a function defined on a domain Ω in $\mathbb{R}^n$. Denote the boundary of Ω by $\partial\Omega$. It is said that u has compact support in Ω if it is nonzero only on a bounded subdomain Ω' of Ω with Ω' lying at some positive distance from $\partial\Omega$. Figure 3.1 illustrates this idea for the cases $n = 1$ and $n = 2$.

Lemma 3.1. *Let $u = u(x_1,\ldots,x_n)$ be a smooth function of n variables having compact support in some bounded domain Ω of $\mathbb{R}^n$. Fix a positive integer m. Then by the chain rule for partial differentiation,*

$$\frac{\partial u^m(x_1,\ldots,x_n)}{\partial x_k} = m u^{m-1}(x_1,\ldots,x_n)\frac{\partial u(x_1,\ldots,x_n)}{\partial x_k} \qquad (k = 1,\ldots,n).$$

Both sides can be integrated w.r.t. the kth argument over the interval $(-\infty, x_k)$:

$$u^m(x_1,\ldots,\xi,\ldots,x_n)\Big|_{-\infty}^{x_k} = m\int_{-\infty}^{x_k} u^{m-1}(x_1,\ldots,\xi,\ldots,x_n)\frac{\partial u(x_1,\ldots,\xi,\ldots,x_n)}{\partial \xi}\,\mathrm{d}\xi.$$

Since u has compact support however, $\lim_{\xi\to-\infty} u^m(x_1,\ldots,\xi,\ldots,x_n) = 0$ and

$$u^m(x_1,\ldots,x_k,\ldots,x_n) = m\int_{-\infty}^{x_k} u^{m-1}(x_1,\ldots,\xi,\ldots,x_n)\frac{\partial u(x_1,\ldots,\xi,\ldots,x_n)}{\partial \xi}\,\mathrm{d}\xi$$

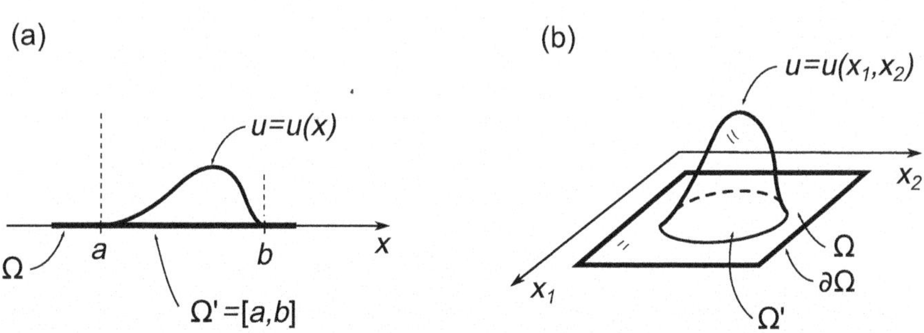

Fig. 3.1 Functions of compact support. (a) One-variable case. (b) Two-variable case.

which henceforth will be written more concisely as

$$u^m = m \int_{-\infty}^{x_k} u^{m-1} u_{x_k} \, \mathrm{d}x_k \, .$$

An application of the triangle inequality gives

$$|u^m| \leq m \int_{-\infty}^{x_k} |u^{m-1} u_{x_k}| \, \mathrm{d}x_k \leq m \int_{-\infty}^{\infty} |u^{m-1} u_{x_k}| \, \mathrm{d}x_k$$

and therefore

$$\max_{x_k} |u^m| \leq m \int_{-\infty}^{\infty} |u^{m-1} u_{x_k}| \, \mathrm{d}x_k \qquad (k = 1, \ldots, n) \, . \tag{3.1}$$

3.3 Ladyzhenskaya's Inequality in Two Dimensions

Let

$$u = u(x_1, x_2)$$

be a smooth function with compact support in some domain $\Omega \subset \mathbb{R}^2$. Writing $dA = \mathrm{d}x_1 \, \mathrm{d}x_2$, one has

$$\begin{aligned} \iint_{-\infty}^{\infty} u^4 \, \mathrm{dA} &= \iint_{-\infty}^{\infty} u^2 u^2 \, \mathrm{dA} \leq \iint_{-\infty}^{\infty} \max_{x_1} u^2 \max_{x_2} u^2 \, \mathrm{d}x_1 \, \mathrm{d}x_2 \\ &= \int_{-\infty}^{\infty} \max_{x_1} |u|^2 \, \mathrm{d}x_2 \int_{-\infty}^{\infty} \max_{x_2} |u|^2 \, \mathrm{d}x_1 \\ &\leq 4 \iint_{-\infty}^{\infty} |u u_{x_1}| \, \mathrm{d}x_1 \, \mathrm{d}x_2 \iint_{-\infty}^{\infty} |u u_{x_2}| \, \mathrm{d}x_1 \, \mathrm{d}x_2 = 4 \prod_{k=1}^{2} \iint_{-\infty}^{\infty} |u u_{x_k}| \, \mathrm{dA} \end{aligned}$$

by (3.1) with $m = 2$. Application of the Schwarz inequality gives

$$\begin{aligned} \iint_{-\infty}^{\infty} u^4 \, \mathrm{dA} &\leq 4 \prod_{k=1}^{2} \left(\iint_{-\infty}^{\infty} u^2 \, \mathrm{dA} \right)^{1/2} \left(\iint_{-\infty}^{\infty} u_{x_k}^2 \, \mathrm{dA} \right)^{1/2} \\ &= 4 \iint_{-\infty}^{\infty} u^2 \, \mathrm{dA} \left(\prod_{k=1}^{2} \iint_{-\infty}^{\infty} u_{x_k}^2 \, \mathrm{dA} \right)^{1/2} . \end{aligned}$$

Finally, by the AM–GM inequality,

$$\iint_{-\infty}^{\infty} u^4 \, \mathrm{d}A \le 4 \iint_{-\infty}^{\infty} u^2 \, \mathrm{d}A \cdot \frac{1}{2} \sum_{k=1}^{2} \iint_{-\infty}^{\infty} u_{x_k}^2 \, \mathrm{d}A$$

and one has Ladyzhenskaya's inequality in two dimensions:

$$\iint_{-\infty}^{\infty} u^4 \, \mathrm{d}A \le 2 \iint_{-\infty}^{\infty} u^2 \, \mathrm{d}A \iint_{-\infty}^{\infty} \sum_{k=1}^{2} u_{x_k}^2 \, \mathrm{d}A \,. \tag{3.2}$$

It can also be expressed as

$$||u||_{L^4(\Omega)} \le 2^{1/4} ||u||_{L^2(\Omega)}^{1/2} \, ||\nabla u||_{L^2(\Omega)}^{1/2} \tag{3.3}$$

where

$$||u||_{L^p(\Omega)}$$

denotes the Lebesgue p-norm of the function u,

$$||u||_{L^p(\Omega)} = \left(\iint_{\Omega} |u|^p \, \mathrm{d}A \right)^{1/p},$$

and in this case ∇ is the nabla operator in two space dimensions.

3.4 Ladyzhenskaya's Inequalities in Three Dimensions

In this section the results

$$||u||_{L^4(\Omega)} \le 2^{1/2} ||u||_{L^2(\Omega)}^{1/4} \, ||\nabla u||_{L^2(\Omega)}^{3/4}, \tag{3.4}$$

$$||u||_{L^6(\Omega)} \le 48^{1/6} \, ||\nabla u||_{L^2(\Omega)}, \tag{3.5}$$

are derived for a smooth function

$$u = u(x_1, x_2, x_3)$$

having compact support in some domain $\Omega \subset \mathbb{R}^3$.

3.4.1 First Result

Writing $\mathrm{d}V = \mathrm{d}x_1 \, \mathrm{d}x_2 \, \mathrm{d}x_3$ one has by (3.2),

$$\iiint_{-\infty}^{\infty} u^4 \, dV = \int_{-\infty}^{\infty} \left(\iint_{-\infty}^{\infty} u^4 \, dx_1 \, dx_2 \right) dx_3$$

$$\leq 2 \int_{-\infty}^{\infty} \left(\iint_{-\infty}^{\infty} u^2 \, dx_1 \, dx_2 \iint_{-\infty}^{\infty} \sum_{k=1}^{2} u_{x_k}^2 \, dx_1 \, dx_2 \right) dx_3 .$$

Next the maximum value of the integral

$$\iint_{-\infty}^{\infty} u^2 \, dx_1 \, dx_2$$

over x_3 is withdrawn from the integral with respect to x_3:

$$\iiint_{-\infty}^{\infty} u^4 \, dV \leq 2 \left(\max_{x_3} \iint_{-\infty}^{\infty} u^2 \, dx_1 \, dx_2 \right) \iiint_{-\infty}^{\infty} \sum_{k=1}^{2} u_{x_k}^2 \, dV .$$

But the maximum value of an integral cannot exceed the integral of the maximum value, and this provides an opportunity to use (3.1):

$$\iiint_{-\infty}^{\infty} u^4 \, dV \leq 2 \left(\iint_{-\infty}^{\infty} \max_{x_3} u^2 \, dx_1 \, dx_2 \right) \iiint_{-\infty}^{\infty} \sum_{k=1}^{2} u_{x_k}^2 \, dV$$

$$\leq 2 \left[\iint_{-\infty}^{\infty} \left(2 \int_{-\infty}^{\infty} |u u_{x_3}| \, dx_3 \right) dx_1 \, dx_2 \right] \iiint_{-\infty}^{\infty} \sum_{k=1}^{2} u_{x_k}^2 \, dV$$

$$= 4 \iiint_{-\infty}^{\infty} |u u_{x_3}| \, dV \iiint_{-\infty}^{\infty} \sum_{k=1}^{2} u_{x_k}^2 \, dV .$$

Now the Schwarz inequality is applied to the first factor on the right:

$$\iiint_{-\infty}^{\infty} u^4 \, dV \leq 4 \left(\iiint_{-\infty}^{\infty} u^2 \, dV \right)^{1/2} \left(\iiint_{-\infty}^{\infty} u_{x_3}^2 \, dV \right)^{1/2} \iiint_{-\infty}^{\infty} \sum_{k=1}^{2} u_{x_k}^2 \, dV .$$

Finally, the two last factors on the right may be increased through the inclusion of additional terms:

$$\iiint_{-\infty}^{\infty} u^4 \, dV \leq 4 \left(\iiint_{-\infty}^{\infty} u^2 \, dV \right)^{1/2} \left(\iiint_{-\infty}^{\infty} \sum_{k=1}^{3} u_{x_k}^2 \, dV \right)^{1/2} \iiint_{-\infty}^{\infty} \sum_{k=1}^{3} u_{x_k}^2 \, dV .$$

The result is Ladyzhenskaya's inequality

$$\iiint_{-\infty}^{\infty} u^4 \, dV \leq 4\left(\iiint_{-\infty}^{\infty} u^2 \, dV\right)^{1/2}\left(\iiint_{-\infty}^{\infty} \sum_{k=1}^{3} u_{x_k}^2 \, dV\right)^{3/2} \tag{3.6}$$

which can also be expressed as (3.4).

3.4.2 Second Result

This time it is temporarily assumed that u is nonnegative (a restriction that will be removed after the final result has been stated). The development starts by writing

$$\iiint_{-\infty}^{\infty} u^6 \, dV = \int_{-\infty}^{\infty}\left[\iint_{-\infty}^{\infty} u^3 u^3 \, dx_2 \, dx_3\right] dx_1 ,$$

replacing the integrand functions on the right by their respective maxima over x_2 and x_3,

$$\begin{aligned}\iiint_{-\infty}^{\infty} u^6 \, dV &\leq \int_{-\infty}^{\infty}\left[\iint_{-\infty}^{\infty} \max_{x_2} u^3 \max_{x_3} u^3 \, dx_2 \, dx_3\right] dx_1 \\ &= \int_{-\infty}^{\infty}\left[\int_{-\infty}^{\infty}\left(\max_{x_2} u^3\right) dx_3 \int_{-\infty}^{\infty}\left(\max_{x_3} u^3\right) dx_2\right] dx_1 ,\end{aligned}$$

and using (3.1) with $m = 3$:

$$\begin{aligned}\iiint_{-\infty}^{\infty} u^6 \, dV &\leq \int_{-\infty}^{\infty}\left[\int_{-\infty}^{\infty}\left(3\int_{-\infty}^{\infty} u^2 |u_{x_2}| \, dx_2\right) dx_3 \int_{-\infty}^{\infty}\left(3\int_{-\infty}^{\infty} u^2 |u_{x_3}| \, dx_3\right) dx_2\right] dx_1 \\ &= 9\int_{-\infty}^{\infty}\left(\prod_{k=2}^{3}\iint_{-\infty}^{\infty} u^2 |u_{x_k}| \, dx_2 \, dx_3\right) dx_1 .\end{aligned}$$

By the Schwarz inequality,

$$\begin{aligned}\iiint_{-\infty}^{\infty} u^6 \, dV &\leq 9\int_{-\infty}^{\infty}\left[\prod_{k=2}^{3}\left(\iint_{-\infty}^{\infty} u^4 \, dx_2 \, dx_3\right)^{1/2}\left(\iint_{-\infty}^{\infty} u_{x_k}^2 \, dx_2 \, dx_3\right)^{1/2}\right] dx_1 \\ &= 9\int_{-\infty}^{\infty}\left[\iint_{-\infty}^{\infty} u^4 \, dx_2 \, dx_3 \prod_{k=2}^{3}\left(\iint_{-\infty}^{\infty} u_{x_k}^2 \, dx_2 \, dx_3\right)^{1/2}\right] dx_1\end{aligned}$$

and one can withdraw the maximum over x_1 of the quantity

$$\iint_{-\infty}^{\infty} u^4 \, dx_2 \, dx_3$$

from the integral with respect to x_1 and apply the Schwarz inequality again:

$$\begin{aligned}\iiint_{-\infty}^{\infty} u^6 \, dV &\le 9\left(\max_{x_1} \iint_{-\infty}^{\infty} u^4 \, dx_2 \, dx_3\right) \int_{-\infty}^{\infty} \prod_{k=2}^{3} \left(\iint_{-\infty}^{\infty} u_{x_k}^2 \, dx_2 \, dx_3\right)^{1/2} dx_1 \\ &\le 9 \max_{x_1} \iint_{-\infty}^{\infty} u^4 \, dx_2 \, dx_3 \prod_{k=2}^{3} \left(\iiint_{-\infty}^{\infty} u_{x_k}^2 \, dV\right)^{1/2}.\end{aligned}$$

By now-familiar techniques,

$$\begin{aligned}\iiint_{-\infty}^{\infty} u^6 \, dV &\le 9 \iint_{-\infty}^{\infty} \max_{x_1} u^4 \, dx_2 \, dx_3 \prod_{k=2}^{3} \left(\iiint_{-\infty}^{\infty} u_{x_k}^2 \, dV\right)^{1/2} \\ &\le 9 \iint_{-\infty}^{\infty} 4 \int_{-\infty}^{\infty} |u^3 u_{x_1}| \, dx_1 \, dx_2 \, dx_3 \prod_{k=2}^{3} \left(\iiint_{-\infty}^{\infty} u_{x_k}^2 \, dV\right)^{1/2} \\ &= 36 \iiint_{-\infty}^{\infty} |u^3 u_{x_1}| \, dV \prod_{k=2}^{3} \left(\iiint_{-\infty}^{\infty} u_{x_k}^2 \, dV\right)^{1/2}\end{aligned}$$

so that

$$\iiint_{-\infty}^{\infty} u^6 \, dV \le 36\left(\iiint_{-\infty}^{\infty} u^6 \, dV\right)^{1/2} \prod_{k=1}^{3} \left(\iiint_{-\infty}^{\infty} u_{x_k}^2 \, dV\right)^{1/2}.$$

Division through by the quantity

$$\left(\iiint_{-\infty}^{\infty} u^6 \, dV\right)^{1/2}$$

gives

$$\begin{aligned}\left(\iiint_{-\infty}^{\infty} u^6 \, dV\right)^{1/2} &\le 36 \prod_{k=1}^{3} \left(\iiint_{-\infty}^{\infty} u_{x_k}^2 \, dV\right)^{1/2} \\ &= 36\left(\prod_{k=1}^{3} \iiint_{-\infty}^{\infty} u_{x_k}^2 \, dV\right)^{1/2}.\end{aligned}$$

Finally, by the AM–GM inequality

$$\left(\iiint\limits_{-\infty}^{\infty} u^6\,\mathrm{d}V\right)^{1/2} \le 36\left[\left(\frac{1}{3}\sum_{k=1}^{3}\iiint\limits_{-\infty}^{\infty} u_{x_k}^2\,\mathrm{d}V\right)^3\right]^{1/2}$$

$$= 36\cdot 3^{-3/2}\left(\iiint\limits_{-\infty}^{\infty}\sum_{k=1}^{3} u_{x_k}^2\,\mathrm{d}V\right)^{3/2}$$

and both sides may be squared:

$$\iiint\limits_{-\infty}^{\infty} u^6\,\mathrm{d}V \le 48\left(\iiint\limits_{-\infty}^{\infty}\sum_{k=1}^{3} u_{x_k}^2\,\mathrm{d}V\right)^3, \tag{3.7}$$

which is (3.5). Although this inequality of Ladyzhenskaya was proved under the assumption that u is nonnegative, it clearly holds without this restriction.

3.5 Further Possible Applications

The inequalities presented above were developed for application to the boundary-value problems of hydrodynamics. Nevertheless one can use them in other problems. Let us briefly consider some of these.

- The first is the theory of elastic membranes and plates with nonlinear elastic support. These problems can be described by the following dimensionless equations:

$$\begin{aligned} \Delta w - w^p &= q, \\ \Delta^2 w + w^p &= q, \end{aligned} \tag{3.8}$$

  where w is a deflection, q is an external transverse load, and Δ is the two-dimensional Laplace operator. A number p can describe an elastic support nonlinearity of power-law type. The value $p = 1$ applies to the Winkler foundation. Obviously, Eqs. (3.8) could be easily generalized for nonlinear plates and shells.
- Another example relates to Korteweg or Cahn–Hilliard fluids, also known as strain gradient fluids (see [9, 26]). The statics of these fluids can be described through the equation

$$\Delta\rho - \frac{\partial W(\rho)}{\partial\rho} + \Omega = 0, \tag{3.9}$$

  where ρ is a mass density, W is a function of ρ, and Ω is a given potential of mass forces [25]. For W a polynomial (e.g., cubic) approximation is used. Note that in (3.9), Δ is the 3D Laplace operator. This model bears a close relation to the recently proposed dilatational strain gradient elasticity [8].
- Within the dilatational strain gradient elasticity under small deformations [7, 8, 14] there exists a strain energy density U defined as follows

$$U = U(\boldsymbol{\varepsilon}, \nabla\nabla\cdot\mathbf{u}), \qquad \boldsymbol{\varepsilon} = \frac{1}{2}\left[\nabla\mathbf{u} + (\nabla\mathbf{u})^{\mathrm{T}}\right], \tag{3.10}$$

where $\mathbf{u} = (u_1, u_2, u_3)$ is a displacement vector and ∇ is the 3D nabla operator [6]. As an example of an isotropic material, the following dependence could be proposed

$$U = \frac{\alpha}{2}\mathrm{dev}\boldsymbol{\varepsilon} : \boldsymbol{\varepsilon} + \frac{\beta}{2m}(\nabla\cdot\mathbf{u})^m + \frac{\gamma}{2}|\nabla\nabla\cdot\mathbf{u}|^2, \tag{3.11}$$

where α, β, and γ are elastic moduli, m is an exponent, and dev is the deviatoric part of a second-order tensor [6]. Equation (3.11) has similar to (3.9) power-law nonlinearity with respect to the dilatation $e = \nabla\cdot\mathbf{u}$. In order to analyze the well-posedness of the corresponding boundary-value problem one can use Ladyzhenskaya's inequality comparing norms of e^m and ∇e.
- Strain gradient fluids provide examples of so-called strain gradient continua. Within this model there exists a strain energy density dependent on the first- and higher-order gradients of the displacements. Recently the model was used to describe beam-lattice metamaterials; see [5]. For such materials one can also expect power-law type nonlinearities as in the case of plates and shells. Ladyzhenskaya's inequality could be useful in studying the well-posedness of these models.

In general, Ladyzhenskaya's inequality could prove useful in any nonlinear problem where a nonlinearity is described by a power-law-type dependence involving an unknown function u where higher-order derivatives of u constitute linear terms in the governing equations. In other words it could be useful for quasilinear systems. Despite the apparent simplicity of these problems, let us note the complexity of the Navier–Stokes equations, whose nonlinearity seems to be rather simple but produces many interesting phenomena such as turbulence, chaos, cascades of instabilities, and so forth.

3.6 Conclusions

The inequalities discussed in this article were published by their originator O.A. Ladyzhenskaya in 1958 and included in her classic book [11]. Olga Ladyzhenskaya belongs to the St. Petersburg school of mechanics and mathematics originated by the famous Leonhard Euler and later by Pafnuty Chebyshev. A characteristic of many Russian mathematicians has been the application of pure mathematics to problems of mechanics and physics. Ladyszenskaya is known not only for her achievements in mathematics, notably in the theory of linear and nonlinear partial differential equations, but also for her results in hydromechanics. Nikita F. Morozov, professor and member of the Russian Academy of Sciences, belongs to the same school and represents an outstanding example of a Russian scientist in the field of mathematical elasticity. For a summary of his most important results, see the following references [16, 21, 22, 23, 24].

References

1. R.A. Adams and J.J.F. Fournier. *Sobolev Spaces*, volume 140 of *Pure and Applied Mathematics*. Academic Press, Amsterdam, 2nd edition, 2003.
2. P.G. Ciarlet. *Mathematical Elasticity. Vol. I: Three-Dimensional Elasticity*. North-Holland, Amsterdam, 1988.
3. P.G. Ciarlet. *Mathematical Elasticity. Vol. III. Theory of Shells*. Elsevier, Amsterdam, 2000.
4. M. Cloud, B. Drachman, and L. Lebedev. *Inequalites: With Applications to Engineering*. Springer, New York, 2013.
5. F. dell'Isola and D.J. Steigmann. *Discrete and Continuum Models for Complex Metamaterials*. Cambridge University Press, Cambridge, 2020.
6. V.A. Eremeyev, M.J. Cloud, L.P. Lebedev. *Applications of Tensor Analysis in Continuum Mechanics*. World Scientific, New Jersey, 2018.
7. V.A. Eremeyev, S.A. Lurie, Y.O. Solyaev, F. dell'Isola. On the well posedness of static boundary value problem within the linear dilatational strain gradient elasticity. *Zeitschrift für angewandte Mathematik und Physik*, 71(6):1–16, 2020.
8. V.A. Eremeyev, A. Cazzani and F. dell'Isola. On nonlinear dilatational strain gradient elasticity. *Continuum Mechanics and Thermodynamics*, 33(4):1429–1463, 2021.
9. D.J. Korteweg. Sur la forme que prennent les équations des mouvements des fluides si l'on tient compte des forces capillaires par des variations de densité. *Archives Néerlandaises des sciences exactes et naturelles*, Sér. II(6):1–24, 1901.
10. O.A. Ladyzhenskaya. *The Boundary Value Problems of Mathematical Physics*, volume 49 of *Applied Mathematical Sciences*. Springer, New York, 1985.
11. O.A. Ladyzhenskaya. *The Mathematical Theory of Viscous Incompressible Flow*. Gordon and Breach, New York, 1963.
12. L.P. Lebedev, I.I. Vorovich, and G.M.L. Gladwell. *Functional Analysis. Applications in Mechanics and Inverse Problems*, volume 100 of *Solid Mechanics and Its Applications*. Kluwer, New York, 2nd edition, 2002.
13. J.L. Lions and E. Magenes. *Non-Homogeneous Boundary Value Problems and Applications. Vol. 1*. Springer, Berlin, 1972.
14. S.A. Lurie, A.L. Kalamkarov, Y.O. Solyaev, A.V. Volkov. Dilatation gradient elasticity theory. *European Journal of Mechanics – A/Solids* 88:104,258, 2021.
15. V. Maz'ya. *Sobolev Spaces: with Applications to Elliptic Partial Differential Equations*, volume 342 of *Grundlehren der mathematischen Wissenschaften*. Springer, Berlin, 2nd edition, 2011.
16. S.G. Mikhlin, N.F. Morozov, and M.V. Paukshto. *Integral Equations of the Elasticity Theory*. Nauka, Stuttgart, 1995.
17. N.F. Morozov. On the non-linear theory of thin plates (in Russ.). *Doklady Akad. Nauk SSSR*, 114:968–971, 1957.
18. N.F. Morozov. The uniqueness of the symmetrical solution in the problem of the large deflection of a symmetrically loaded circular plate. *Soviet Physics Doklady*, 3:1275–1278, 1958.
19. N.F. Morozov. On the existence of a non-symmetric solution in the problem of large deflections of a circular plate with a symmetric load (in Russ.). *Izv. Vysh. Uchebn. Zaved. Matematika*, 21:126–129, 1961.
20. N.F. Morozov. The existence of a smooth solution to the problem of the non linear vibrations of a thin plate (in Russ.). *USSR Computational Mathematics and Mathematical Physics*, 6(4):228—232, 1966.
21. N.F. Morozov. *Selected Problems of Two-dimensional Theory of Elasticity (in Russ.)*. Leningrad University, Leningrad, 1978.
22. N.F. Morozov. *Mathematical Question of the Thory of Cracks (in Russ.)*. Nauka, Moscow, 1984.
23. N.F. Morozov and M.V. Paukshto. *Discrete and Hybrid Modelas of Fracture (in Russ.)*. St. Petersburg University, St. Petersburg, 1984.
24. N.F. Morozov and Yu.V. Petrov. *Dynamics of Fracture*. Springer, Berlin, 2000.

25. G. Rosi, I. Giorgio, and V.A Eremeyev. Propagation of linear compression waves through plane interfacial layers and mass adsorption in second gradient fluids. *ZAMM - Journal of Applied Mathematics and Mechanics / Zeitschrift für Angewandte Mathematik und Mechanik*, 93(12):914–927, 2013.
26. P. Seppecher. *Les Fluides de Cahn-Hilliard*. Mémoire d'habilitation à diriger des recherches, Université du Sud Toulon, 1996.
27. I.I. Vorovich. *Nonlinear Theory of Shallow Shells*, volume 133 of *Applied Mathematical Sciences*. Springer, New York, 1999.

Chapter 4
Mechanical Behaviour of Nonwovens: Continuous Approach with Parametric Finite-element Modelling

Vincenzo Cucumazzo and Vadim V. Silberschmidt

Abstract In the last decade, various numerical models, continuous and discontinuous, were proposed in the literature with an aim to simulate deformation and damage behaviour of nonwoven materials. However, these models offer only partial solutions as they are mostly limited to specific problems and neglect important design parameters. Due to their broad use, hot-calendered nonwovens were considered in this study. These materials have a complex heterogeneous microstructure that comprises three mechanically distinct domains – a fibrous matrix, bond areas and interface regions. Variability in design parameters such as type of polymer, fabric planar density, type of fibre, orientation distribution of fibres and bond pattern makes it difficult to develop a universal model that can accommodate them. So, a parametric numerical tool (*FabricGEN*) was developed to model and simulate deformation and damage behaviours of calendered nonwovens. Various experimental techniques were employed to assess the deformation, damage and failure mechanisms as well as the microstructural properties of the selected calendered fabrics. The numerical framework was based on the experimental observations using a continuous modelling approach. The numerical tool enabled to compute effective elastic-plastic properties of fabric domains along the fabric's three principal directions, based on the fabric's microstructure. The numerical models demonstrated good predictive capabilities in simulating deformation and damage behaviours under tensile loading of medium- and high-density fabrics. Finally, a parametric study was conducted to gain insight into the effects design parameters had on the mechanical response of the fabrics.

Key words: Calendered nonwovens, Deformation and damage, Parametric modelling, Finite-element analysis

Vincenzo Cucumazzo · Vadim V. Silberschmidt
Mechanics of Advanced Materials Research Group, Wolfson School of Mechanical, Electrical and Manufacturing Engineering, Loughborough University, Loughborough, Leicestershire LE11 3TU, UK,
e-mail: v.cucumazzo@gmail.com,v.silberschmidt@lboro.ac.uk

H. Altenbach et al. (eds.), *Advances in Solid and Fracture Mechanics*,
Advanced Structured Materials 180,
https://doi.org/10.1007/978-3-031-18393-5_4

4.1 Introduction

It may be argued whether or not *nonwoven* materials can be traced back to ancient times. If felt is considered a nonwoven material, it would have been the case. Indeed, felt is known as the oldest textile material, first used around 6000 B.C. [1]. However, Millman et al. [2] reported that the term *nonwoven* can be found in a British patent issued in 1853; further confirmation was also provided by Tanchis [3]. Today, nonwovens have a well-defined identity. They are engineered materials made of fibres that are randomly-oriented or aligned in preferential directions, consolidated with thermal, mechanical or chemical bonding techniques [4]. The definition excludes paper, woven, knitted, tufted or felted products. Due to their high versatility as well as cheaper and faster manufacturing processes (compared to woven materials), nonwovens have experienced a rapid growing demand in various sectors. They cover a large spectrum of applications [5]. According to EDANA [4], the nonwoven market is largely dominated by the hygiene sector, followed by the personal-care and construction ones. At the current stage of technological advancement, the strongest materials are manufactured as structures made of fibres with a high aspect ratio and small diameter (in the range of nm to μm). Nonwoven manufacturing companies also have a great interest to produce lighter products with high resistance.

Mechanical properties and, therefore, mechanical performance of nonwoven fabrics is strongly influenced by two main factors. The first one is the raw material used to manufacture their constituent fibres. In fact, the material, structure and composition of fibres affect the dimensions, structure and properties of the final fabric. The second one is the manufacturing process employed to produce them, with spunbonding being the most widely used one (Fig. 4.1a). Spun-bonding is a one-step process starting from the raw polymer (or a blend of polymers), in the form of granules/chips, to the final product [6]. The melted polymer is extruded through tiny nozzles (10–25 μm in diameter) and then through a stretching channel to increase the degree of crystallinity [7]. The produced filaments are laid onto a conveyor belt and partially re-oriented along a preferential direction called *machine direction* – MD.

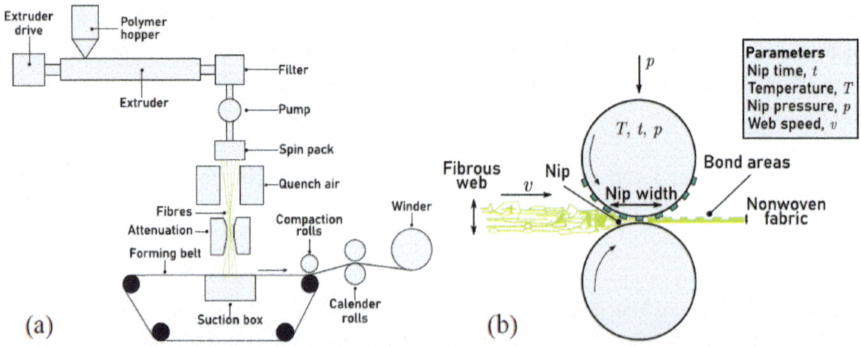

Fig. 4.1 Spun-bonding process (a) and hot-calendering bonding (b) [8]

At this stage, the resulting fibrous structure has poor mechanical strength as it is composed of loose filaments. Enhancement of the mechanical properties of the web is obtained through the bonding process, such as *hot calendering* (Fig. 4.1b). In the process, the fibrous mat passes through two rolls (*calenders*), one or both heated up to about the melting temperature of the polymer used, termed *calendering temperature*. One roll has a customised embossed bonding pattern, with raised parts having the desired shape (e.g., diamond, elliptical, rectangular, etc.), size and pattern, which is pressed at a certain pressure against the loose and disordered fibrous mat. The resulting nonwoven fabric – *calendered nonwoven* – is a heterogeneous material, composed of three mechanically and microstructurally distinct regions: a *fibrous matrix* (FM), *bond areas* (BAs) and *interface regions* (IRs). The first is a set of fibres partially oriented along MD. This allows to define two additional orthogonal directions in the three-dimensional space; that is, *cross* and *through-thickness direction* – CD and TD, respectively. The bond areas, formed by the pressed raised parts of the roll, are strong and well-consolidated regions, made of highly-compressed fibres, impart strength to the fabric. Finally, the interface regions are transitional zones between the bond areas and the fibrous matrix, which are well known in the literature for being potential locations of damage initiation [9, 10, 11, 12].

Nonwovens are engineered to provide specific functions to ensure their fitness-for-purpose, with a life span ranging from seconds to years, depending on the application. It is clear that the development of a numerical tool can enable to predict the mechanical behaviour of these materials, while accounting for a number of design parameters, would streamline the design and verification stages of the product development process.

With this aim, this chapter is structured as follows. Initially, the background for modelling of mechanics of nonwovens using a continuous approach is provided. Then, the proposed parametric finite-element (FE) modelling strategy is explained, with the developed numerical scheme and tool implemented based on experimental observations and findings. Information about micro-structure and design variables of the selected analysed nonwoven fabrics is provided. Computation and assessment of effective local material properties is carried out. A round-up of case studies, each with a specific FE model, is presented and discussed to investigate the effects of design variables on mechanical response of fabrics. These include the loading direction, planar density, fabric size and bond pattern. Finally, the results obtained from FE simulations are compared with experimental data to evaluate the reliability and efficiency of the proposed numerical modelling approach.

4.2 Background

Before the advent of computers, the study of mechanical behaviour of fibrous networks in the continuous domain relied entirely on the development of analytical formulations. General analytical solutions were obtained to predict the mechanical behaviour of generic fibrous networks, with a major focus on paper.

Pioneering work was conducted by Cox [13], who developed the first analytical set of equations to describe the deformation behaviour of fibrous networks based on the theory of elasticity. Cox presented a network model, in which fibres were assumed to extend from one side of the network to the other and carry only an axial load, neglecting their mutual interaction. The concept of *orientation distribution of fibres* was introduced for the first time. Based on this notion, the elastic stiffnesses of the fibrous mat were derived. From the knowledge of the force acting on the fibre, the fabric's stiffness matrix was determined.

The first example of the application of continuum mechanics to nonwovens goes back to Backer and Petterson [14], who developed the first continuum model for nonwovens based on the theory of orthotropic elasticity in fibrous materials (Cox's model). Interest in fibrous materials proceeded with the development of discrete numerical models, while the use of a continuous approach remained rather stagnant.

Curiskis and Carnaby [15] questioned the extent, to which the *continuum theory* was applicable to fibrous networks. They treated bundles of fibres as homogeneous continuum medium, assuming them as an untwisted assembly of fibres, exhibiting a preferential direction. The authors highlighted that the limitations associated with any continuum idealisation were dictated by the statistical reproducibility of the initial fibre geometry and associated packing arrangements as well as strain gradients.

A similar approach was used by Bais-Singh and Goswami [16] to predict the biaxial tensile behaviour of spun-bond nonwovens based on the mechanical behaviour of its constituent fibres, in particular, their orientation distribution. Analytical solutions were obtained employing the theory of laminated composites. The nonwoven fabric was idealised as a number of layers of fibres. Fibres within one layer were assumed to be straight and oriented along a fixed direction. Moreover, they were capable of carrying only the axial load. The elastic response was fairly captured both in MD and CD. To the best of our knowledge, continuum models tend to provide more accurate predictions along directions with higher stiffness (e.g., MD) compared to their counterparts (e.g., CD).

Development of analytical solutions progressed, moving from 2D to 3D fabrics. Narter et al. [17] extended Cox's model to 3D anisotropic fibrous fabrics (applicable to through-air bonded nonwovens). The resulting simulations were able to predict the elastic moduli in various directions of the fibrous webs and 15 spherical harmonic coefficients of the *orientation distribution function* (ODF). Furthermore, a direct dependence between the Young's moduli, E_i, and the *fibre orientation distribution density* (FODD) along the same axis was found. Some assumptions were made in this framework, i.e., the bonds were considered as rigid.

Kim and Pourdeyhimi [18] adopted a unit-cell approach to study structural parameters such as the ODF of fibres, the bond area strain and shear deformation of the unit cell, occurring during controlled-displacement tests. The role of the bonding temperature on the fabric's deformation behaviour was assessed in this study.

Rawal [19], based on the work of Adanur and Liao [20], proposed a theoretical micromechanical model for predicting the in-plane tensile mechanical response of thermally-bonded and needle-punched nonwovens, based on the geometric and mechanical properties of their constituent fibres. Fibre orientation and curliness were incorporated into the model and investigated. In the case of thermally-bonded nonwovens, bonds were modelled as rigid links between fibres.

Later, the author extended his work to through-air bonded nonwovens, proposing a detailed analytical study on the extension, compression, bending and shear responses of fabrics [21]. In this work, a realistic ODF was obtained using digital image processing. In a more recent study, the author analysed the Poisson's effect in needle-punched nonwovens [22].

Hou et al. [23] developed one of the first 2D-continuous FE models of a low-density (20 g/m^2) calendered nonwoven to predict its deformation behaviour under uniaxial tension. The classical composite-based theory was employed to generate the continuous FE model at macroscopic level. The material was modelled as orthotropic. Due to difficulties in obtaining the local mechanical properties of the bond area, their stiffness was assumed to be three times that of the fibrous matrix, taking the stiffness in TD identical to that in CD.

In a similar fashion, Demirci et al. [24] developed a continuous shell-based FE model to simulate the tensile response of various calendered nonwovens composed of (core/sheath) bi-component fibres. The numerical model attempted to predict the viscous-elastic-plastic behaviour of the material undergoing in-plane uniaxial-tensile loading first, and, later, out-of-plane loading [25]. A fair agreement was observed between the numerical and experimental results, although the model was not able to predict damage and failure behaviour due to the absence of a failure criterion. The numerical scheme also required the use of two separated pieces of software (MATLAB® and MSC Marc®) and could not generate automatic simulations in the same way this research aims to.

Silberstein et al. [26] developed a constitutive material model capable of capturing the elastic-plastic behaviour of electrospun nonwovens, stretched uniaxially under monotonic and cyclic loadings. The constitutive model captured the membrane stress-strain behaviour as a function of the properties of fibres and fabric. The behaviour of a single fibre was incorporated into the macroscopic membrane model of the fabric. The effect of the fibre alignment and bending as well as network's consolidation was also modelled. The proposed model was idealised as a layered network using an RVE approach. The FE model was composed of several RVEs, each characterised by a different initial orientation. The elastic-plastic response of the material, including loading and unloading, was predicted with a fair agreement with the experimental data. However, the model could not predict damage and failure behaviour of the nonwoven fabric. Additionally, the size effect was not investigated (the maximum fabric width was 3 mm).

A constitutive material model was developed by Ridruejo et al. [27] to account for deformation, damage and failure behaviour of nonwoven fabrics. The model provided the constitutive response of the fabric in the meso-domain and could account for damage localisation. The constitutive behaviour of the nonwoven fabric was expressed in terms of fibre volume fraction, stress carried by the fibres as a function of the fibre stretch and fraction of fibres forming an angle with the reference direction. Additionally, an "orientation index", quantifying the degree of orientation of fibres within the fabric, was also included in the model. The damage model was implemented in such a way that the load carried by damaged fibres could increase at later stages of the deformation process, accounting for the re-loading of fibres once they are aligned to the loading direction. A Weibull probability function was used to account for the stochastic nature of damage in bonds. The model reproduced rather accurately the main deformation and fracture micromechanisms.

This numerical scheme was updated to implement a constitutive material model for simulation of mechanical behaviour of thick needle-punched nonwovens [28]. The model incorporated non-affine deformation effects, anisotropic connectivity induced by entanglement points, fibre uncurling and re-orientation as well as fibre disentanglement and pull-out from the knots. Within the limits of a continuous model, re-orientation of fibres, strain/stress non-uniformity and softening mechanisms were accurately captured from the simulations. Numerical simulations showed very good agreement with the experimental results.

More recent works carried out by the same authors focused on the ballistic performance of hybrid laminates, made of a front layer of needle-punched nonwoven and several layers of woven fabric. First, they demonstrated that the hybrid laminate exhibited enhanced mechanical properties compared to those of the individual material [29]. Second, the previously developed numerical model could also account for high-strain-rate and out-of-plane loading. The obtained numerical results, expressed in terms of residual vs. initial velocity of the spherical projectile, were in line with the experimental observations.

The same model and approach were employed to simulate the impact behaviour of a hybrid metal/nonwoven fabric [30]. Steel plates were modelled with a standard elastic-plastic constitutive law at macroscale. The numerical model demonstrated high predictive capabilities, also in this context. However, in their extensive work, the authors did not provide any application to low-density fabrics, which would be of great interest.

Damage and failure behaviours of nonwovens, specifically through-air bonded ones, was also studied by Chen and Silberstein [31]. In their work, a micromechanical damage model was presented based on the work conducted by Ridruejo et al. [27]. The model was built by linking local damage events (bond failure) to macroscopic behaviour of the fabric. In this model, a non-linear term was introduced to describe the non-affine deformation of fibres at the bond level. A bond was allowed to fail when its traction load exceeded a critical value. This local information was used to update the global damage state employing a classical continuum-damage-mechanics framework. The model could predict rather accurately the global mechanical response of nonwoven fabrics with various planar densities.

Raina and Linder [32] developed a method starting from a modified affine network model and incorporating fibre undulation and re-orientation phenomena typical for nonwoven materials. The developed model was able to capture a non-linear and anisotropic response of the fabric, observed experimentally. The fibres' unfolding mechanism was captured by modifying the linear elastic response of individual fibres in the lower stretch regime. Re-orientation of fibres, instead, was modelled as a rigid-body rotation without producing any strain under the influence of external loadings.

A developed homogenisation technique was later combined with an *enhanced deformation gradient* to model failure phenomena in nonwoven materials with a random network microstructure at finite deformations [33]. This allowed capturing the anisotropic and nonlinear bulk material responses with propagating cracks in the failing nonwoven at finite strains. Failure, in the form of cracks, was incorporated locally as displacement jumps, called *strong discontinuities*.

A stochastic approach was used to develop a constitutive model for predicting the random mechanical response of isotropic thin fibre networks of arbitrary size [34]. The model was based on *stochastic volume elements* (SVEs) with the introduction of stochastic constitutive-law parameters. An empirical relationship between the equivalent stress and strain was used to describe the mechanical response of each SVE forming the fibrous network. The softening part of the SVE's mechanical response was modelled with a quadratic expression as a function of the equivalent and failure stresses and strains. The authors stated that the proposed continuous model significantly reduced the computational cost compared to a discrete one. Simulation of a 24 mm × 24 mm continuous fabric took 2 min on a modern 128 GB RAM supercomputer, while a corresponding discontinuous fibrous network took 2 days [34]. This conclusion confirms that although the discrete modelling approach allows a detailed capture of complex mechanical behaviours of fabrics, it makes product development prohibitive due to the overwhelming computational costs required to simulate and design products of larger sizes.

Considering a period of 70 years, it can be asserted that few continuous numerical models are reported in the literature, and most of them do not account for damage and failure behaviour. This is particularly true for hot-calendered nonwovens. Furthermore, the traditional modelling approach used could not account for direct changes in design parameters (material or geometric) as a parametric one would do. The state-of-the-art offers only partial solutions to specific problems without providing a more universal model.

This situation paves the way for the development and implementation of a parametric numerical scheme and tool that would enable prediction of mechanical behaviour of calendered nonwoven materials based on specified material and geometric characteristics. The tool should allow designers and manufacturers of nonwovens to design and assess a novel nonwoven fabric with minimal effort and cost, by-passing experimental characterisation, which, sometimes, can be impractical and insufficient for a comprehensive understanding of complex deformation and damage phenomena.

4.3 Parametric Finite-element Modelling Strategy

A numerical scheme was developed to model, simulate and predict the deformation and damage behaviours of mono- and bi-component fibre calendered nonwovens. In order to be able to model different types of fabrics with various design variables (geometric and material properties), the formulation of a parametric approach was necessary. A stand-alone software, named *FabricGEN*, was developed and interfaced through a graphical user interface (GUI) to the FE software MSC Marc®, which functioned as solver. Implementation of the algorithm was performed using an object-oriented programming approach in Python™ environment. *FabricGEN* was interfaced to the FE environment via specific *application programming interface* (API) functions suitably created. The parametric modelling strategy adopted is illustrated in Fig. 4.2.

FabricGEN constitutes the core of the numerical scheme and is used to generate and transfer inputs to the FE package (Table 4.1). Gaining knowledge of local mechanical properties of heterogeneous materials, such as calendered nonwovens, is challenging and, in many cases, not viable. Therefore, the numerical tool was developed to compute local mechanical properties of fibrous matrix, bond areas and interface regions based on some properties (physical, material and geometric) of fibres and fabric that can be either found in the literature, obtained experimentally or from manufacturers. Generation of the FE model and simulation of a specified calendered fabric necessitate the knowledge of geometric properties of each domain, boundary conditions and selection of a suitable mesh. The result of this process is a shell-based FE model simulating a calendered nonwoven under uniaxial tension. The model is able to predict the mechanical performance of the specified fabric and provide information (e.g., distribution of stresses, strains and damage) not accessi-

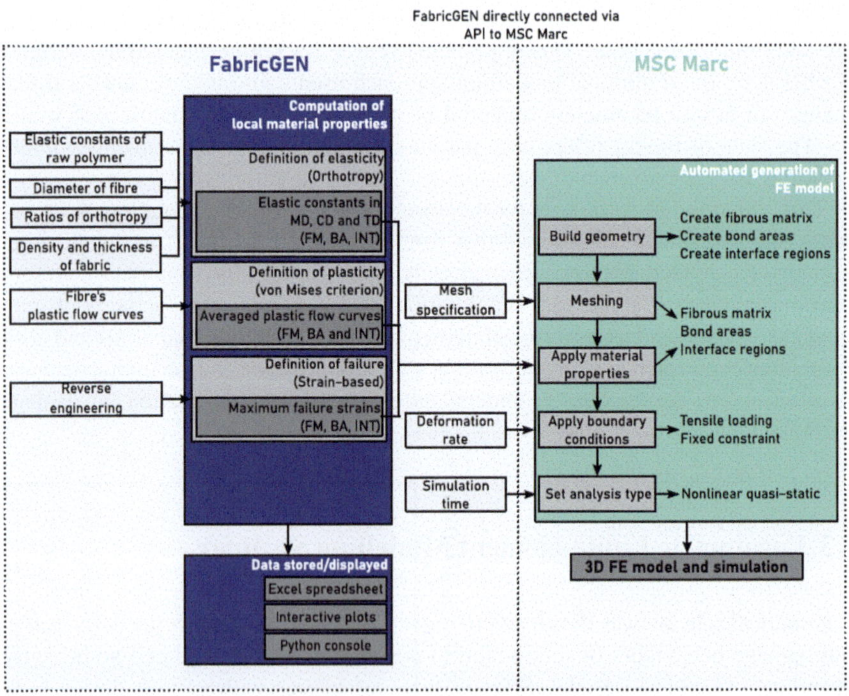

Fig. 4.2 Parametric modelling scheme [8]

Table 4.1 Inputs and outputs of numerical scheme

Input (*FabricGEN*)			Output (*FabricGEN*)	Output (MSC Marc®)
Fabric and fibre properties			Elastic-plastic properties of fibrous matrix, bond areas and interface regions	Elastic-plastic response up to failure
Fabric planar density, γ_{fab}	Sheath material in fibre, $s_{\%}$		3D-shell-based FE model	Stress and strain distributions
Ratios of orthotropy, ζ_{MD} and ζ_{CD}	Fibre diameter, $\overline{d}_{\text{fib}}$			Damage distribution
Elastic properties				
Fibre density, ρ_{core} and ρ_{sheath}	Young's modulus, E_{core} and E_{sheath}			
Poisson's ratio, ν_{core} and ν_{sheath}	Shear modulus, G_{core} and G_{sheath}			
Elastic-plastic properties of fibre				
Young's modulus of fibre, E_{fib}		Plastic flow curve of fibre, σ_{fib}		
Geometric properties				
Fibrous matrix	*Bond pattern*	*Interface region*		
Width, W	Shape	Thickness, t_{INT}		
Height, H	Bond width, a	Centre x-coordinate, x_C		
Thickness, t_{FM}	Bond height, b	Centre y-coordinate, y_C		
	Thickness, t_{BA}	Width, c_x		
	Orientation, θ	Height, c_y		
	Horizontal bond spacing, s_x			
	Vertical bond spacing, s_y			
	Bond mutual angle, $\Delta\theta$			
Meshing				
Fibrous matrix	*Bond pattern*	*Interface region*		
FE size and type				
Simulation parameters				
Deformation rate, $\dot{u}$	Simulation time, t	Loading direction, ϑ		

ble through experimentation. All the parameters are explained and illustrated in the next sections.

FabricGEN is composed of two sections, one dedicated to material properties and another to model and simulation parameters (see Figs. 4.3a and 4.3b, respectively). In the first section, fibre and fabric properties of a specific nonwoven are required. Elastic constants and physical properties of the polymeric materials, used in core and sheath parts of bi-component fibres, are also a necessary input of the numerical scheme. The second section deals with geometric and simulation parameters required to generate the FE model and simulation. The size of the fabric, bond

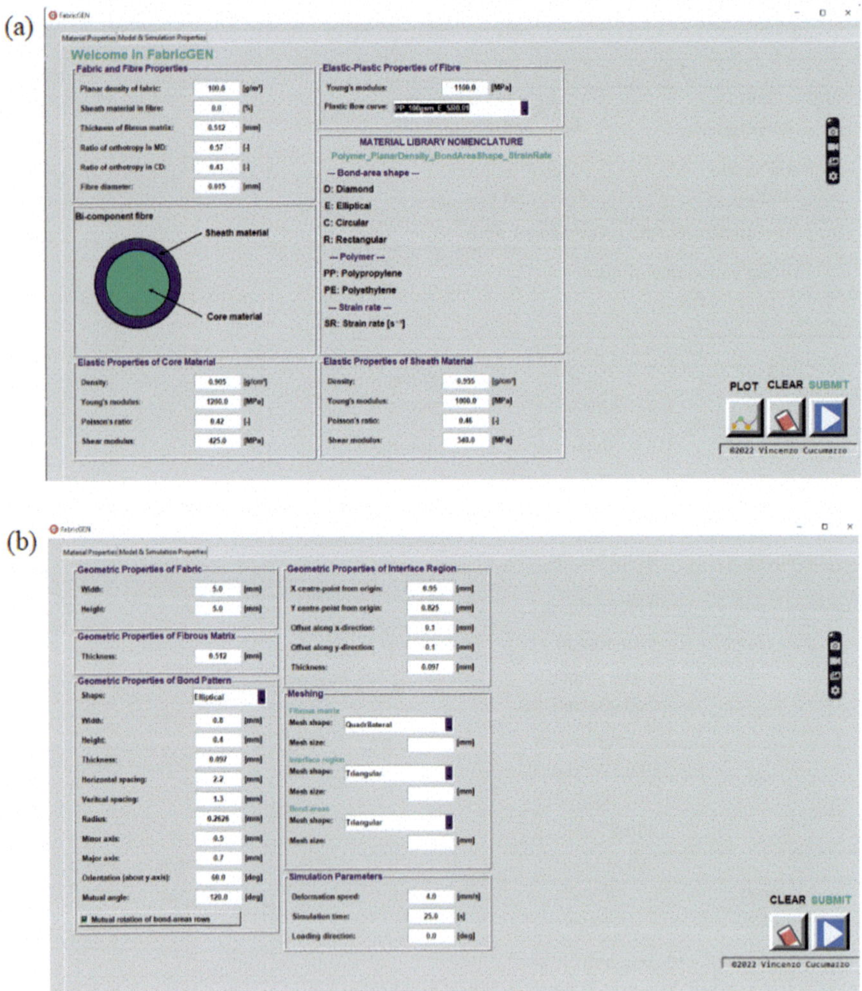

Fig. 4.3 Sections of *FabricGEN*: (a) materials properties; (b) model and simulation properties [8]

area pattern (bond area shape, size, orientation, spacing in x and y direction and orientation angle between two consecutive rows of bonds) as well as the interface region's width and thickness of each domain are used to create the model. Finally, information about meshing, deformation rate (according to the fabric length and the strain rate of the plastic flow curve chosen) and loading direction must be provided.

4.4 Multiscale Experimental Characterisation

4.4.1 Calendered Fabrics

Due to their popularity, the nonwoven materials considered in this work were produced with spun-bonding and bonded with the hot-calendering technique. Variability in material and structural parameters such as polymeric resin, fabric's planar density, type and orientation distribution of fibres as well as bond pattern were considered. With this in mind, three types of calendered nonwoven fabrics were selected. The low-density fabrics, with a planar density of 25 g/m^2, was manufactured with bi-component core-sheath fibres made of 70% polypropylene (PP) and 30% high-density polyethylene (HDPE). The medium- and high-density ones, with a planar density of 40 and 100 g/m^2, respectively, were produced with mono-component fibres made of 100% PP. The 100 g/m^2 fabric was a bi-layer nonwoven. Based on their planar density, the selected fabrics are designated as *F25*, *F40* and *F100* throughout this study. The properties of fabrics and fibres of the nonwovens studied are reported in Table 4.2.

4.4.2 Experimental Methods

Development of a numerical model to simulate and predict mechanical behaviour of calendered fabrics requires information about their microstructural and mechanical characteristics. A number of experimental methods were employed to characterise the selected nonwoven fabrics at various scales. Each method provided specific outputs, which not only served as inputs for the developed numerical scheme but also helped improve and validate the obtained numerical models (Tables 4.3 and 4.4). Detailed analysis and results can be found in [35].

Microstructural characterisation provided fundamental input parameters of the selected fabrics. These are reported in Table 4.5 and used to generate FE models below. A schematic view of a generic model of calendered fabric, including geometric parameters, is illustrated in Fig. 4.4.

Table 4.2 Fabric and fibre properties of *F25*, *F40* and *F100* fabrics

	Properties of fabric and fibre			
	Fabric planar density, γ_{fab} (g/m^2)		Sheath material of fibre, $s_{\%}$ (%)	
F25 (70/30 PP/PE)	25		30	
F40 (PP)	40		0	
F100 (PP)	100		0	
	Elastic properties of polymers			
	Fibre density, ρ^{fib}_{core} and ρ^{fib}_{core} (g/cm^3)	Young's modulus, E_{core} and E_{sheath} (MPa)	Poisson's ratio, ν_{core} and ν_{sheath}	Shear modulus, G_{core} and G_{sheath} (MPa)
F25 (70/30 PP/PE)	0.905 / 0.955	1200 / 1000	0.42 / 046	425 / 340
F40 (PP)	0.905	1200	0.42	425
F100 (PP)	0.905	1200	0.42	425

4.5 Generation of Calendered-fabric Model

Considering the fabric's structure (Fig. 4.4), the development of the model begins with the generation of the fabric itself (see algorithm in Table 4.6). The fabric is represented by a rectangular region defined by its sides (straight lines) with its initial point, A, coinciding with the origin of coordinates. In this context, the fabric implicitly represents the fibrous matrix. The model lies in the *x*–*y* plane, with the width and height of the fabric defining its size. Thicknesses, instead, impart three-dimensionality to the model. Then, the bond pattern is generated as needed, including interface regions. Depending on the loading direction and, therefore, rotation of the bond pattern, the algorithm trims elements of the model outside the fabric's domain. Finally, once the three material domains are defined, boundary conditions and materials properties are suitably applied to the model. It is clear that the parametric nature of the algorithm allows the generation of various models that, for instance, have different bond patterns (Fig. 4.5).

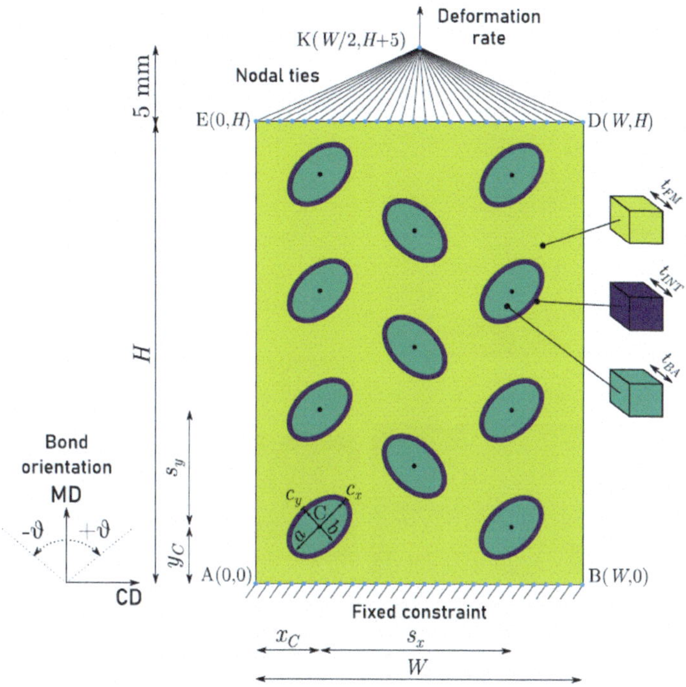

Fig. 4.4 General scheme of model with notation of geometric parameters

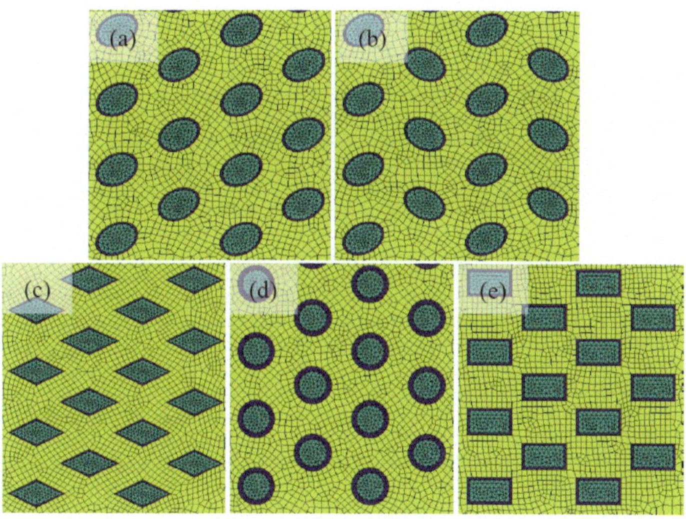

Fig. 4.5 Examples of models for various bond patterns

Table 4.3 Experimental methods: microstructural characterisation [35]

Technique	Experimental output	Numerical input
Scanning Electron Microscopy (SEM)	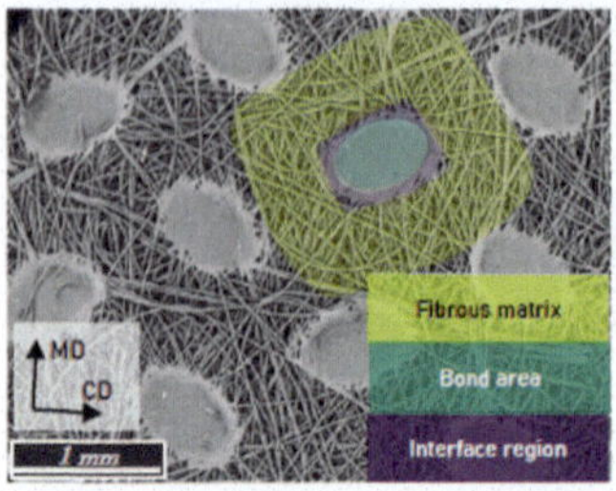	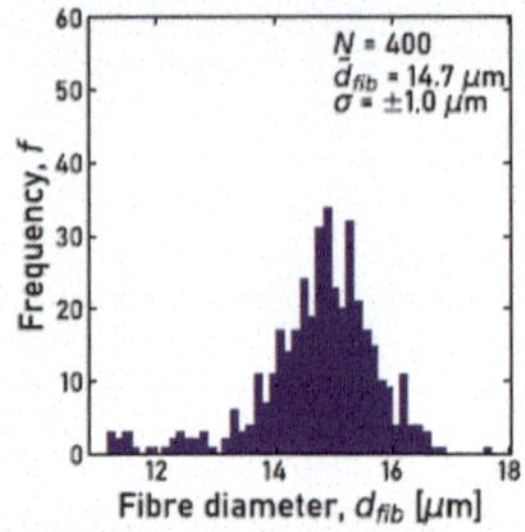 • (Mean) Fibre diameter • Bond orientation
X-ray Micro Computed Tomography (CT)		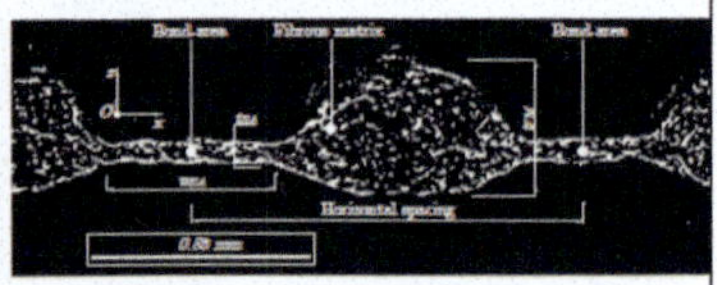 • Bond width, height and thickness • Vertical and horizontal spacing between bonds • Fibrous matrix thickness
Parametric block-based CAD algorithm	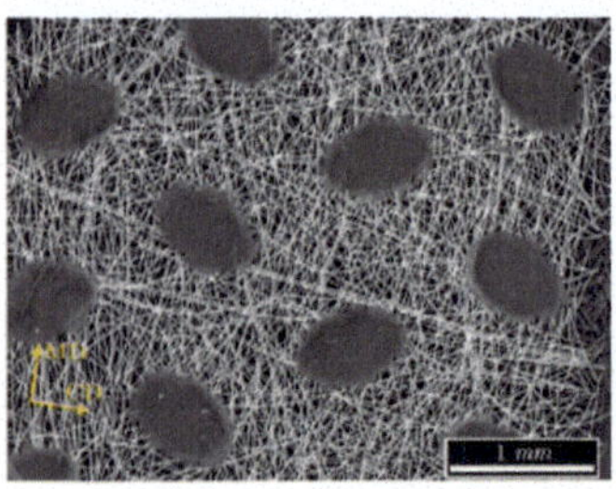	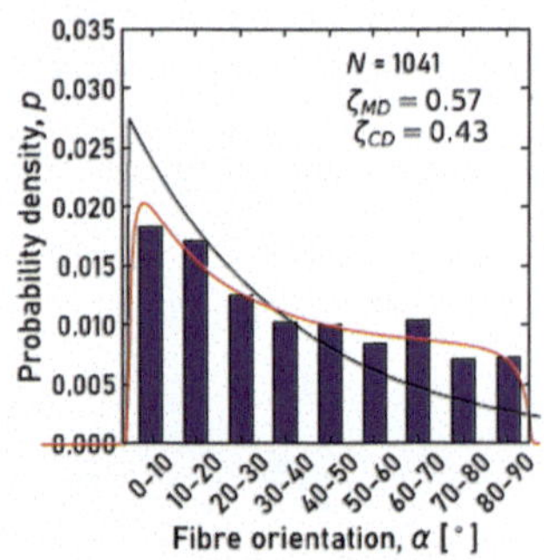 • Orientation distribution of fibres • Ratios of orthotropy in MD and CD

4.5.1 Computation of Local Mechanical Properties

As discussed, calendered nonwovens are heterogeneous materials composed of three mechanically differently domains – fibrous matrix, bond areas and interface regions. According to the experimental results, bond areas are stiffer than the fibrous matrix. Therefore, it was appropriate to assume that the material properties of the nonwoven could be described using the classic theory of mechanics of composites.

The BAs could be treated as the reinforcement of the composite, whereas the FM as its matrix. The latter is porous and compressible in TD, but stiff in MD.

Table 4.4 Experimental methods: mechanical characterisation [35]

Technique	Experimental output	Numerical input
Single-fibre tensile test	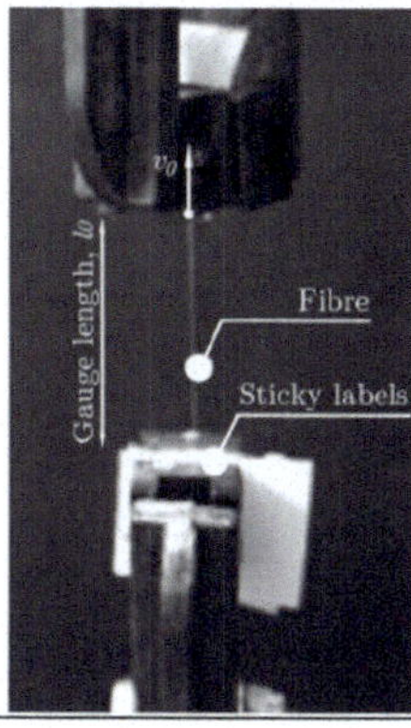	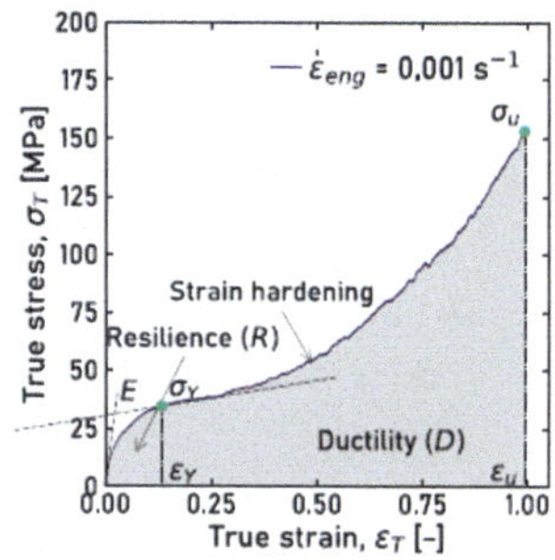 • Fibre's plastic flow curves • Fibre's Young's modulus
Fabric uniaxial tensile test	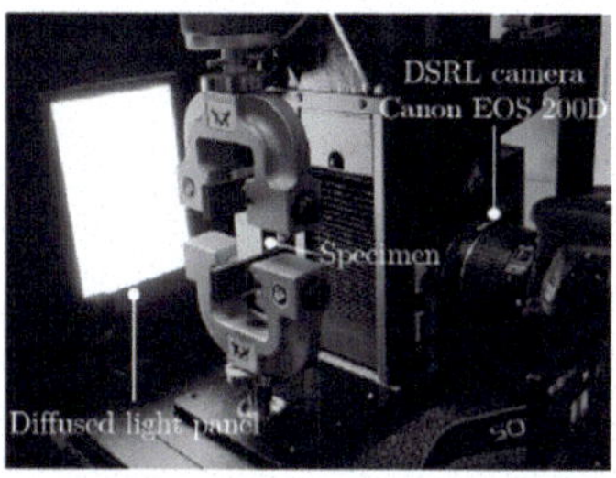	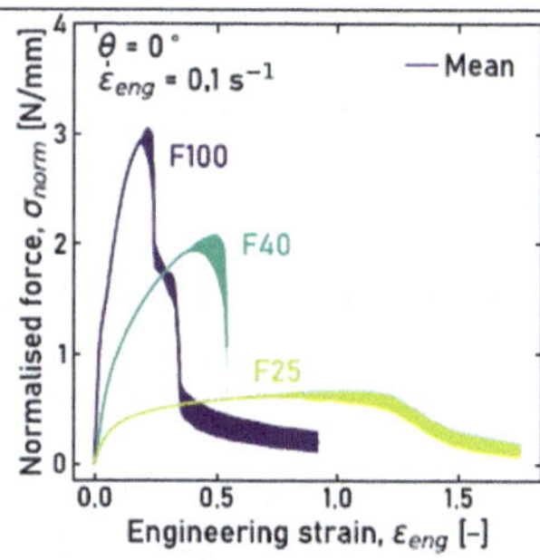 • Fabric's width and height • Fabric's Young's modulus, ultimate strength and anisotropy parameters • Deformation, damage and failure mechanisms, including effect of design parameters • Boundary conditions • Refinement and validation of numerical models

The latter are solid and continuous regions, almost incompressible. By contrast, the interface region, where most imperfections take place, exhibits an intermediate behaviour between those of the fibrous matrix and bond areas. So, the mechanical properties of the material should be computed individually for each region (domain). Besides, due to the orthotropic nature of the material, these properties are direction-dependent and, therefore, should be considered in three directions.

An attempt was made to compute the *local elastic-plastic properties* of the fibrous matrix, bond area and interface region based on the properties of the fabric and its constituent fibres. Mechanical properties were obtained under the assumption that the material was orthotropic, defined its three principal directions.

Table 4.5 Geometric properties of *F25*, *F40* and *F100* fabrics

Geometric properties	*F25*	*F40*	*F100*
W (mm)	25	25	25
H (mm)	40	40	40
t_{FM} (mm)	0.35	0.4	0.512
a (mm)	0.70	0.78	0.70
b (mm)	0.50	0.42	0.46
ϑ (°)	±60°	±60°	±60°
t_{BA} (mm)	0.023	0.056	0.097
c_x (mm)	0.05	0.05	0.05
c_y (mm)	0.05	0.05	0.05
t_{INT} (mm)	0.186	0.228	0.304
x_C (mm)	1.0	0.85	0.95
y_C (mm)	0.75	0.875	0.825
s_x (mm)	2.3	2.2	2.1
s_y (mm)	1.4	1.3	1.3
Ratio of orthotropy, ζ_{MD} and ζ_{CD}	0.63 / 0.37	0.54 / 0.46	0.57 / 0.43
Fibre diameter, d_{fib} (mm)	0.017	0.0145	0.015

After the acquisition of the parameters needed, the algorithm started with calculating the number of fibres per 1 mm^2 of the fabric area, using the orthotropic parameters and the nonwoven's planar density. Since fibres acted as trusses, and ζ_{MD} and ζ_{CD} quantify the level of orthotropy of the fabric, randomly-oriented fibres could be represented with truss elements oriented along MD and CD. In other words, the anisotropic microstructure was converted into an orthotropic one.

As observed in the experiments, the deformation behaviour of fibres and fabrics was characterised by two main stages: *elastic* and *plastic* with strain hardening. The first stage of the deformation process was *elastic*. Therefore, elastic constants of the fibrous matrix and bond areas were first calculated. The properties of the third domain, the interface region, were obtained by averaging those of the other two regions. In order to develop a continuum model, effective mechanical properties must be considered. The Young's modulus of fibrous matrix and bond areas (indicated with i) for a given loading direction, E_i^θ, was computed as a function of the Young's modulus of the fibre (E_{fib}), the number of fibres in 1 mm^2 (N_{fib}), the cross-sectional area of the fibre (A_{fib}), the orthotropy ratio in MD (ζ_{MD}), the anisotropy coefficient applied to the elastic domain (Λ_E^θ) for the specified loading direction, and the thickness of the domain considered (t_i):

$$E_i^\theta = \frac{E_{fib} N_{fib} \zeta_{MD} A_{fib} \Lambda_E^\theta}{t_i}. \tag{4.1}$$

Table 4.6 Steps for generation of model based on input parameters

Algorithm 1 Generation of fabric with elliptical bonds

Parameters	Step
1: $f = f(W, H, t_{\text{FM}})$	▷ Generate fabric
2: *BA_shape = elliptical*	▷ Select shape of bond/interface
3: $f = f(\theta)$	▷ Set bond/interface orientation (clockwise) about MD
4: *BA_mutual_angle = True*	▷ Consider 'shift angle' between two consecutive bond rows and alternately place bonds/interfaces belonging to even rows at half s_x
5: $f = f(\Delta\theta)$	▷ Set 'shift angle'
$f = f(\Delta\theta + \vartheta)$	▷ Set 'shift angle' considering loading direction
6: $f = f(x_C, y_C)$	▷ Define centre of 1st bond/interface in relation to point A
7: $f = f(a, b, t_{\text{BA}})$	▷ Define bond size
8: $f = f(a, b, c_x, c_y, t_{\text{INT}})$	▷ Expand bond in x and y directions to create interfaces based on specified offset
9: $f = f(W, H, s_x, s_y)$	▷ Create bond pattern by copying bonds/interfaces spaced in x and y directions based on specified fabric size
10:	▷ Trim bond pattern outside fabric's domain
11:	▷ Mesh fibrous matrix, bond areas and interface regions
12: $f = f(E_i, G_i, \nu_i, \sigma_i)$	▷ Assign computed material properties to each domain
13: $f = f(\dot{u}, u, t)$	▷ Apply boundary conditions (strain rate, fixed fabric's bottom edge and simulation time)

Differently, the Young's modulus of the fibrous matrix in TD is given by $E_{\text{FM}}^{\text{TD}} = E_{\text{FM}}^{\text{MD}}/R_T$, where R_T is a large number (around 120) that relates the mechanical stress in MD to that in TD. This was found by Demirci [24] from compression tests and, basically, indicates negligible compressive behaviour of the matrix in TD. Transversal elastic constants, such as the Young's modulus in TD of bond areas ($E_{\text{BA}}^{\text{TD}}$), were calculated using Halpin-Tsai equations [36]

$$E_{\text{BA}}^{\text{TD}} = \frac{1 + 2\eta_1 C_{\text{core}} E_{\text{sheath}}}{1 - \eta_1 C_{\text{core}}}, \tag{4.2}$$

where η_1 is given as

$$\eta_1 = \frac{E_{\text{core}}/E_{\text{sheath}} - 1}{E_{\text{core}}/E_{\text{sheath}} + 2}, \tag{4.3}$$

where C_{core}, C_{sheath}, E_{core} and E_{sheath} are the volumetric fractions and the Young's moduli of core and sheath materials, respectively. These formulations indicate that bonds were modelled as composite materials, in which the core material represents the fibre, and the sheath the matrix. The shear modulus in MD of the bond area, $G_{\text{BA}}^{\text{MD}}$, was calculated with a formula similar to (4.2):

$$G_{\text{BA}}^{\text{MD}} = \frac{1 + 2\eta_2 C_{\text{core}} G_{\text{sheath}}}{1 - \eta_2 C_{\text{core}}}, \tag{4.4}$$

$$\eta_2 = \frac{G_{\text{core}}/G_{\text{sheath}} - 1}{G_{\text{core}}/G_{\text{sheath}} + 1}, \tag{4.5}$$

G_{core} and G_{sheath} are the shear moduli of core and sheath materials, respectively. The shear moduli of the fibrous matrix along the direction θ were calculated assuming a linear proportionality between the Young's and shear moduli of bond areas ($E_{\text{BA}}^{\text{MD}}$ and $G_{\text{BA}}^{\text{MD}}$, respectively) and the Young's modulus of the matrix (E_{FM}^{θ})

$$G_{\text{FM}}^{\theta} = \frac{G_{\text{BA}}^{\text{MD}} E_{\text{FM}}^{\theta}}{E_{\text{BA}}^{\text{MD}}}, \tag{4.6}$$

Due to its porous nature, the fibrous matrix is characterised by low flexural stiffness, which can reasonably be assumed negligible. Therefore, the in-plane Poisson's ratio of the fibrous matrix and bond areas, $\nu_{MD/CD}$, was assumed to be equivalent to that of the fibre. This latter was computed using by applying the rule of mixtures (RoM)

$$\nu_{\text{fib}} = \nu_{\text{core}} C_{\text{core}} + \nu_{\text{sheath}} C_{\text{sheath}}, \tag{4.7}$$

where ν_{core} and ν_{sheath} are the Poisson's ratios of the core and sheath materials, respectively. The out-of-plane Poisson's ratios were assumed to be zero as a result of compressibility of the fibrous matrix.

Finally, the thickness of the bond area was obtained assuming it as solid without voids with the following relationship:

$$t_{\text{BA}} = \frac{m_{1\text{mm}^2}}{\rho_{\text{fib}}}, \tag{4.8}$$

where $m_{1\text{mm}^2}$ is the mass per 1 mm^2 of the fabric and ρ_{fib} is given by the RoM in the form $\rho_{\text{fib}} = \rho_{\text{sheath}} C_{\text{sheath}} + \rho_{\text{core}} C_{\text{core}}$. Here, ρ_{sheath} and ρ_{core} are the sheath and core densities, respectively.

The local elastic constants in MD, CD and TD for the selected nonwoven fabrics were computed (Table 4.7). From the results, it is possible to note that the bond area was characterised by high values of Young's moduli as compared to the other regions. This effect was more marked in TD, due to their high stiffness along that direction. Similarly, shear moduli were found to be rather high in the bond-area domain as it can sustain shear stresses.

In order to define the material's *plastic behaviour* in the FE environment, plastic-flow curves of the fibrous matrix, bond areas and interface regions (σ_i^{θ}) are required.

Table 4.7 Computed orthotropic elastic constants of selected fabrics

Elastic constants	Fibrous matrix			Bond area			Interface region		
	F25	*F40*	*F100*	*F25*	*F40*	*F100*	*F25*	*F40*	*F100*
E_{MD} (MPa)	29.18	46.36	100.04	375.84	420.19	463.54	202.51	233.21	563.58
E_{CD} (MPa)	16.41	39.55	75.46	211.41	357.94	349.68	113.91	198.74	212.57
E_{TD} (MPa)	0.24	0.38	0.83	568.62	1200	1200	284.43	600.19	600.41
$\nu_{\mathrm{MD,CD}}$	0.432	0.42	0.42	0.432	0.42	0.42	0.432	0.42	0.42
$\nu_{\mathrm{CD,TD}}$	0	0	0	0.432	0.42	0.42	0.21	0.21	0.21
$\nu_{\mathrm{TD,MD}}$	0	0	0	0.432	0.42	0.42	0.21	0.21	0.21
$G_{\mathrm{MD,CD}}$ (MPa)	15.42	46.96	91.72	198.67	425.00	425.00	107.04	235.98	258.36
$G_{\mathrm{CD,TD}}$ (MPa)	8.67	40.00	69.19	111.75	362.03	320.61	60.21	201.01	389.80
$G_{\mathrm{TD,MD}}$ (MPa)	0.13	0.39	0.76	300.58	1213.73	1100.23	150.35	607.06	550.50

The following expression was used to compute a *flow curve* in the direction θ:

$$\sigma_i^{\theta} = \frac{\sigma_{\mathrm{fib}} N_{\mathrm{fib}} \zeta_{\mathrm{MD}} A_{\mathrm{fib}} \Lambda_{\sigma_u}^{\theta}}{t_i}, \tag{4.9}$$

where σ_{fib} is the plastic flow stress of fibre and $\Lambda_{\sigma_u}^{\theta}$ is the anisotropy coefficient applied to plastic domain for the loading direction considered. The plastic flow stress of bond areas in TD was assumed to be linearly proportional to that in MD according to the following relation:

$$\sigma_{\mathrm{BA}}^{\mathrm{TD}} = \frac{\sigma_{\mathrm{BA}}^{\mathrm{MD}} E_{\mathrm{BA}}^{\mathrm{TD}}}{E_{\mathrm{BA}}^{\mathrm{MD}}}, \tag{4.10}$$

where $\sigma_{\mathrm{BA}}^{\mathrm{TD}}$ and $\sigma_{\mathrm{BA}}^{\mathrm{MD}}$ are the plastic flow stresses; $E_{\mathrm{BA}}^{\mathrm{MD}}$ and $E_{\mathrm{BA}}^{\mathrm{TD}}$ are the Young's moduli in MD and TD, respectively.

A material database containing plastic-flow curves for several types of fibres was implemented. For instance, the nomenclature "PP_PE_70_30_25gsm_E_SR0.1" denotes a fibre extracted from the *F25* fabric made of 70% polypropylene and 30% polyethylene with elliptical bond areas tested at a strain rate of 0.1 s^{-1} (Fig. 4.6).

Fig. 4.6 Database of plastic-flow curves of fibres

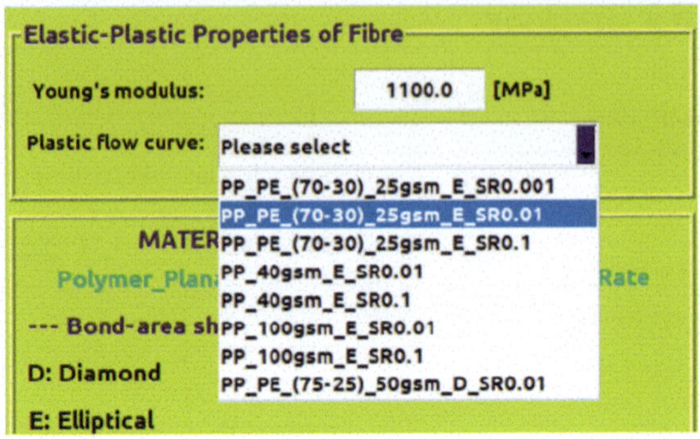

The definition of plasticity in the FE software required only one plastic-flow curve for each region of the fabric. In order to achieve this, *average flow curves* were computed using *FabricGEN* (Fig. 4.7). In this way, it was possible to assign these plastic properties directly to the domains in the FE environment. The magnitude of the curves reflects the mechanical behaviour of each material's domain observed in the experiments. It should be recalled that the plastic-flow curve of the interface region was an average between that of the fibrous matrix and bond area.

It can be noted that the level of flow stress in the fibrous matrix was significantly lower compared to that in the bond area. In particular, the stress was significantly higher along TD as the material is nearly incompressible. Poor mechanical performance was exhibited by the fibrous matrix in TD. These curves are employed to define the von Mises yield criterion in the numerical scheme.

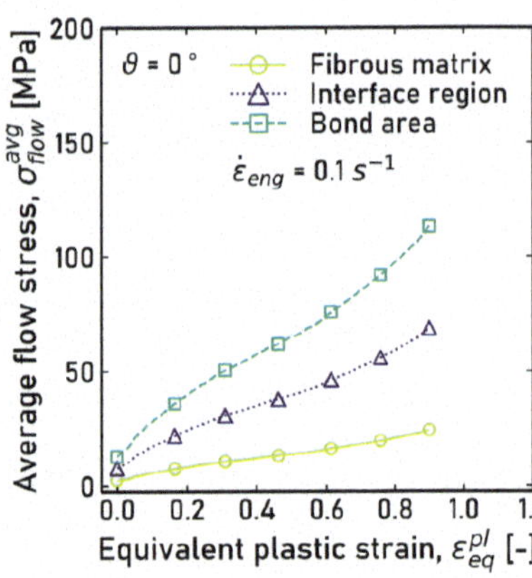

Fig. 4.7 Computed average plastic-flow curves of fibrous matrix, bond areas and interface regions for *F100* fabric at strain rate 0.1 s^{-1}

4.5.2 Meshing

Realistic simulations of complex deformation and damage behaviours of a nonwoven fabric require appropriate implementation of a FE type to reproduce mechanical characteristics of the three domains. Due to geometric (large region to discretise) and computational (reduction in number of elements) reasons, the fibrous matrix was modelled with quadrilateral shell elements (element type 139 in MSC Marc®), characterised by six nodal displacement variables. Due to their complex curvilinear shape and small surface area, bond areas and interface regions were modelled using three-node thin-triangular shell elements (element type 138 in MSC Marc®), with characteristics of element type 139.

To assess mesh independence, a mesh-sensitivity analysis was carried out. Due to its larger size (as compared to bond areas and interface regions), the fibrous matrix was discretised with a larger element size. The bond areas and interface regions were discretised with a finer mesh to better approximate these small regions.

Tensile behaviour of an *F100* fabric of size 25 mm × 40 mm was simulated using a 3.60 GHz (RAM 128 GB) Intel Xeon CPU PC with 6 cores. Levels of ultimate strength and computational time were used as reference variables to determine an appropriate mesh size for the model while maintaining a satisfactory level of accuracy and reasonable computational effort. The obtained results (Fig. 4.8), demonstrated that the ultimate strength converged asymptotically to the experimental value, providing a good approximation in Case B with the computational cost roughly three times lower than that in Case C. Thus, it could be concluded that the mesh size of Case B was sufficiently dense and not overly demanding in terms of computing resources; therefore, it was used in subsequent simulations.

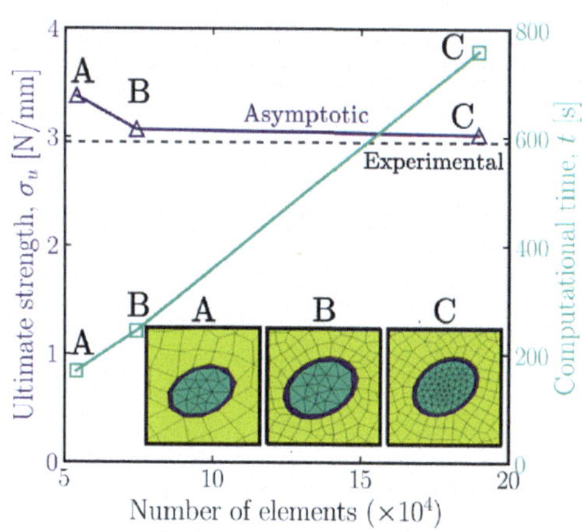

Fig. 4.8 Mesh-convergence analysis

4.5.3 Failure Criterion

Experimental observations showed that calendered fabrics are more sensitive to strain rather than stress. Therefore, a maximum-strain-based failure criterion was chosen to model the failure behaviour. The criterion is formulated in 6 failure indices as follows

$$\left(\frac{\varepsilon_1}{e_{xt}}\right), \left(\frac{\varepsilon_2}{e_{yt}}\right), \left(\frac{\varepsilon_3}{e_{zt}}\right) = 1, d_\varepsilon = 1 \tag{4.11}$$

ε_1, ε_2, ε_3 are the components of strain at any integration point of an element. e_{xt}, e_{yt}, e_{zt} are the maximum allowable longitudinal tensile strains in 3 directions (assumed to be the same) and $d_\varepsilon = 1$ is the damage variable associated with the failure mode of an element, depending upon the strains applied. An element-deletion approach was used to remove the elements from the model based on the value of damage variables as calculated with (4.11). An element was assumed to fail and then removed from the model when the damage condition, $d_\varepsilon = 1$, was satisfied at a given integration point.

With the presence of a three-domain microstructure, the failure behaviour varied with the domain in question. A reverse-engineering approach was used to identify the local maximum allowable failure strains of each phase. A number of simulations were run until a fair match was attained between numerical and experimental results, while still retaining the physics observed in the experiments. The magnitude of the maximum allowable failure strains was assigned in the following ascending order: interface regions, fibrous matrix and bond areas (Table 4.8). In other words, damage initiated at interfaces then propagated through the fibrous matrix with failure of fibres, and, finally, bond area started failing.

Table 4.8 Maximum allowable strains for each domain in various loading directions in *F40* and *F100* models

	0°		30°		45°		60°		90°	
	F40	*F100*	*F40*	*F100*	*F40*	*F100*	*F40*	*F100*	*F40*	*F100*
Fibrous matrix	0.15	0.08	0.23	0.09	0.22	0.08	0.17	0.10	0.18	0.08
Bond area	0.20	0.15	0.25	0.10	0.25	0.10	0.25	0.15	0.20	0.15
Interface region	0.08	0.08	0.10	0.05	0.08	0.08	0.08	0.08	0.08	0.08

4.6 Results and Discussion

4.6.1 Developed Models and Validation

In this section, the developed FE models and simulations are introduced and validated with experimental observations. The full model of *F100* fabric and a detailed view of it is shown in Fig. 4.9. The model was generated to simulate uniaxial tensile loading along MD. Due to the negligible role of the strain rate in the mechanical response of the fabric, all the simulations were performed keeping the strain rate at $0.1\ s^{-1}$.

The simulation results (Fig. 4.10a) present the true-strain distribution at 25% extension. As mentioned, it is more meaningful to assess strain distributions in the material rather than those of stress. A qualitative comparison with the experiment (Fig. 4.10b) indicates the evident similarity with the numerical output (e.g., necking behaviour and model's deformed configuration), although damage occurred somewhat prematurely in simulations. It should be noted that the experimental evidence chosen is one of several obtained and, therefore, not directly comparable to the numerical results. The latter indicate that a higher level of strain was localised at the model's mid-height, spreading within the fabric declining at the top and bottom of the fabric. The strain tended to reach a peak value at interfaces, causing localised damage of the structure with consequent failure. This damage mechanism was observed in the experimental studies. In the models, failure occurred near the fixed boundary conditions (top right end and bottom left end) due to the specific configuration of the bond pattern (symmetry), in particular, near the bonds located close to the boundaries. In these areas, where there was no bond, damage initiated prematurely, leading to failure of the fibrous matrix.

A more in-depth study was conducted to investigate the effect of each phase on the Cauchy stress. A nodal path was designed by considering nodes in the fibrous matrix, the interface region and the bond area. Based on the node arrange-

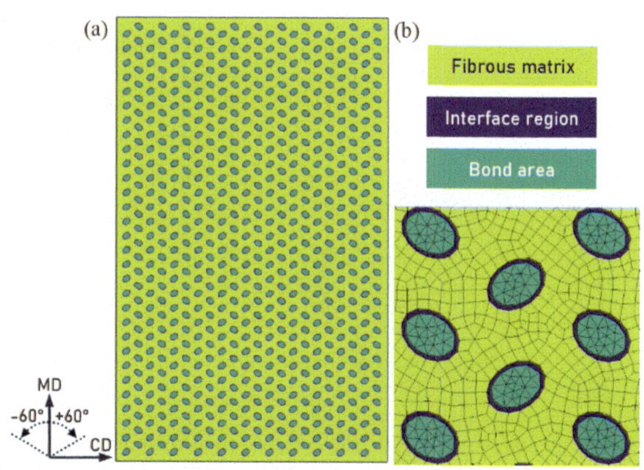

Fig. 4.9 FE model of *F100* fabric in MD: (a) full model; (b) detailed view

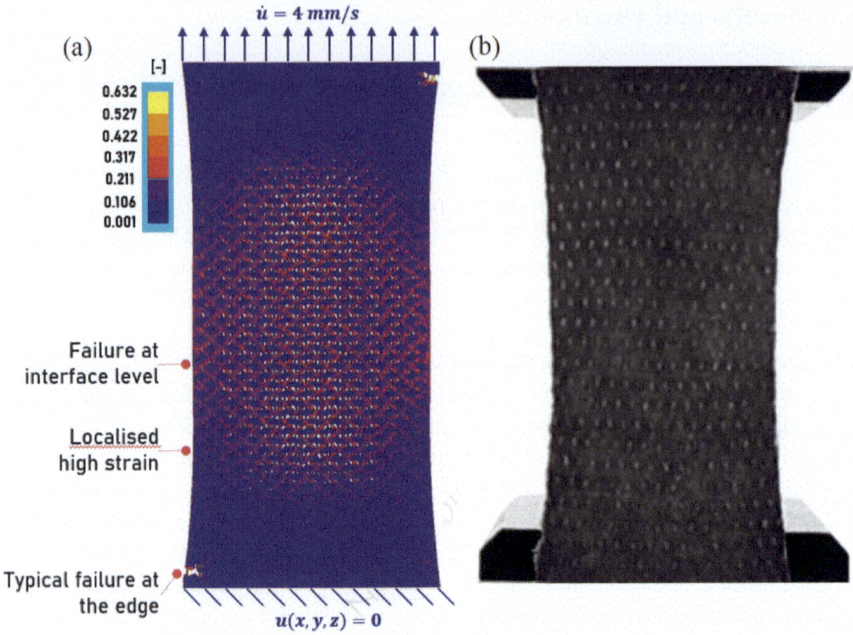

Fig. 4.10 Computed distribution of total true strain (a) and experimental test (b) of *F100* fabric stretched in MD (0°) at 25% extension

ment, an attempt was made to position the path orthogonally to the loading direction (Fig. 4.11a). For the selected nodes, axial stress values were extracted at 20% extension of the fabric and plotted against their distance along the selected path (Fig. 4.11b). A low stress level was observed in the fibrous matrix, with a sudden

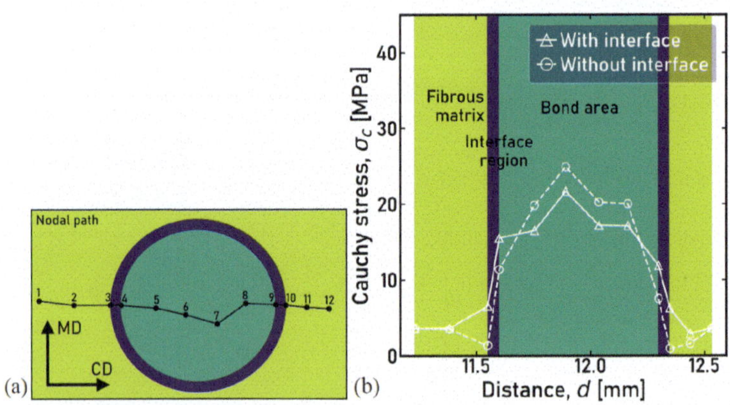

Fig. 4.11 Nodal path in *F100* fabric (a) and Cauchy-stress distribution (b) at 20% extension of fabric in MD

rise at the bond periphery, indicating a geometric discontinuity. As expected, the stress reached its peak value in the middle of the bond area as this region is characterised by high stiffness. A bell-shaped function was identified, which resembles that of fibre stress distribution in a pull-out test of a fibre embedded into a matrix [37]. Typically, the interface region is hard to clearly define geometrically and measure in calendered nonwovens. Perhaps, a definition could be given in terms of other parameters, for instance, stress. Two simulations were performed, one with the interface region and one without it. The presence of the interface region seems to relieve the stress inside the bond area and smoothen the transition between the interface and the fibrous matrix. It appears that the interface region had little effect on the overall mechanical performance of the nonwoven fabric, even locally. However, the presence of the interface is important for damage initiation in continuous models of calendered fabrics.

Similar simulations were performed to predict the mechanical performance of *F40* and *F25* fabrics. In these cases, distributions of true strains and stresses were computed. The results showed that, despite the *F40* fabric was made of the same polymer as *F100*, it exhibited a different behaviour, even locally (Fig. 4.12). This

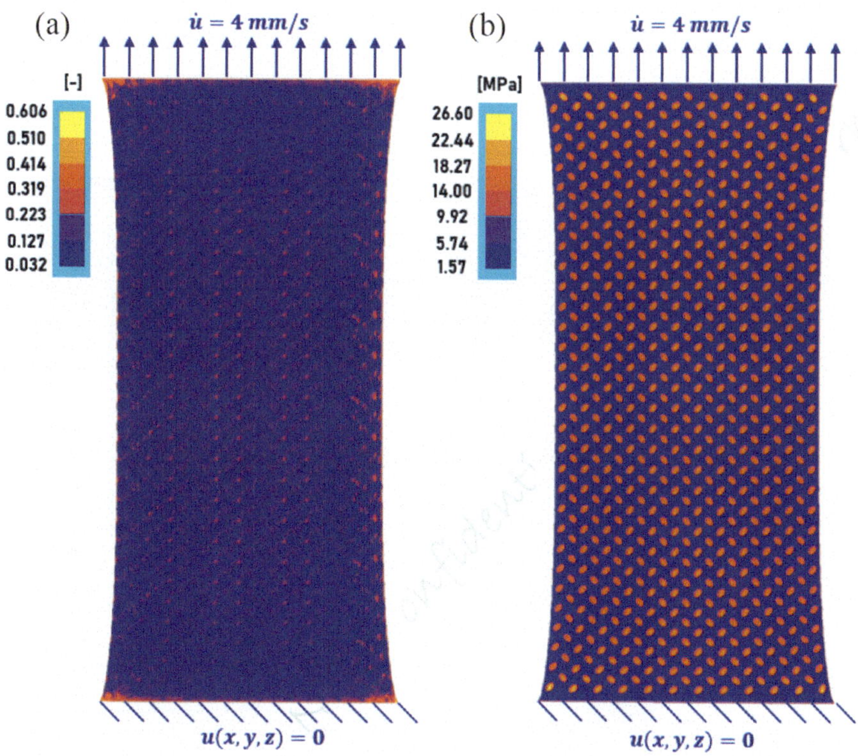

Fig. 4.12 Distributions of total true strains (a) and stresses (b) for *F40* fabric stretched in MD (0°) at 25% extension

effect proved a considerable importance of an account for various parameters for prediction of mechanical behaviour of a nonwoven material. In the *F40* fabric, the strain was more uniformly distributed as compared to *F100*. The reason may be linked to the ratio between the fibre-matrix thickness, t_{FM}, and that of bond areas, t_{BA}: this ratio was higher in the *F40* and *F25* fabrics than that in *F100*. This hypothesis was supported by the character of the strain distribution in the *F25* fabric (Fig. 4.13). As expected, stress localised at bond areas, which acted as stress concentration zones due to their high stiffness. The *F25* fabric exhibited a more marked necking effect as compared to its counterparts.

In order to compare the numerical results with experimental observations, the force-extension graphs were used (Fig. 4.14); the force was normalised with the specimen width. The obtained results revealed that the model could successfully simulate the mechanical behaviour of the *F40* and *F100* fabrics. The elastic part was appreciably captured for these fabrics together with the plastic one. However, as the fabric density decreased (*F25* fabric), the model could not mimic the behaviour of the material realistically. In particular, the numerical model could describe adequately neither the elastic nor the plastic response of the material. Furthermore, simulations of the *F25* fabric were affected by severe convergence issues due to large material deformations. It is well known that continuous models, used in the literature, have limitations in simulating the realistic behaviour of low-density

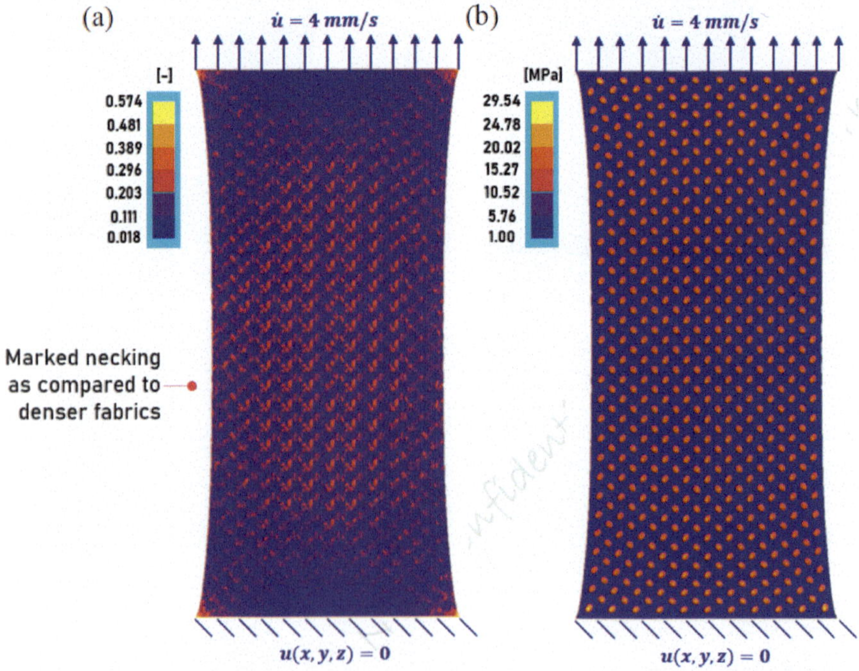

Fig. 4.13 Distributions of total true strains (a) and stresses (b) for *F25* fabric stretched in MD (0°) at 25% extension

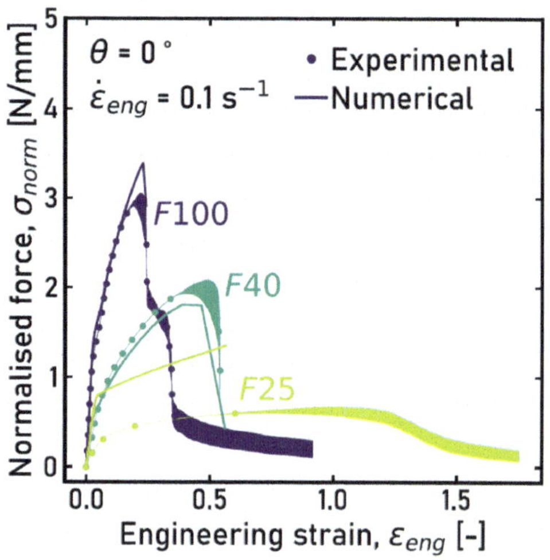

Fig. 4.14 Experimental and numerical normalised force *vs.* engineering strain curves of *F25*, *F40* and *F100* fabrics stretched in MD at strain rate 0.1 s^{-1}

nonwovens. The structure of these fibrous networks greatly differs from that of traditional solids, and porosity, for instance, becomes an important factor that is not accounted for in this research framework. However, an attempt was made to include these fabrics in this study and try to elucidate the limitations of a continuous modelling approach. The fibre-re-orientation mechanism is more prominent in these fabrics due to a lower fibre content as compared to the high-density ones. In light of this result, the next sections deal only with the *F40* and *F100* fabrics.

4.6.2 Effect of Load Direction

As demonstrated in [35], the mechanical performance of nonwoven materials highly depends on the distribution and orientation of their fibres. Hence, the *loading direction* is a key factor to consider. This phenomenon was partially studied experimentally by testing fabrics in MD and CD, but not numerically. Numerical simulations were performed to assess the reliability of the developed numerical scheme in predicting the mechanical response of fabrics stretched in various loading directions (0°, 30°, 45°, 60° and 90° with respect to MD), see Fig. 4.15. The results demonstrated a good agreement with the experimental data. The elastic-plastic response was well captured in all the cases, while ultimate strength was slightly overestimated. A higher level of error was observed in the mechanical response in CD ($\vartheta = 90°$). This discrepancy could be attributed to a lower stiffness exhibited by the material in CD as compared to the other directions – recalling the case of the *F25* fabric. Difficulties in capturing the mechanical response in CD were also reported in other works [24], [38].

The effect of the loading direction was also analysed in the *F40* fabric. The results reported in Fig. 4.15b depict a fair agreement with the experimental data, marked

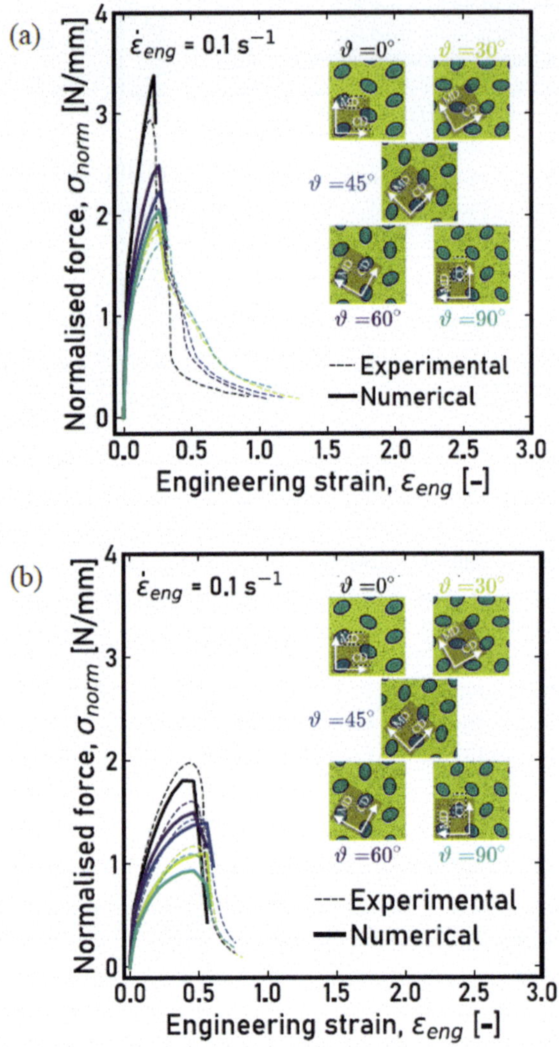

Fig. 4.15 Experimental and numerical normalised force *vs.* engineering strain curves of *F40* (a) and *F100* (b) fabrics stretched in various loading directions at strain rate 0.1 s^{-1}

by some underestimation of the latter. The elastic-plastic response was reasonably captured. It is evident that a decrease in planar density led to a decrease in the accuracy of the predictive capability of the numerical scheme.

4.6.3 Effect of Fabric Size

Hou [39] demonstrated experimentally that the *size* of a nonwoven fabric can affect its mechanical behaviour. In order to account for this effect, three case studies are proposed in this section. The width and height of the generated FE models were changed while maintaining the width-to-height aspect ratio at 0.625 (Table 4.9). The deformation speed was adjusted according to the gauge length (model height) in order to simulate the tensile behaviour of fabrics at the same strain rate of 0.1 s^{-1}.

Results illustrated in Fig. 4.16 indicate an increase in toughness as the fabric size grew. This expected result is important from a structural perspective, especially for the design of nonwovens for ballistic applications or, in general, in cases where high resistance is required. Less obvious was the relation between the ultimate force initiating the fabric failure, F_u, and the area of the fabric A_{fab}. One reason for size-induced strengthening, especially prominent in the *F40* fabric, is the reduction of the effect of areas near the specimen vertices. In these areas, the type of specimen's fixture (reflected in the respective boundary condition) caused a loss of uniaxiality and, subsequently, higher total deformations. As a result (see above), the damage initiated in these areas. Finally, the stiffness was observed to increase with an increase in the fabric size, confirming Hou's findings, [39].

Table 4.9 Geometric parameters in various models of *F40* and *F100* fabrics

Model	Fabric size, $W \times H$ (mm)	Fabric area, A_{fab} (mm^2)	Centre coordinates of first bond area, (x_C, y_C) (mm)	Deformation rate, $\dot{u}$ (mm/s)
A	6.25×10	62.5	(0.925, 0.755)	1.0
B	12.5×10	250	(0.75, 0.90)	2.0
C	25×40	1000	(0.95, 0.825)	4.0

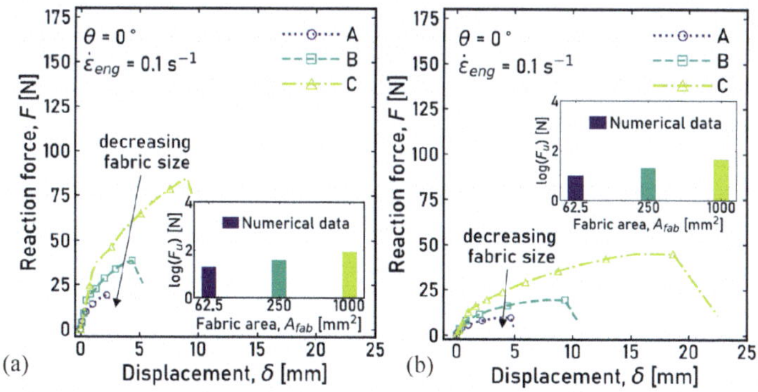

Fig. 4.16 Force-displacement curves of *F40* (b) and *F100* (b) in MD for various models (fabric sizes)

4.6.4 Effect of Bond Pattern

With the confirmed predictive capability of the developed numerical scheme, this section aims to simulate and assess the effect of bond pattern on the mechanical performance of calendered nonwovens. A *bond pattern* can be defined by four main parameters: *bond-pattern orientation*, *spacing* between bond areas, bond area's *shape* and *orientation*. The FE models of the *F40* and *F100* fabrics were identical to those presented in Sect. 4.6.1 but had a variation in bond area spacing, shape and orientation. Below, three case studies are proposed for each parameter studied.

The *spacing* between bond areas can affect stiffness along a specific direction. Spacings between bond areas in the x and y directions, s_x and s_y, respectively, were varied according to Table 4.10. The reference case, Sp-B, was analysed earlier. It should be noted that a variation in spacing entails a variation in the centre coordinates of the first bond area generated in the model. As a result, the bond pattern is defined by a unique arrangement for a given fabric size. Contrary to expectations, the obtained results shown in Fig. 4.17 indicate that an increase or decrease in bond spacing had no effect on the mechanical performance of the *F100* fabric. In the case of the *F40* fabric, in contrast, the mechanical response was greatly affected by the bond spacing. In particular, a smaller bond spacing led to a higher fabric strength and lower stretchability. This behaviour developed progressively as the bond spacing decreased. The discrepancy in the mechanical behaviour between the two fabrics could be attributed to different ratios between thicknesses of the fibrous matrix and bond areas. In fact, the higher the ratio, the higher the freedom of movement offered to bond areas within the fibrous matrix thickness. These differences in the mechanical behaviour of (an apparently similar) nonwoven reflect the complex nature of these materials.

The second factor defining the bond pattern is the *bond area shape*. Three bond shapes were studied, *elliptical*, *rectangular* and *diamond-like* with their area kept at 0.28 mm^2 (Fig. 4.18). The orientation of bond areas was set to 90° with respect to the machine and loading directions.

The reference case here is Sh-A, the *F100* fabric generated with all the geometric and material properties identical to the reference case introduced in Sect. 4.6.1. Two other cases are identical to Sh-A, but with the bond areas having a different shape.

Table 4.10 Case studies for variation of bond area spacing

Case study	Spacing in x–direction, s_x (mm)	Spacing in y–direction, s_y (mm)	Bond area centre coordinates, (x_C, y_C) (mm)
Sp-A	1.5	0.7	(0.875, 0.75)
Sp-B	2.2	1.3	(0.95, 0.825)
Sp-C	4.4	2.6	(1.5, 1.15)

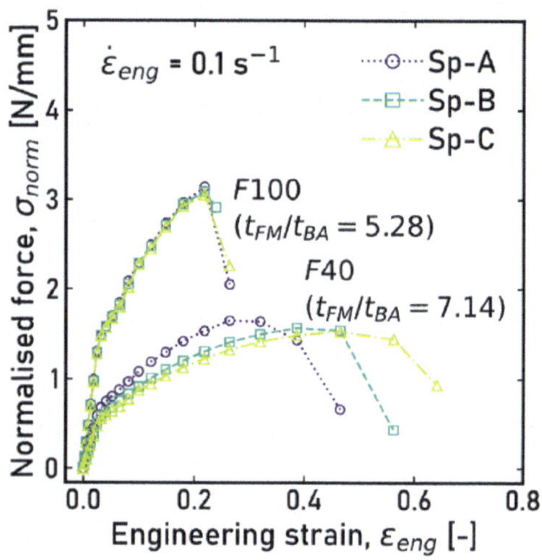

Fig. 4.17 Normalised force vs. engineering strain curves of *F40* and *F100* fabrics with various bond area spacings

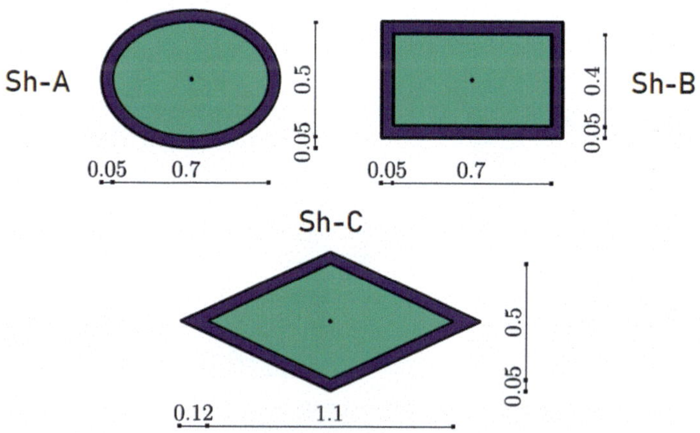

Fig. 4.18 Bond area shapes in different case studies

No evidence about the influence of the bond area shape on the mechanical response of the material was found (Fig. 4.19). In fact, only the bonded-area fraction should matter in this respect.

Since the variation in the bond area shape had no effect on the global response of the material, an attempt was made to analyse its local mechanical response, in particular the stress distributions around the bond area and inside it. The selected bond areas were located roughly in the middle of the specimen, specifically at the 34th row and 7th column of the bond pattern. The *F100* fabrics were stretched in MD, and with results collected at 20% fabric extension, as shown in Fig. 4.20. As expected, the peak stress was observed in the central part of the bond area. In Sh-A, the elliptical bond deformed into a circle and some elements at interface failed and were deleted as the failure condition was met. Among the three cases, this one displayed the lowest

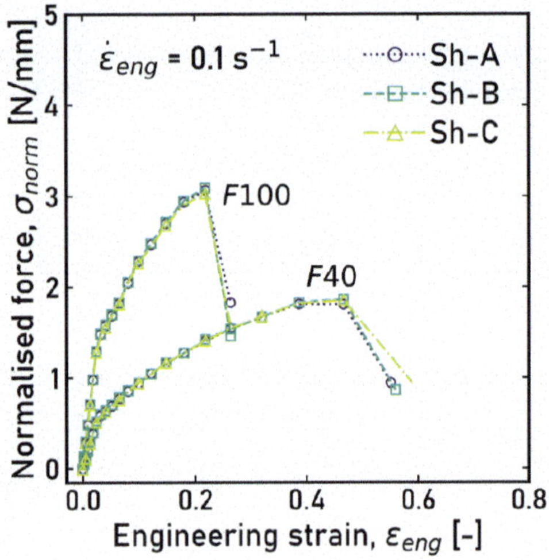

Fig. 4.19 Normalised force *vs.* engineering strain curves of *F40* and *F100* fabrics with various bond area shapes

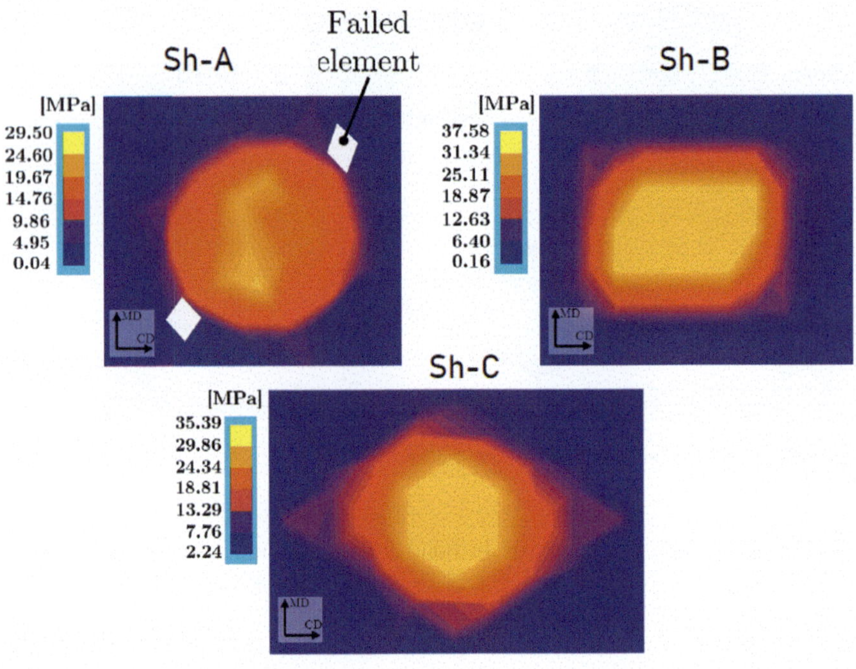

Fig. 4.20 Cauchy-stress distribution in various bond areas of *F100* fabrics at 20% extension

stress-bearing capacity. Rectangular bond areas seem to withstand a higher level of stress as compared to the other two cases. Therefore, according to the results, it can be concluded that in terms of stress the most suitable bond shape is the rectangular one. The diamond-like bond shape provided an intermediate scenario.

The roundup of case studies completes with the investigation of the effect of the *orientation of bond areas* on mechanical performance of *F40* and *F100* fabrics. Three case studies were formulated with the bond areas rotated by 0°, 45° and 60° with respect to MD. The models were generated under the same conditions as in Sh-A. The fabrics were stretched uniaxially up to failure and then results were collected.

The computed normalised force *vs.* engineering strain curves of the *F40* and *F100* fabrics for various orientations of bond areas (Fig. 4.21) revealed that the difference in the mechanical response among the models was rather marginal in this case. A reasonable explanation could be that the shapes of bonds were symmetric along all directions and, therefore, there was no predominant direction that acted as a lever to generate moments and/or forces.

4.7 Conclusions

This work attempted to advance the knowledge in the field of mechanics of nonwoven materials, particularly the spun-bonded ones, produced with the hot-calendering technique. Its contribution in relation to the current state-of-the-art can be summarised as follows:

- Characterisation, modelling and prediction of deformation and damage behaviours of calendered nonwovens at macroscopic level through multiscale mechanical testing and introduction of a third domain – interface region – zone of damage localisation and initiation
- Extension of the study to low-density nonwovens and verification of the limitations of the developed numerical scheme in this context.

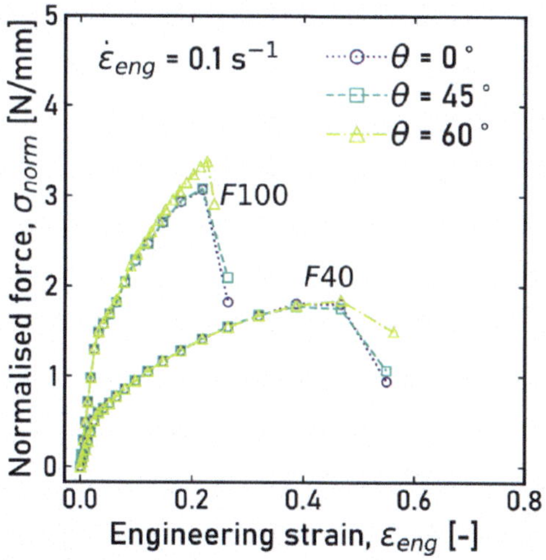

Fig. 4.21 Normalised force *vs.* engineering strain curves of *F40* and *F100* fabrics with various bond area orientations

- Development of an algorithm to accurately assess and quantify the level of anisotropy if fabrics' microstructure, also accounting for curliness of fibres.
- Development of a stand-alone software with a graphical user interface that facilitates its use and links the numerical scheme to the FE environment. In addition, it allows automatic and efficient generation of shell-based FE models based on user-defined material and geometric properties of calendered fabrics (e.g., bond pattern, fabric size, interface-region size, fibre diameter, etc.), including studies of the effects of design parameters and estimation of local distributions of strains, stresses and damage as well as mechanical anisotropy. The tool can facilitate the design and optimisation of nonwoven products by reducing trial-and-error stages of product development. It can help manufacturers to lower the cost and time of bringing new products to the market, and potentially be extended to model other materials, thanks to its general nature.
- Identification of maximum allowable strains in various loading directions for each fabric domain and development of a local damage criterion based on experimental observations.

References

1. M.E. Burkett, A.H.A. Gallery. *The Art of the Felt Maker*. Abbot Hall Art Gallery, 1979.
2. G.E. Millman, J.P. Figh, H.D Wellington. The Expanding Market for Nonwovens. *American Dyestuff Reporter* 58(7):32–39, 1969.
3. G. Tanchis. *The Nonwovens – Reference Books of Textile Technologies*. Fondazione ACIMIT, 2008.
4. EDANA (2018) What are nonwovens? https://www.edana.org/nw-related-industry/what-are-nonwovens
5. W.W. Albrecht, H. Fuchs, W. Kittelmann. *Nonwoven Fabrics: Raw Materials, Manufacture, Applications, Characteristics, Testing Processes*. Weinheim: Wiley-VCH, 2003.
6. S.K. Batra, B. Pourdeyhimi. *Introduction to Nonwovens Technology*. Destech Publications, 2012.
7. R. Jubera, A. Ridruejo, C. González, J. LLorca. Mechanical behavior and deformation micromechanisms of polypropylene nonwoven fabrics as a function of temperature and strain rate. *Mechanics of Materials* 74:14-25, 2014.
8. V. Cucumazzo. Deformation and Damage Behaviour of Calendered Nonwovens: Experimental and Numerical Analyses. Loughborough University, 2020
9. R.K. Dharmadhikary, T.F. Gilmore, H.A. Davis, S.K. Batra. Thermal bonding of nonwoven fabrics. *Textile Progress* 26(2):1-37, 1995.
10. K. Smith, A.A. Ogale, R. Maugans, L.K. Walsh, R.M. Patel. Effects of bond roll pattern and temperature on the microstructure and properties of polyethylene nonwovens. *Textile Research Journal* 73(10):845-853, 2003.
11. J. Rittenhouse, R. Wijeratne, E.B. Orler, D.A. Dillard, R.B. Moore, R. De Vita. Effect of areal density and fiber orientation on the deformation of thermomechanical bonds in a nonwoven fabric. *Polymer Engineering and Science* 59(2):311-322, 2019.
12. R.S. Wijeratne, R. De Vita, J.A. Rittenhouse, E.B. Orler. Biaxial properties of individual bonds in thermomechanically bonded nonwoven fabrics. *Textile Research Journal* 89(5):698-710, 2019.
13. H.L. Cox. The elasticity and strength of paper and other fibrous materials. *British Journal of Applied Physics* 3(3):72-79, 1952.

14. S. Backer, D.R. Petterson. Some principles of nonwoven fabrics. *Textile Research Journal* 30(9):704-711, 1960.
15. J.I. Curiskis, G.A. Carnaby. Continuum mechanics of the fiber bundle. *Textile Reseach Journal* 55(6):334-344, 1985.
16. S. Bais-Singh, B.C. Goswami. Predicting the biaxial tensile deformation behavior of spunbonded nonwovens. *Textile Research Journal* 68(3):219-227, 1998.
17. M.A. Narter, S.K. Batra, D.R. Buchanan. Micromechanics of three-dimensional fibrewebs: constitutive equations. *Proceedings of the Royal Society A: Mathematical, Physical and Engineering Sciences* 455(1989):3543-3563, 1999.
18. H.S. Kim, A. Deshpande, B. Pourdeyhimi, A.S. Abhiraman, P. Desai. Characterizing structural changes in point-bonded nonwoven fabrics during load-deformation experiments. *Textile Research Journal* 71(2):157-164, 2001.
19. A. Rawal. A modified micromechanical model for the prediction of tensile behavior of nonwoven structures. *Journal of Industrial Textiles* 36(2):133–149, 2006.
20. S. Adanur, T. Liao. Fiber arrangement characteristics and their effects on nonwoven tensile behavior. *Textile Research Journal* 69(11):816-824, 1999.
21. A. Rawal, S. Lomov, T. Ngo, I. Verpoest. Mechanical behavior of thru-air bonded nonwoven structures. *Textile Research Journal* 77(6):417-431, 2007.
22. A. Rawal, S. Sharma, V. Kumar, K. Rao, H. Saraswat, N.K. Jangir, R. Kumar, D. Hietel, M. Dauner. Micromechanical analysis of nonwoven materials with tunable out-of-plane auxetic behavior. *Mechanics of Materials* 129:236-245, 2019.
23. X. Hou, M. Acar, V.V. Silberschmidt. 2D finite element analysis of thermally bonded nonwoven materials: Continuous and discontinuous models. *Computational Materials Science* 46(3):700-707, 2009.
24. E. Demirci, M. Acar, B. Pourdeyhimi, V.V. Silberschmidt. Finite element modelling of thermally bonded bicomponent fibre nonwovens: Tensile behaviour. *Computational Materials Science* 50(4): 1286-1291, 2011.
25. E. Demirci, F. Farukh, M. Acar, B. Pourdeyhimi, V.V. Silberschmidt. Numerical analysis of dynamic out-of-plane loading of nonwovens. *Journal of Physics: Conference Series* 451: 012021, 2013.
26. M.N. Silberstein, C.L. Pai, G.C. Rutledge, M.C. Boyce. Elastic–plastic behavior of nonwoven fibrous mats. *Journal of the Mechanics and Physics of Solids* 60(2):295-318, 2012.
27. A. Ridruejo, C. González, J. Llorca. A constitutive model for the in-plane mechanical behavior of nonwoven fabrics. *International Journal of Solids and Structures* 49(17):2215-2229, 2012.
28. F. Martínez-Hergueta, A. Ridruejo, C. González, J. LLorca. A multiscale micromechanical model of needlepunched nonwoven fabrics. *International Journal of Solids and Structures* 96:81-91, 2016.
29. F. Martínez-Hergueta, A. Ridruejo, C. González, J. Llorca. Numerical simulation of the ballistic response of needle-punched nonwoven fabrics. *International Journal of Solids and Structures* 106-107:56–67, 2017.
30. J. Vila-Ortega, A. Ridruejo, F. Martínez-Hergueta. Multiscale numerical optimisation of hybrid metal/nonwoven shields for ballistic protection. *International Journal of Impact Engineering* 138:103478, 2020.
31. N. Chen, M.N. Silberstein. A micromechanics-based damage model for non-woven fiber networks. *International Journal of Solids and Structures* 160:18-31, 2020.
32. A. Raina, C. Linder. A homogenization approach for nonwoven materials based on fiber undulations and reorientation. *Journal of the Mechanics and Physics of Solids* 65:12-34, 2014.
33. A. Raina, C. Linder. A micromechanical model with strong discontinuities for failure in nonwovens at finite deformations. *International Journal of Solids and Structures* 75-76:247-259, 2015.
34. R. Mansour, A. Kulachenko, W. Chen, M. Olsson. Stochastic constitutive model of isotropic thin fiber networks based on stochastic volume elements. *Materials* 12(3):538, 2019.

35. V. Cucumazzo, E. Demirci, B. Pourdeyhimi, V.V. Silberschmidt. Anisotropic mechanical behaviour of calendered nonwoven fabrics: Strain-rate dependency. *Journal of Composite Materials* 55(13):1783–1798,2020
36. T.W. Chou. Microstructural Design of Fiber Composites. Cambridge University Press, 1992.
37. J.K. Kim, Y.W. Mai. *Engineered Interfaces in Fiber Reinforced Composites*. Elsevier, 1998.
38. F. Farukh. E. Demirci, M. Acar, B. Pourdeyhimi, V.V. Silberschmidt. Large deformation of thermally bonded random fibrous networks: Microstructural changes and damage. *Journal of Materials Science* 49(11):4081-4092, 2014.
39. X. Hou. Experimental and Numerical Analysis of Deformation of Low-density Thermally Bonded Nonwovens. Loughborough University, 2010.

Chapter 5
Free Vibrations of a Cylindrical Shell Closed with the Cap

Sergei B. Filippov, Grigory A. Nesterchuk, and Andrei L. Smirnov

Abstract Low-frequencies and vibration modes of the construction consisted of a closed circular cylindrical shell joined with the circular plate are obtained by means of numerical and asymptotic methods. Three types of vibrations are analyzed. The frequencies of vibrations of the first type are close to plate frequencies. The frequencies of the second type vibrations slightly differ from the frequencies of the cylindrical shell. Two different asymptotic approaches are used to get approximate asymptotic formulas for plate-like and shell-like vibration frequencies. In the approximate analysis the third type vibrations of the structure may be considered as beam vibrations with circular cross-sectional area. The optimal ratio of the plate and shell thicknesses that provides the maximum value of the fundamental frequency of the structure with the given mass is evaluated. The asymptotic and numerical results obtained with FEA are in good agreement.

Key words: Plate/shell coupled structure, Free vibrations, Asymptotic method, Optimal thickness

5.1 Introduction

The equations of the theory of thin shells contain the dimensionless shell thickness h_s as a small parameter. Hence, asymptotic methods [1, 8, 9, 12] may be applied to solve these equations, in particular, approximate solutions may be obtained by means of Vishik-Lyusternik algorithm [1, 13], when solutions are represented in the form of a sum of slowly varying functions and edge effect integrals. In such a

Sergei B. Filippov · Grigory A. Nesterchuk · Andrei L. Smirnov
St. Petersburg State University, 7/9 University Embankment, St. Petersburg, 199034, Russian Federation
e-mail: s_b_filippov@mail.ru,g.nesterchuk@spbu.ru,a.l.smirnov@spbu.ru

H. Altenbach et al. (eds.), *Advances in Solid and Fracture Mechanics*,
Advanced Structured Materials 180,
https://doi.org/10.1007/978-3-031-18393-5_5

way the initial singularly perturbed system of differential equations is reduced to an approximate system of the smaller order [12].

Vibrations of joined shells are analyzed in many papers, including [2]. Asymptotic approach was used in [3, 4, 5], but the low-frequency vibrations of a shell closed with the plate essentially differ from low-frequency vibrations of joined shells. The reason for this is as follows. The non-dimensional fundamental frequency of the plate joined with the shell f_* is of the order of h_p, where $h_p \ll 1$ is the non-dimensional plate thickness and the fundamental mode is axisymmetric. On the other hand, the non-dimensional lowest frequency of axisymmetric vibrations of joined shells of revolution, $f_* \sim 1$ [1]. Therefore, the asymptotic analysis of joined shells of revolution cannot be transferred to structures containing thin plates.

In [6, 7] new asymptotic technique was developed to find low-frequencies and vibration modes of a cylindrical shell joined with annular plates. In the current research the same asymptotic method is applied to find approximate solution of the eigenvalue problem describing free vibrations of a cylindrical shell stiffened by a circular plate at the end. The analysis of vibrations of joined thin-walled elements of such kind is of particular interest since they are often used in actual designs (see Fig. 5.1).

5.2 Numerical Results

A numerical study of the lowest natural vibration frequencies of a structure consisting of a circular cylindrical thin shell, rigidly clamped at one end and joined with a circular thin plate at the other, was carried out in the COMSOL Multiphysics® Version 5.6 finite element package. We considered a copper shell of medium length with the following parameters: E = 110 GPa — the Young modulus, $\nu = 0.35$ — Poisson's ratio, $\rho = 8960$ kg/m^3 — density, $R = 1$ m, $l = L/R = 4$ — shell length, $h = H/R = 0.01$ — characteristic thickness of the structure.

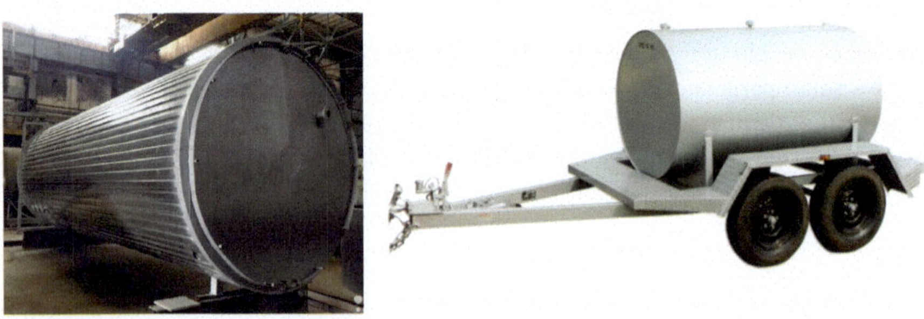

Fig. 5.1 Actual structures: cylindrical shell with the attached end plate.

The convergence of the method on different grids was considered. The best convergence is provided by mapping into rectangular elements (mapped mesh) with characteristic element sizes in the interval $[0.008, 0.08]$. A further decrease in the size of the element has little effect on the results, since the relative change in the knowledge of frequencies does not exceed 1%. As a result, the lower part of the spectrum of natural dimensionless frequencies $\Omega = (\omega^2 \cdot 12\rho\sigma R^2/(Eh^2))^{1/4}$ of the structure under consideration, where $\sigma = 1 - \nu^2$, and f, $\omega = 2\pi f$ are the frequency and cyclic frequency of vibrations, respectively, is shown in Fig. 5.2

Natural vibration frequencies may be divided into three groups of shell-like, plate-like and beam-like frequencies. The corresponding eigenmodes are shown in Fig. 5.3.

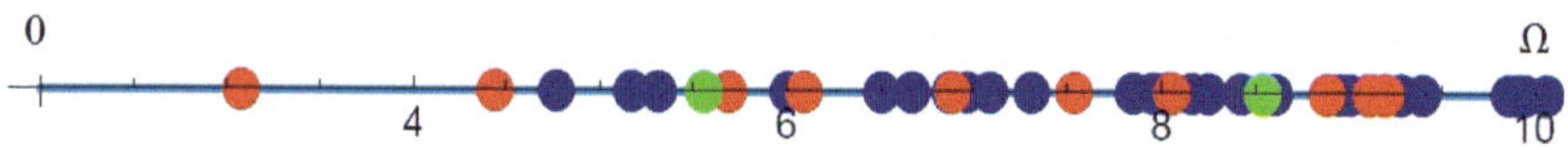

Fig. 5.2 Lower natural frequencies of a cylindrical shell with the attached end plate. Dimensionless shell-like frequencies are marked in blue, plate-like frequencies — in red, and beam-like frequencies — in green.

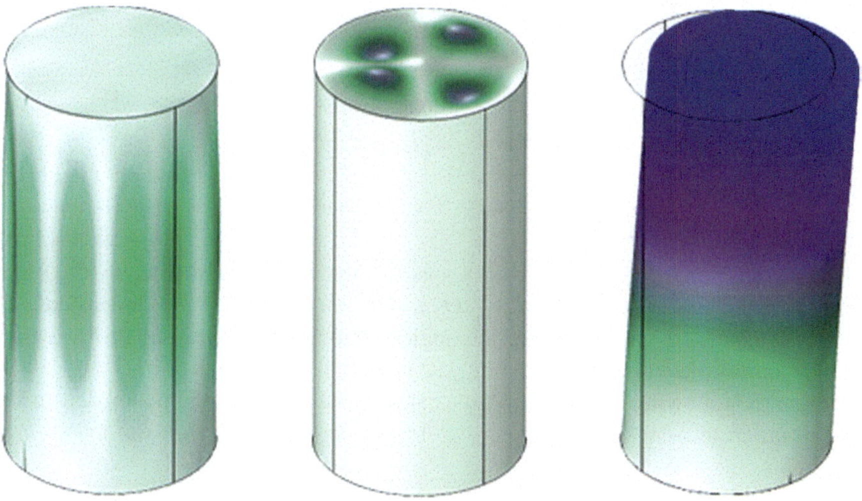

a) b) c)

Fig. 5.3 Vibration modes: a) shell-like, b) plate-like, c) beam-like.

5.3 Basic Equations

We consider small free low-frequency vibrations of a thin cylindrical shell stiffened by the circular plate. The plate is located on the parallel $s = l$ of the shell, where s is the dimensionless axial coordinate on the shell middle surface. The radius of the cylindrical shell R is taken as the characteristic length (Fig. 5.4).

After the separation of variables non-dimensional differential equations describing free vibrations of a cylindrical shell [6, 10] are written in the form

$$\begin{aligned}
&T_1' + mS + \lambda u = 0, \quad S' - mT_2 + Q_2 + 2H' + \lambda v = 0,\\
&Q_1' + mQ_2 - T_2 + \lambda w = 0, \quad Q_1 = M_1' + 2mH, \quad Q_2 = -mM_2,\\
&M_1 = \mu^4(\vartheta_1' + \nu m\vartheta_2), \quad M_2 = \mu^4(m\vartheta_2 + \nu\vartheta_1'), \quad H = \mu^4(1-\nu)\vartheta_2',\\
&T_1 = u' + \nu(w + mv), \quad T_2 = w + mv + \nu u', \quad 2S = (1-\nu)(v' - mu),\\
&\vartheta_1 = -w', \quad \vartheta_2 = mw + v,
\end{aligned} \tag{5.1}$$

where ($'$) denotes the derivative with respect to the coordinate $s \in [0,l]$, m is the circumferential wave number, u, v and w are components of the displacement vector, $T_1, T_2, S, Q_1, Q_2, M_1, M_2, H$ are the dimensionless stress-resultants, ϑ_1 and ϑ_2 are the angles of rotation of the normal, $\lambda = \omega^2\sigma\rho R^2 E^{-1}$ is the frequency parameter, $\mu^4 = h_s^2/12$ is a small parameter, h_s is the dimensionless shell thickness

The plate bending equations [6, 10] are

$$\begin{aligned}
&(s_pQ_{1p})' + mQ_{2p} + \lambda s_p w_p = 0,\\
&s_pQ_{1p} = (s_pM_{1p})' - M_{2p} + 2mH_p, \quad s_pQ_{2p} = -mM_{2p} + 2H_p,\\
&s_pM_{1p} = \mu_p^4[s_p\vartheta_{1p}' + \nu(m\vartheta_{2p} + \vartheta_{1p})], \quad s_pM_{2p} = \mu_p^4(m\vartheta_{2p} + \vartheta_{1p} + \nu s_p\vartheta_{1p}'),\\
&H_p = \mu_p^4 s_p(1-\nu)\vartheta_{2p}', \quad \vartheta_{1p} = -w_p', \quad s_p\vartheta_{2p} = mw_p.
\end{aligned} \tag{5.2}$$

Here ($'$) denotes the derivative with respect to the radial coordinate, $s_p \in [0,1]$, w_p is the transverse deflection, $Q_{1p}, Q_{2p}, M_{1p}, M_{2p}, H_p$ are the dimensionless stress-resultants, ϑ_{1p} and ϑ_{2p} are the angles of rotation of the normal, h_p is the dimensionless plate thickness, $\mu_p^4 = h_p^2/12$ is a small parameter.

From the following equations [6] one can find the tangential (in plane) deformation of the plate

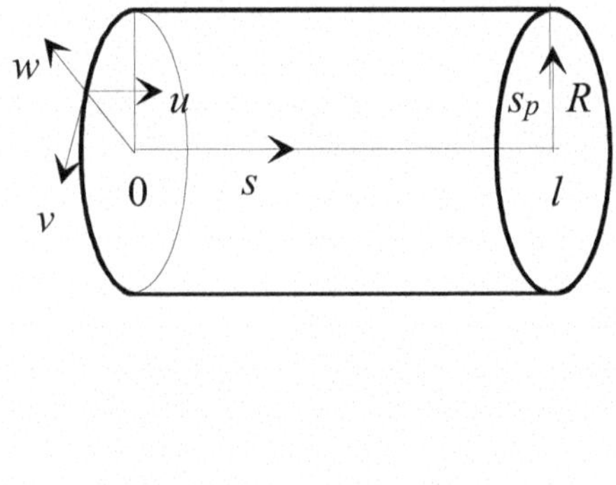

Fig. 5.4 Shell with the attached end plate

$$
\begin{aligned}
&(s_pT_{1p})' - T_{2p} + mS_p + \lambda s_p u_p = 0, \quad s_pS'_p + 2S_p - mT_{2p} + \lambda v_p = 0,\\
&s_pT_{1p} = s_pu'_p + \nu(mv_p + u_p), \quad s_pT_{2p} = u_p + mv_p + \nu s_pu'_p,\\
&2s_pS_p = (1-\nu)(s_pv'_p - mu_p - v_p),
\end{aligned}
\tag{5.3}
$$

where u_p and v_p are the tangential components of the displacement, T_{1p}, T_{2p}, S_p are the dimensionless stress-resultants.

If the shell and plate made of the same material, then the following 8 continuity conditions at the circumference $s = l$, $s_p = 1$ are to be satisfied

$$
\begin{aligned}
&w = u_p, \quad u = -w_p, \quad v = v_p, \quad \vartheta_1 = \vartheta_{1p},\\
&h_sQ_1 = -h_pT_{1p}, \quad h_sT_1 = h_pQ_{1p}, \quad h_sS = -h_pS_p, \quad h_sM_1 = -h_pM_{1p}.
\end{aligned}
\tag{5.4}
$$

At the shell edge $s = 0$ four homogeneous boundary conditions should be introduced. As an example, we consider a shell with the clamped edge, i.e.

$$
u = w = v = \vartheta_1 = 0 \quad \text{for} \quad s = 0. \tag{5.5}
$$

If for $\lambda = \lambda_k$ Eqs. (5.1)-(5.3) have a nontrivial solution satisfying boundary conditions (5.4)-(5.5) then λ_k is an eigenvalue of the boundary eigenvalue problem (5.1)-(5.5). The minimal positive eigenvalue λ_1 corresponds to the fundamental frequency f_1.

5.4 Asymptotic Solution for the Plate-like Vibrations

Let us assume that $\mu_p \sim \mu$, $\lambda \sim \mu^4$ and seek the approximate solution of system (5.1) as a sum of the membrane solutions and the edge effect functions:

$$
y = \mu^{I_0(y)} y_0 + \mu^{I_1(y)+1} y_1 + \mu^{I_1(y)} y_2, \tag{5.6}
$$

Here y denotes any unknown function, $I(y)$ and $I_1(y)$ are the intensity indices. The appropriate choice of the intensity indices permits to satisfy the boundary conditions in the first and subsequent approximations, and assures the existence of nontrivial solutions of the corresponding eigenvalue problems. Usually, the evaluation of the intensity indices is based on one of the exhaustive search methods and/or intuitive reasons.

Functions u_0, v_0, T_{10} and S_0 satisfy plane stress equations derived from Eqs. (5.1) assuming $\mu = 0$ and neglecting the small terms λu, λv and λw:

$$
T'_{10} + mS_0 = 0, \quad S'_0 = 0, \quad 2S_0 = (1-\nu)(v'_0 - mu_0), \quad T_{10} = \sigma u'_0. \tag{5.7}
$$

The edge effect functions y_1 and y_2 have the form

$$
y_1 = \sum_{j=1}^{2} D_j \hat{y}_j \exp(r_j s/\mu), \quad y_2 = \sum_{j=3}^{4} D_j \hat{y}_j \exp[r_j(s-l)/\mu]. \tag{5.8}
$$

Here $D_j \sim 1$ are arbitrary constants,

$$r_{1,2} = g(-1 \pm i), \quad r_{3,4} = g(1 \pm i), \quad g = \sigma^{1/4}/\sqrt{2}, \quad i^2 = -1.$$

In particular, $\hat{\omega}_j = 1$, $j = 1,2,3,4$. Functions y_1 decreases rapidly as s increases from 0 up to l and function y_2 decreases as s decreases from l to 0. Assuming $l \gg \mu$, it follows that $y_1(l) \ll 1$ and $y_2(0) \ll 1$. The intensity indices are given in Table 5.1.

We suppose that

$$w_p \sim \vartheta_{1p} \sim 1, \quad M_{1p} \sim Q_{1p} \sim \mu^4, \quad u_p \sim v_p \sim T_{1p} \sim S_p \sim \mu^3. \tag{5.9}$$

Substituting solutions (5.6) and (5.9) into Eqs. (5.1)-(5.3), continuity conditions (5.4) and boundary conditions (5.5) the eigenvalue problem (5.1)-(5.5) in the first approximation may be split into five separated problems:

1. The eigenvalue problem for equations (5.2), describing the transverse bending deformation of the plate with boundary conditions

$$w_p = \vartheta_{1p} = 0, \quad s_p = 1. \tag{5.10}$$

2. The linear algebraic equations

$$w_2(l) = 0, \quad h_s M_{12}(l) = -h_p M_{1p}(1) \tag{5.11}$$

for the unknown constants D_3, D_4. Solving these equations we obtain the edge effect functions at the parallel $s = l$.

3. Non-homogeneous boundary value problem for membrane shell equations (5.7) with boundary conditions

$$u_0(0) = v_0(0) = 0, \quad T_{10}(l) = 0, \quad h_s[S_0(l) + S_2(l)] = -h_p S_p(1). \tag{5.12}$$

4. The linear algebraic equations

$$w_1(0) = -w_0(0), \quad \vartheta_{11}(0) = 0. \tag{5.13}$$

for the unknown constants D_1 and D_2.

5. Non-homogeneous boundary value problem for plate equations (5.3) with boundary conditions

$$v_p(1) = v_0(l), \quad h_p T_{1p}(1) = -h_s Q_{12}(l).$$

Table 5.1 Intensity indices for the vibration mode of the first (plate-like) type

	Functions							
Indices	u	v	w	ϑ	T_1	S	M_1	Q_1
I_0	3	3	3	3	3	3	7	7
I_1	3	4	2	1	4	3	4	3

Thus, the approximate solution of the eigenvalue problem (5.1)-(5.5) is reduced to the solution of five more simple problems.

First, one has to solve eigenvalue problem 1 for the plate bending deformation. Then, solution of Eqs. (5.11) is used to find edge effect functions at the parallel $s = s_k$. Further, one can obtain solution of plane stress problem 3, using boundary conditions (5.12) and find the edge effect integrals at parallels $s = 0$ and $s = l$ with the help of Eqs. (5.13). Finally, boundary value problem 5, describing tangential displacements of the plate, may be solved.

We consider only problem 1, since solving it one finds frequency parameter λ and the main term of the vibration mode. Displacements of the cylindrical shell and the tangential displacement of the plate are very small compared to the plate deflection (see relation (5.9) and Table 5.1).

To obtain the improved first approximation solution, problem 1 may be refined by replacing boundary condition (5.10) with more accurate condition. From relations

$$w_2(l) = 0, \quad \vartheta_{1p}(1) = \vartheta_{12}(l), \quad h_s M_{12}(l) = -h_p M_{1p}(1)$$

we find

$$D_3 + D_4 = 0, \quad \vartheta_{1p}(1) = -\mu(r_3 D_3 + r_4 D_4), \quad h_p M_{1p}(1) = \mu^4 h(r_3^2 D_3 + r_4^2 D_4). \tag{5.14}$$

It follows from (5.14) that

$$M_{1p}(1) = -\sqrt{2} g \mu^3 \frac{h_s}{h_p} \vartheta_{1p}(1).$$

The refined value of frequency parameter λ is obtained using the following condition

$$w_p(1) = 0, \quad M_{1p}(1) = -\sqrt{2} g \mu^3 \frac{h_s}{h_p} \vartheta_{1p}(1) \tag{5.15}$$

instead of the boundary condition (5.10).

Equations (5.2) describing free transverse vibrations of a circular plate are reduced to the following equation

$$\Delta^2 w - \beta^4 w = 0, \quad \beta^4 = \frac{\lambda}{\mu_p^4}, \tag{5.16}$$

where

$$\Delta = \frac{1}{s_p} \frac{d}{ds_p} \left(s_p \frac{d}{ds_p} \right) - \frac{m^2}{s_p^2}.$$

The exact solution of Eq. (5.16) has the form

$$w = C_1 J_m(\beta s_p) + C_2 I_m(\beta s_p), \tag{5.17}$$

where C_1 and C_2 are the arbitrary constants, J_m is the Bessel function and I_m is the modified Bessel function. The substitution of solution (5.17) into boundary conditions (5.15) provides a system of linear algebraic equations for the unknowns C_1 and C_2. The system has nontrivial solutions if its determinant vanishes:

$$J_m(\beta)I_{m-1}(\beta) - J_{m-1}(\beta)I_m(\beta) - \frac{2\beta J_m(\beta)I_m(\beta)}{1-\nu-k} = 0, \tag{5.18}$$

where

$$k = \frac{\sqrt{2}\sigma^{1/4}}{\mu\delta^3}, \quad \delta = \frac{h_p}{h_s}.$$

For the plate joined with the shell the roots $\beta(m,n)$, $m = 0,1,2,\ldots$, $n = 1,2,\ldots$ of Eq. (5.18) are in the improved first approximation the plate vibration frequency parameters. Vibrations are axisymmetric for $m = 0$. The case $k = 0$ corresponds to vibrations of a plate with the simply supported edge. If $k \to \infty$ then the roots of Eq. (5.18) tend to the roots of equation

$$J_m(\beta)I_{m-1}(\beta) - J_{m-1}(\beta)I_m(\beta) = 0, \tag{5.19}$$

and the roots of Eq. (5.19) for the plate with the clamped edge are in the first approximation the plate vibration frequency parameters.

Consider the structure with the following parameters $h_p = h_s = h = 0.01$, $\nu = 0.35$. The first roots of Eqs. (5.19) and (5.18) and values for frequency parameter obtained by Finite Elements Analysis (FEA) for the plate joined with the cylindrical shell are given in Table 5.2. The last column contains values of β for a plate with the simply supported edge. The dependence of angular frequency on the frequency parameter is given by formula

$$\omega^2 = \beta^4 \frac{h^2 E}{12\sigma\rho R^2}. \tag{5.20}$$

Table 5.2 Lower plate vibrations frequency parameters β

m	n	Eq. (5.19)	Eq. (5.18)	FEA	$k = 0$
0	1	3.196	3.086	3.070	2.238
1	1	4.611	4.460	4.422	3.736
2	1	5.906	5.722	5.658	5.067
0	2	6.306	6.111	6.058	5.457
3	1	7.144	6.932	6.835	6.326
1	2	7.799	7.570	7.480	6.967
4	1	8.347	8.111	7.978	7.543
2	2	9.197	8.939	8.805	8.377
0	3	9.439	9.176	9.002	8.615

5.5 Asymptotic Solution for the Shell-like Vibrations

The transverse flexural stiffness of the plate is essentially smaller than its tangential (in plane) stiffness. Neglecting the flexural stiffness of the plate we obtain $Q_{1p} = M_{1p} = 0$. Then from (5.4) two boundary conditions at the shell edge $s = l$ are found as:

$$T_1 = 0, \quad M_1 = 0 \quad \text{for} \quad s = l. \tag{5.21}$$

If the tangential stiffness of the plate in plane is infinitely large, then $u_p = v_p = 0$ and

$$v = 0, \quad w = 0 \quad \text{for} \quad s = l. \tag{5.22}$$

Conditions (5.21) and (5.22), corresponding to the simply supported shell edge, are the simplest join conditions between a shell and a plate.

If the ratio $\delta = h_p/h_s$ is large then the flexural stiffness of the plate cannot be ignored and conditions (5.21) need to be revised. On the other hand, for small values of δ it is unjustified to suppose that the tangential plate stiffness is very large and use conditions (5.22). The comparison of frequencies obtained by asymptotic and numerical methods shows, that in the case $\delta \sim 1$ the use of conditions (5.21) and (5.22) in the asymptotic analysis provides good results.

For the low-frequency second (shell-like) type vibrations the vibration mode of cylindrical shell stiffened by the circular plate is similar to the vibration mode of the non-stiffened shell. In both cases the circumferential wave number m is large. We seek approximate solution of the eigenvalue problem (5.1)-(5.5) as a sum of the semi-membrane solution and edge effect functions [6]:

$$y = m^{I_0(y)} y_0 + m^{I_1(y)} (y_1 + y_2), \quad k = 1, 2, \dots, n+1. \tag{5.23}$$

Here y denotes any unknown function, $I(y)$ and $I_1(y)$ are the intensity indices. Table 5.3 lists values of the intensity indices for the problem under consideration

The function v_0 satisfy the semi-membrane equation

$$\frac{d^4 v_0}{ds^4} - \alpha_s^4 v_0 = 0, \tag{5.24}$$

where

$$\alpha_s^4 = \frac{\lambda m^4 - \mu^4 m^8}{\sigma},$$

Table 5.3 Intensity indices for the vibration mode of the first (plate-like) type

	Functions							
Indices	u	v	w	ϑ	T_1	S	M_1	Q_1
I_0	-2	-1	0	0	-2	-3	-6	-6
I_1	-4	-5	-2	0	-4	-3	-6	-4

and

$$w_0 = -v_0, \quad u_0 = \frac{\mathrm{d}v_0}{\mathrm{d}s}, \quad T_{10} = \frac{\mathrm{d}^2 v_0}{\mathrm{d}s^2}, \quad S_0 = -\sigma \frac{\mathrm{d}^3 v_0}{\mathrm{d}s^3}. \tag{5.25}$$

The edge effect functions have the form (5.8).

Boundary conditions for Eqs. (5.24) may be obtained by splitting boundary conditions into main and additional conditions [12]. Neglecting the small terms in the main conditions we obtain the boundary conditions for Eqs. (5.24).

At the edge $s = 0$ the main conditions are

$$v = u = 0.$$

The separation of the boundary conditions (5.21) and (5.22) gives the following main conditions at the edge $s = l$

$$v = T_1 = 0.$$

Taking into account intensity indices and formulas (5.25) we obtain boundary conditions for Eq. (5.24):

$$v_0(0) = \frac{\mathrm{d}v_0}{\mathrm{d}s}(0) = 0, \quad v_0(l) = \frac{\mathrm{d}^2 v_0}{\mathrm{d}s^2}(l) = 0. \tag{5.26}$$

The boundary value problem (5.24), (5.26) also describes the vibrations of a beam with the clamped edge $s = 0$ and the simply supported edge $s = l$. The solution of this problem is well known (see [11]). The eigenvalues for problem (5.24), (5.26) are given by formula $\alpha_{sn} = \kappa_n / l$, where κ_n are the roots of the equation

$$\tan \kappa = \tanh \kappa. \tag{5.27}$$

The minimal positive root of Eq. (5.27) is $\kappa_1 = 3.927$.

The frequency parameter

$$\lambda(m,n) = \frac{\sigma \kappa_n^4}{m^4 l^4} + \mu^4 m^4 \tag{5.28}$$

attains its minimal value, corresponding to the fundamental frequency, if $n = 1$ and m is close to m_0 where

$$m_0^4 = \frac{\sqrt{\sigma} \kappa_1^2}{\mu^2 l^2}.$$

Consider the cylindrical shell with the length $l = 4$ and thickness $h_s = 0.01$, which is joined with the plate of same thickness $h_p = h_s$. In Table 5.4 the values of parameter

$$\eta = (12\lambda)^{1/4} / \sqrt{h} \tag{5.29}$$

are given for $\nu = 0.35$. The third and fourth columns list the results of FEA obtained for the shell-plate structure and for the shell with simply supported edge respectively. The last column contains values of η due to (5.28).

Table 5.4 Frequency parameters for the shell-like vibrations.

n	m	FEA, constr.	FEA, shell	Eq. (5.28)
1	4	4.770	4.746	5.026
1	5	5.173	5.166	5.287
1	3	5.315	5.218	5.991
1	6	6.010	6.007	6.086
2	5	6.508	6.465	6.901
2	6	6.669	6.652	6.760
1	2	6.872	6.388	8.848
2	4	7.089	6.962	8.083
1	7	6.968	6.965	7.030
2	7	7.311	7.302	7.293
1	8	7.956	7.955	8.012

The attached plate makes the boundary condition more rigid, than the simply supported edge. However, for large wave number m the influence of a plate on shell frequencies quickly decreases from 7.8% at $m = 2$ up to fractions of a percent. For fixed m the plate influence slightly increases with n. The asymptotic formula gives a good approximation for frequencies at $m \geq m_0$.

5.6 Vibrations of the Third (Beam-like) Type

Among the lower eigenfrequencies of the shell, there may also be frequencies of beam-like vibrations, at which the structure vibrates like a cantilever beam with a load at the end (see Fig. 5.3). The equation of transverse vibrations of a beam with one end clamped and loaded by a concentrated mass on the other end has the form

$$w''''(s/l) - \alpha^4 w(s/l) = 0, \quad \alpha^4 = \frac{\rho S l^4 R^4}{EJ}\omega^2, \quad 0 \leq s \leq l, \tag{5.30}$$

and the boundary conditions are

$$w(0) = w'(0) = 0, \quad w''(1) = 0, \quad EJw'''(1)/l^3 = -m\omega^2 w(1).$$

Here J is the moment of inertia of the shell section about its diameter, $m = \pi R^2 h_p \rho$ is the plate mass, $S = 2\pi R h_s$ is the shell cross-sectional area. Substituting the solution of Eq. (5.30)

$$w(s/l) = C_1 \sin(\alpha s/l) + C_2 \cos(\alpha s/l) + C_3 \sinh(\alpha s/l) + C_4 \cosh(\alpha s/l)$$

into the boundary conditions and equating the determinant of the homogeneous linear system to zero, we obtain an equation to find the parameter α

$$\gamma\alpha(\cos(\alpha)\sinh(\alpha)-\cosh(\alpha)\sin(\alpha))+\cosh(\alpha)\cos(\alpha)+1=0, \tag{5.31}$$

where $\gamma=\frac{h_p}{2lh_s}$

Results of analytical and finite element analysis for different shell lengths are given in Table 5.5. For shells of small relative length, Eq. (5.31) gives strongly overestimated frequencies, but the accuracy of the analytical formula increases with increasing shell length. Note that as the shell length increases, the beam frequency rapidly decreases and at relative lengths shell over 8.4 (FEM) (8.7 (analytical)) it becomes the lowest natural frequency of the structure.

5.7 Spectrum Optimization by Thickness Variation

The results obtained for the spectrum may be used in solving optimization problems. The most common optimization problem for a spectrum is the maximization the value of the lowest natural frequency of vibrations when changing parameters of the system, either geometric or physical.

We confine ourselves to considering the effect of the wall thickness on natural frequencies, provided that the mass of the structure is preserved. Since the lowest eigenfrequencies of the structure are plate frequencies, we will try to increase the thickness of the plate so that its lowest frequency would coincide with the lowest shell frequency.

For the shell and the plate with thicknesses h and $h_p=\delta h$ relatively, for angular frequency ω formula similar to (5.20) is valid

$$\omega^2=\beta_\delta^4\frac{h^2E}{12\sigma\rho R^2}, \quad \beta_\delta^4=(\beta\sqrt{\delta})^4. \tag{5.32}$$

The second line of Table 5.6 shows the first root β of Eq. (5.8), and the third line — the value of β_δ. The fourth line contains β_δ values corresponding to the fundamen-

Table 5.5 Frequency parameter α vs shell length l for beam-like vibrations of a shell

l	Analytical	FEA	Error
4	6.411	5.561	15.3%
8	3.357	3.203	4.8%
12	2.277	2.223	2.4%
16	1.723	1.701	1.3%
20	1.387	1.375	0.9%

Table 5.6 Frequency parameter for the fundamental frequency vs. plate thickness.

δ	1.0	1.5	2	2.5	3	3.5	4
β	3.086	2.906	2.715	2.565	2.461	2.393	2.348
β_δ	3.086	3.560	3.840	4.055	4.263	4.477	4.697
FEA	3.069	3.537	3.815	4.028	4.231	4.439	4.648

tal frequency found by means of FEA. The difference between the asymptotic and numerical results is less than 1.1%.

The condition of conservation of mass of the structure allows us to determine the shell thickness h_s, which decreases with increasing the plate thickness. Indeed, from the condition of the preservation of the volume

$$2\pi hl + \pi h = 2\pi h_s l + \pi h\delta$$

the shell thickness $h_s = (1+(1-\delta)/2/l)h = \delta_s h$ is determined. For $\delta = 1$, the plate and shell thicknesses are equal $h_s = h_p = h$.

Figure 5.5 on the left shows the δ dependences of the frequency parameter Ω for the lowest plate frequency (at $m = 0$), two lowest shell frequencies (at $m = 4$ and $m = 5$) and the first beam frequency for $h = 0.01$. The frequency parameter Ω is related to parameters α, β and η as follows:

$$\Omega = \begin{cases} \beta_\delta \text{ for plate-like vibrations (see (5.32))} \\ \eta \text{ for shell-like vibrations (see (5.29))} \\ \alpha/l(3\sigma(4/h_s^2+1)^{1/4} \text{ for beam-like vibrations (see (5.30))} \end{cases}$$

For $\delta = 1$, the plate frequency turns out to be lower. As δ increases, h_s decreases and the plate mass increases, hence, both shell frequencies and beam frequencies decrease. As the length L of the shell increases, the lowest beam frequency decreases and for $L > 8$ it becomes the lowest frequency of the structure.

When changing parameter δ, the frequencies may become multiples. If plate-like and shell-like frequencies collide, then the distortions of the corresponding vibration modes are negligible. When the plate-like and beam-like frequencies are close the vibration mode is a superposition of beam and plate modes, which prevents the determination of the vibration type. The collision of the first beam and second plate frequencies explains the nonmonotonicity of the green line in Fig. 5.5.

Table 5.7 shows the values of the optimal thicknesses of structural elements and the corresponding frequency parameter found analytically and with the finite element method.

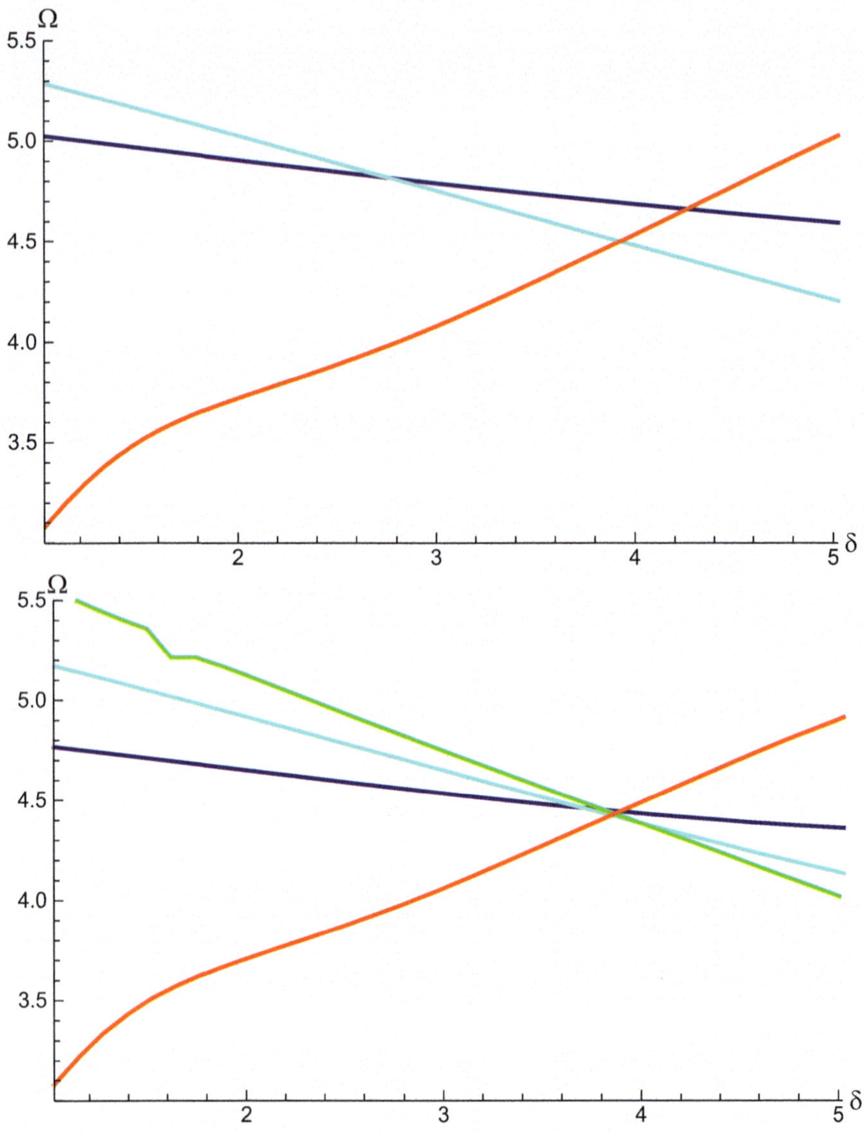

Fig. 5.5 Dependence of the frequency parameter for the lowest natural frequencies on the thickness of the plate while maintaining the total mass of the structure. Plate frequency — red line, shell frequencies for $m = 4$ — dark blue line and for $m = 5$ — light blue line, beam frequency — green line. Analytical results on the top of figure, numerical (FEM) — on the bottom.

Table 5.7 Optimal thicknesses of structural elements and the corresponding frequency parameter

	Analytical	FEA
Ω	4.50	4.43
h_p	3.90	3.83
h_s	0.64	0.65

5.8 Conclusions

Natural frequencies and vibration modes of the structure consisting of a circular cylindrical thin shell joined with a circular thin plate were studied numerically using FEM. The application of asymptotic methods to the analysis of the low-frequency vibrations of the structure permits to obtain simple approximate formulas for natural frequencies and vibration modes.

The comparison of numerical and asymptotic results shows that for the 9 first plate-like frequencies the relative error of asymptotic formula is less than 2%. In the case of shell-like vibrations for the lowest 6 frequencies the maximal difference between numerical and asymptotic values does not exceed 11% and decreases with the shell thickness. Analytical formula for beam-like vibrations of a shell with the length $l = 4$ has approximately the same accuracy that increases with the shell length. Hence, before starting the design approximate formulas may be used to estimate optimal parameters of the structure.

References

1. S.M. Bauer, S.B. Filippov, A.L. Smirnov, P.E. Tovstik and R. Vaillancourt. *Asymptotic Methods in Mechanics of Solids*. Springer International Publishing, Switzerland, 2015.
2. D. Bushnell. Stress, stability and complex, branched shells of revolution. *Comput. Struct.*, 4(2):399-435, 1974.
3. S.B. Filippov. Low-frequency vibration of cylindrical shells. Part II: Connected shells. In: *Asymptotic Methods in Mechanics (Ed. by R. Vaillancourt and A.L. Smirnov)*, CRM Proceedings & Lecture Notes, Vol. 3, AMS, pp. 205-216, 1993.
4. S.B. Filippov and N.V. Naumova. Axisymmetric vibrations of thin shells of revolution joint at a small angle. *Technische Mechanik*, 18(4):285-290, 1998.
5. S.B. Filippov. *Theory of Connected and Stiffened Shells (in Russ.)*. St. Petersburg State University Press, St. Petersburg, 1999.
6. S.B. Filippov. Optimal Design of Stiffened Cylindrical Shells Based on an Asymptotic Approach. *Technische Mechanik*, 24(3-4):221-230, 2004.
7. S.B. Filippov. Asymptotic approximations for frequencies and vibration modes of cylindrical shell stiffened by annular plates. In: *Analysis of Shells, Plates, and Beams. A State of the Art Report (Ed. by H. Altenbach, N. Chinchaladze, R. Kienzler and W.H. Müller)*, Springer's Series Advanced Structured Materials, Vol. 123, pp. 123-140, 2020.
8. A.L. Gol'denveizer, V.B. Lidsky and P.E. Tovstik. *Free Vibrations of Thin Elastic Shells (in Russ.)*. Nauka, Moscow, 1979.
9. A.L. Gol'denveizer. Asymptotic Method in Theory of Shells. *Advances in Mechanics*, 5(1/2):137-182, 1982.

10. W. Soedel. *Vibrations of Shells and Plates*. 3rd ed, CRC Press, 2004.
11. S.P. Timoshenko. *Vibration Problems in Engineering*, Van Nostrand, New York, 1955.
12. P.E. Tovstik, A.L. Smirnov. *Asymptotic Methods in the Buckling Theory of Elastic Shells*. World Scientific Publishing Co Ltd., Singapore, New Jersey, London, Hong Kong, 2001.
13. M.I. Vishik and L.A. Lyusternik. Regular degeneration and boundary layer for linear differential equations with small parameter. *Uspekhi Matematicheskikh Nauk*, 12(5(77)):3-122, 1957.

Chapter 6
Indentation of an Absolutely Rigid Thin Inclusion into One of the Crack Faces in an Elastic Plane Under Slippage at the Ends

Vahram N. Hakobyan, Harutyun A. Amirjanyan, Lilit L. Dashtoyan, and Avetik V. Sahakyan

Abstract The article considers the stress state of an elastic plane with a crack of finite length, when an absolutely rigid thin inclusion of the same length is pressed into one of edges, under the action of a concentrated force. For the contacting side of the inclusion, it is assumed that in its middle part, there is adhesion to the matrix, and slippage occurs along the edges, which is described by the law of dry friction. The problem is mathematically formulated as a system of singular integral equations, the behavior of the unknown functions in the vicinity of the ends of the inclusion-crack and at the separation points of the adhesion and slip zones is studied. The governing system of integral equations is solved by the method of mechanical quadratures. The laws of distribution of contact stresses, as well as the lengths of the adhesion and slip zones, depending on the coefficient of friction, Poisson's ratio of the half-plane material, and the angle of inclination of the external force, are found.

Key words: Elasticity, Mixed boundary value problem, Thin inclusion, Crack, Contact model, Adhesion and slip zones, Quadrature rules

6.1 Introduction

The study of the stress-strain state in the vicinity of the ends of stress concentrators of various types and the development of methods for reducing the level of their harmful effect on the performance and reliability of structures and their parts has always been and remains one of the most important areas of contact and mixed problems of the theory of elasticity and fracture mechanics. A lot of work has been

Vahram N. Hakobyan · Harutyun A. Amirjanyan · Lilit L. Dashtoyan · Avetik V. Sahakyan
Institute of Mechanics, National Academy of Sciences of Republic Armenia, 24B Baghramyan Ave., Yerevan 0019, Republic of Armenia,
e-mail: vhakobyan@sci.am, amirjanyan@gmail.com, lilit.dashtoyan@sci.am, avetik.sahakyansci.am

H. Altenbach et al. (eds.), *Advances in Solid and Fracture Mechanics*,
Advanced Structured Materials 180,
https://doi.org/10.1007/978-3-031-18393-5_6

fulfilled in this direction, among which we note the well-known monographs [1, 2, 3, 4, 5], which contain quite a lot of fundamental results. We also point to the monograph [6] and works [7, 8], where exact solutions were obtained for a number of problems for a composite plane and space with interfacial cracks on one of the faces of which an absolutely rigid stamp is pressed under various contact conditions. In [9, 10], closed solutions were obtained for a number of contact problems for an elastic homogeneous and orthotropic plane with finite cracks, on one or both sides of which absolutely rigid stamps act simultaneously, the base length of which is less than the length of the crack. In these works, it was assumed that either the stamp is fully coupled with the crack edge, or there is a smooth contact between them.

It is known that in 1945, L.A. Galin proposed a model of contact between a stamp and an elastic half-plane, which assumed adhesion in the middle part of the contact zone and slippage, according to the Coulomb dry friction law, along its edges [1, 11]. Using a conformal mapping, L.A. Galin constructed an approximate solution in a closed form. Many researchers were interested in this model. The solution of the Galin problem with the use of the Fuchs-class equation [12], as well as its reduction to the vector Riemann problem [13], is known. In [14], using the Wiener-Hopf method, the solution of this problem is reduced to an infinite system of algebraic equations. The work [15] is devoted to solving the Galin problem by the method of discrete singularities (method of mechanical quadratures), the numerical results obtained in [14] and [15] coincide.

This paper, within the framework of the model of the Galin problem, considers the stress state of an elastic homogeneous plane with a crack of finite length, into one of the edges of which an absolutely rigid thin stamp of the same length is pressed into. The same problem in a symmetrical formulation was presented at the conference and published in [16]. However, there are only descriptions of the governing equations and no solution.

6.2 Statement of the Problem and Derivation of the Governing System of Integral Equation

Let a homogeneous elastic plane related to the Cartesian coordinate system xy, on the interval $(-a,a)$ of the abscissa axis contains an absolutely rigid thin inclusion, which is not previously coupled to the matrix, and is deformed under the action of a concentrated force P_0 applied to the midpoint of the upper face. It is assumed that the contact of the inclusion with the matrix is described by the model of the contact problem by Galin [1], i.e. it is considered that on a certain interval (b,c) of the middle part of the lower side of the inclusion there is adhesion with the matrix, and on the end intervals $(-a,b,)$ and (c,a) slippage occurs, obeying the law of dry friction (Fig. 6.1). The problem is to determine the dimensions of the adhesion and sliding zones, the contact stresses under the inclusion, and the concentration factors of destructive stresses at the endpoints of the stress concentrator of the mixed type inclusion-crack.

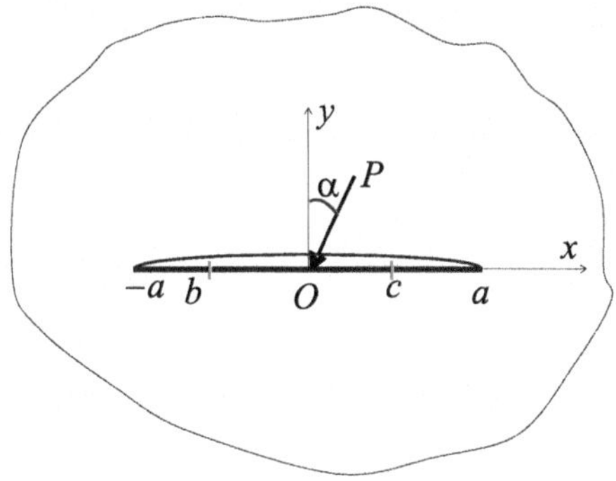

Fig. 6.1 A rigid thin inclusion inside a crack

Dividing the plane along the Ox axis into two half-planes and providing the stress and displacement components related to the points of the upper and lower half-planes with superscripts (1) and (2), respectively, at the intervals $(-a,a)$ of each half-plane the following conditions have occurred:

$$\begin{cases} \sigma_y^{(1)}(x,0) = 0 \\ \tau_{xy}^{(1)}(x,0) = 0 \end{cases} \quad (-a < x < a) \tag{6.1}$$

$$\begin{cases} u_2'(x,0) = 0 & (b < x < c) \\ v_2'(x,0) = 0 & (-a < x < a) \end{cases} \tag{6.2}$$

$$\tau_{xy}^{(2)}(x,0) = -f\,\mathrm{sgn}(x)\,\sigma_y^{(2)}(x,0) \quad x \in (-a,b) \bigcup (c,a). \tag{6.3}$$

Here, $u_j(x,y)$ and $v_j(x,y)$ $(j = 1,2)$ are the components of displacement, which satisfy the Lamé equations in the half-planes $y \geq 0$ and $y \leq 0$, and f is the friction coefficient.

Using discontinuous solutions for an elastic plane with a cut [6], which allow us to express all components of the displacement vector and the stress tensor in terms of stress and displacement jumps on the interval $(-a,a)$, we write out the expressions necessary to satisfy (6.1) and (6.2),

$$\sigma_y^{(1)}(x,0) = \frac{1}{2}\left\{\sigma(x) + \frac{\theta}{\pi}\int_{-a}^{a}\frac{\tau(s)\,\mathrm{d}s}{s-x} + \frac{\lambda}{\pi}\int_{-a}^{a}\frac{V'(s)\,\mathrm{d}s}{s-x}\right\},$$

$$\tau_{xy}^{(1)}(x,0) = \frac{1}{2}\left\{\tau(x) - \frac{\theta}{\pi}\int_{-a}^{a}\frac{\sigma(s)\,\mathrm{d}s}{s-x} + \frac{\lambda}{\pi}\int_{-a}^{a}\frac{U'(s)\,\mathrm{d}s}{s-x}\right\},$$

$$u_2'(x,0) = -\frac{1}{2}\left\{U'(x)+\frac{1}{2\pi\theta_0}\int_{-a}^{a}\frac{\tau(s)}{s-x}ds-\frac{\theta}{\pi}\int_{-a}^{a}\frac{V'(s)}{s-x}ds\right\},$$

$$\upsilon_2'(x,0) = -\frac{1}{2}\left\{V'(x)+\frac{1}{2\pi\theta_0}\int_{-a}^{a}\frac{\sigma(s)}{s-x}ds+\frac{\theta}{\pi}\int_{-a}^{a}\frac{U'(s)}{s-x}ds\right\}.$$

The following notations are introduced here:

$$\begin{aligned}\sigma(x) &= \sigma_y^{(1)}(x,0)-\sigma_y^{(2)}(x,0),\ \ \tau(x)=\tau_{xy}^{(1)}(x,0)-\tau_{xy}^{(2)}(x,0),\\ U'(x) &= u_1'(x,0)-u'_2(x,0),\ \ V'(x)=\upsilon_1'(x,0)-\upsilon'_2(x,0),\end{aligned}\tag{6.4}$$

$$\theta=\frac{1-2\nu}{2(1-\nu)},\quad \theta_0=\frac{2(1-\nu)}{3-4\nu}\mu,\quad \lambda=\frac{\mu}{1-\nu},$$

μ is the shear modulus, and ν is the Poisson's ratio of the plane material.

Satisfying the conditions of the boundary problem (6.1)-(6.2), taking into account condition (6.3), we come to the following system of singular integral equations:

$$\begin{aligned}
&\sigma(x)+\frac{\theta}{\pi}\left[f\int_{-a}^{b}\frac{\sigma(s)}{s-x}ds+\int_{b}^{c}\frac{\tau(s)}{s-x}ds-f\int_{c}^{a}\frac{\sigma(s)}{s-x}ds\right]+\frac{\lambda}{\pi}\int_{-a}^{a}\frac{V'(s)}{s-x}ds=0;\\
&\hspace{25em}(|x|<a),\\
&f\sigma(x)-\frac{\theta}{\pi}\int_{-a}^{a}\frac{\sigma(s)}{s-x}ds+\frac{\lambda}{\pi}\int_{-a}^{a}\frac{U'(s)}{s-x}ds=0;\ (-a<x<b),\\
&\tau(x)-\frac{\theta}{\pi}\int_{-a}^{a}\frac{\sigma(s)}{s-x}ds+\frac{\lambda}{\pi}\int_{-a}^{a}\frac{U'(s)}{s-x}ds=0;\ (b<x<c),\\
&-f\sigma(x)-\frac{\theta}{\pi}\int_{-a}^{a}\frac{\sigma(s)}{s-x}ds+\frac{\lambda}{\pi}\int_{-a}^{a}\frac{U'(s)}{s-x}ds=0;\ (c<x<a),\\
&U'(x)+\frac{1}{2\pi\theta_0}\left[f\int_{-a}^{b}\frac{\sigma(s)}{s-x}ds+\int_{b}^{c}\frac{\tau(s)}{s-x}ds-f\int_{c}^{a}\frac{\sigma(s)}{s-x}ds\right]-\frac{\theta}{\pi}\int_{-a}^{a}\frac{V'(s)}{s-x}ds=0;\\
&\hspace{25em}(b<x<c),\\
&V'(x)+\frac{1}{2\pi\theta_0}\int_{-a}^{a}\frac{\sigma(s)}{s-x}ds+\frac{\theta}{\pi}\int_{-a}^{a}\frac{U'(s)}{s-x}ds=0;\ (|x|<a).
\end{aligned}\tag{6.5}$$

The obtained system should be considered under the conditions of closing of the ends of the crack and equilibrium of the stamp, which in the notation (6.4) have the form:

$$\int_{-a}^{a} V'(s)\mathrm{d}s = 0; \quad \int_{-a}^{a} U'(s)\mathrm{d}s = 0; \quad \int_{-a}^{a} \sigma(s)\mathrm{d}s = P\cos\alpha; \quad \int_{-a}^{a} \tau(s)\mathrm{d}s = P\sin\alpha. \tag{6.6}$$

The presence in the resulting system (6.5) of equations with different definition domains enforces us on each of the intervals $(-a,b)$, (b,c), and (c,a) consider the unknown functions as independent unknowns, thereby tripling their number.

However, due to conditions (6.3), the number of unknown functions is reduced to ten. Dividing the domains of definition of the first and last equations of system (6.5) into three subdomains and considering them as separate equations, we bring the number of equations of system (6.5) to ten. As a result, we obtain a system of ten singular integral equations with respect to the functions

$$\{\varphi_1,\varphi_2,\varphi_3\}^{(ol,i,or)} = \{U',V',\sigma\}\;; \quad \varphi_4^{(i)} = \tau, \tag{6.7}$$

defined on the intervals $(-a,b)$, (b,c) and (c,a) and, in accordance with the interval of definition, marked with superscripts ol (outer-left), i (internal) and or (outer-right) . Four of the equations will be defined on the interval (b,c) and three each on the outer intervals $(-a,b)$ and (c,a).

Dividing the stress components by the shear modulus and reducing each of the intervals for determining the equations to the interval (-1,1), we pass to dimensionless quantities, while retaining their designations. As a result, we obtain a system of ten singular integral equations having a form similar to the following:

$$\begin{aligned}\varphi_3^{ol}(\eta) &+ \frac{\theta}{\pi}\left[f\int_{-1}^{1}\frac{\varphi_3^{ol}(\xi)}{\xi-\eta}\mathrm{d}\xi + \int_{-1}^{1}\frac{\varphi_4^{i}(\xi)}{\xi-z_0}\mathrm{d}\xi - f\int_{-1}^{1}\frac{\varphi_3^{or}(\xi)}{\xi-z_1}\mathrm{d}\xi\right] \\ &+ \frac{1}{2\pi(1-\nu)}\left[\int_{-1}^{1}\frac{\varphi_2^{ol}(\xi)}{\xi-\eta}\mathrm{d}\xi + \int_{-1}^{1}\frac{\varphi_2^{i}(\xi)}{\xi-z_0}\mathrm{d}\xi + \int_{-1}^{1}\frac{\varphi_2^{or}(\xi)}{\xi-z_1}\mathrm{d}\xi\right] = 0, \\ &\qquad (-1<\eta<1)\end{aligned} \tag{6.8}$$

where

$$b^* = \frac{b}{a} < 1, \qquad c^* = \frac{c}{a} < 1,$$

$$z_0 = \frac{b^*+1}{c^*-b^*}\eta - \frac{1+c^*}{c^*-b^*},$$

$$z_1 = \frac{b^*+1}{1-c^*}\eta - \frac{2+c^*-b^*}{1-c^*}.$$

The solution to the obtained system we will seek in the class of functions that have power-law behavior in the vicinity of the ends of the integration interval. Using the well-known results of N. Muskhelishvili on the behavior of the Cauchy-type integral at the ends of the integration line [17], we investigate the behavior of equations in the vicinity of the points ± 1. In a result we obtain the equations to determine the singularity exponents. The corresponding equations for external intervals don't have a unique solution and the singularity exponents will have three various values. Based on this, we represent the functions $\varphi_i^{(ol)}, \varphi_i^{(or)}$ $(i = 1,2,3)$ as linear combinations of new six unknowns $\psi_i^{(ol)}, \psi_i^{(or)}$ $(i = 1,2,3)$, whose behavior at the ends is uniquely determined:

$$\varphi_1^{(ol)}(s) = \psi_1^{(ol)}(s) + \frac{-f\sqrt{1-f^2\theta^2} + (1+f^2)\theta}{2(1-\theta)}\psi_2^{(ol)}(s) + \frac{f\sqrt{1-f^2\theta^2} + (1+f^2)\theta}{2(1-\theta)}\psi_3^{(ol)}(s),$$

$$\varphi_2^{(ol)}(s) = -\frac{\sqrt{1-f^2\theta^2}}{2(1-\theta)}\psi_2^{(ol)}(s) + \frac{\sqrt{1-f^2\theta^2}}{2(1-\theta)}\psi_3^{(ol)}(s),$$

$$\varphi_3^{(ol)}(s) = \psi_2^{(ol)}(s) + \psi_3^{(ol)}(s),$$

$$\varphi_1^{(or)}(s) = \psi_1^{(or)}(s) + \frac{f\sqrt{1-f^2\theta^2} + (1+f^2)\theta}{2(1-\theta)}\psi_2^{(or)}(s) + \frac{-f\sqrt{1-f^2\theta^2} + (1+f^2)\theta}{2(1-\theta)}\psi_3^{(or)}(s),$$

$$\varphi_2^{(or)}(s) = -\frac{\sqrt{1-f^2\theta^2}}{2(1-\theta)}\psi_2^{(or)}(s) + \frac{\sqrt{1-f^2\theta^2}}{2(1-\theta)}\psi_3^{(or)}(s),$$

$$\varphi_3^{(or)}(s) = \psi_2^{(or)}(s) + \psi_3^{(or)}(s).$$

Taking into account that all unknown functions at the points of separation of the adhesion and slip zones take finite values, we consider them as unknown constants, and the solution of the governing system should be looking for in the following form:

$$
\begin{aligned}
\varphi_j^{(i)}(s) &= \frac{1-s}{2}C_{j1} + \frac{1+s}{2}C_{j2} + \varphi_{j*}^{(i)}(s)\left(1-s^2\right) \quad (j=1,2,3),\\
\varphi_4^{(i)}(s) &= \frac{1-s}{2}fC_{31} + \frac{1+s}{2}fC_{32} + \varphi_{4*}^{(i)}(s)\left(1-s^2\right)^{0.5-\gamma_1},\\
\psi_1^{(ol)}(s) &= 2^{0.5}(1+s)^{-0.5}\left(C_{11} - fC_{21} - \frac{\theta\left(1+f^2\right)}{2(1-\theta)}C_{31}\right)\\
&\quad + \psi_{1*}^{(ol)}(s)(1-s)^{0.5-\gamma_1}(1+s)^{-0.5},\\
\psi_2^{(ol)}(s) &= 2^{0.5-\gamma_2}(1+s)^{-0.5+\gamma_2}\left(\frac{1}{2}C_{31} - \frac{(1-\theta)}{\sqrt{1+f^2\theta^2}}C_{21}\right)\\
&\quad + \psi_{2*}^{(ol)}(s)(1-s)^{0.5-\gamma_1}(1+s)^{-0.5+\gamma_2},\\
\psi_3^{(ol)}(s) &= 2^{0.5-\gamma_3}(1+s)^{-0.5+\gamma_3}\left(\frac{1}{2}C_{31} + \frac{(1-\theta)}{\sqrt{1+f^2\theta^2}}C_{21}\right)\\
&\quad + \psi_{3*}^{(ol)}(s)(1-s)^{0.5-\gamma_1}(1+s)^{-0.5+\gamma_3},\\
\psi_1^{(or)}(s) &= 2^{0.5}(1-s)^{-0.5}\left(C_{12} + fC_{22} - \frac{\theta\left(1+f^2\right)}{2(1-\theta)}C_{32}\right)\\
&\quad + \psi_{1*}^{(or)}(s)(1-s)^{-0.5}(1+s)^{0.5-\gamma_1},\\
\psi_2^{(or)}(s) &= 2^{0.5+\gamma_3}(1-s)^{-0.5-\gamma_3}\left(\frac{1}{2}C_{32} - \frac{(1-\theta)}{\sqrt{1+f^2\theta^2}}C_{22}\right)\\
&\quad + \psi_{2*}^{(or)}(s)(1-s)^{-0.5-\gamma_3}(1+s)^{-0.5+\gamma_3},\\
\psi_3^{(or)}(s) &= 2^{0.5-\gamma_2}(1-s)^{-0.5+\gamma_2}\left(\frac{1}{2}C_{32} + \frac{(1-\theta)}{\sqrt{1+f^2\theta^2}}C_{22}\right)\\
&\quad + \psi_{3*}^{(or)}(s)(1-s)^{-0.5+\gamma_2}(1+s)^{0.5-\gamma_2}
\end{aligned}
\tag{6.9}
$$

where

$$
\begin{aligned}
\gamma_1 &= \arctan f\theta,\\
\gamma_2 &= \arctan\left(f\theta + \sqrt{1+f^2\theta^2}\right),\\
\gamma_3 &= \arctan\left(f\theta - \sqrt{1+f^2\theta^2}\right).
\end{aligned}
\tag{6.10}
$$

Here, C_{ij} $(i=1,2,3;\ j=1,2)$ are the unknown constants representing the values of the function $\varphi_j^{(i)}$ $(j=1,2,3)$ at the ends of the domain, and the functions with asterisks satisfy the Hölder condition on the segment $[-1,1]$ and take finite (non-zero) values at the ends ± 1.

Thus, the solution of the stated problem is reduced to a system of ten singular integral equations for ten functions with asterisks with appropriate weight functions and six constants. The presence of determined and different weight functions implies the use of the mechanical quadrature method [18] with different nodes and a different number of collocation points, depending on whether the weight function under the singular integral tends to infinity at both ends, at one end, or at none. As a result, the solution of the problem is reduced to a system of linear algebraic equations. Obviously, it is not expedient to write down here the system of linear algebraic equations, to which the system (6.8) of ten singular integral equations and four integral conditions (6.6) is reduced. It should also be noted that in the case under consideration, the expression for normal stresses contains two terms tending to infinity and, therefore, there is no stress intensity factor in the traditional sense.

6.3 Numerical Analysis

Having obtained representation (6.9), where all the unknown functions act with their weight functions, the solution of the resolving system can be constructed by the method of mechanical quadrature [18]. With the order of approximation n from representations (19), we will have $10n+6$ unknowns: $10n$ values of functions with asterisks at the corresponding nodal points and six constants. According to (19), six weight functions corresponding to the slip zones turn to infinity at one of the ends, and four functions corresponding to the adhesion zone turn to zero at both ends. Therefore, choosing n collocation points for six equations, and $n+1$ collocation points for four, we obtain $10n+4$ equations. To close the resulting system of linear algebraic equations, two of the discretized conditions (6.6) should be added to it. Moreover, one of them should be the third condition, because only it can ensure the heterogeneity of the resulting system of linear equations and the possibility of obtaining a solution. The other two conditions (6.6) are necessary to determine the coordinates of the interface points of the adhesion and slip zones b^* and c^*, which are non-linearly included in the matrix of singular integral algebraic equations.

An analysis of the results obtained for different approximation orders showed that $n = 5$ for the functions $\psi_{k*}^{(ol)}$ and $\psi_{k*}^{(or)}$ $(k = 1,2,3)$, and $n = 10$ for the functions $\varphi_{k*}^{(i)}$ $(k = 1,2,3,4)$, an accuracy of order 10^{-3} is obtained. Such an accuracy is quite sufficient for the graphical presentation of the results, therefore, further calculations were carried out at these values of the approximation orders. Figures 6.2 and 6.3 shows typical curves representing the distribution of normal and tangential contact stresses related to the shear modulus. These curves are qualitatively repeated for various values of the variable parameters of the problem: the friction coefficient, Poisson's ratio, and the angle of inclination of the applied force.

If the curves of the distribution of contact stresses at different values of the parameters of the problem are not very informative in themselves, then the dependence of the values b^* and c^* on the indicated parameters can also tell a lot about the distribution of contact stresses. Figure 6.4 shows the dependence curves of b^* and c^* on the friction coefficient for two values of Poisson's ratio. The solid lines correspond to the value $\nu = 0.1$, and the dotted lines correspond to the value $\nu = 0.2$. As we can

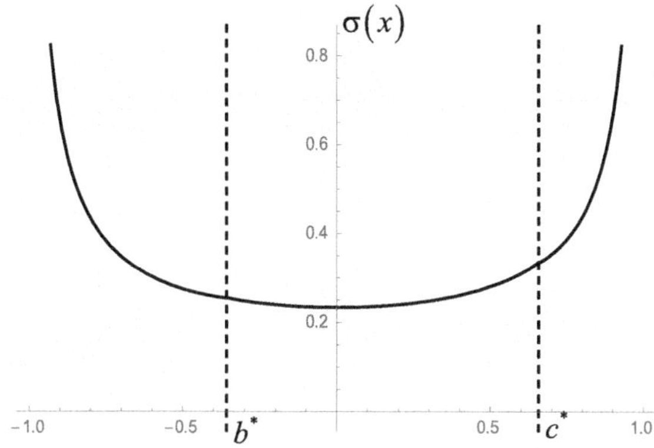

Fig. 6.2 Distribution of dimensionless normal contact stresses ($\alpha = 0.01$, $\nu = 0.2$, $f = 0.1$)

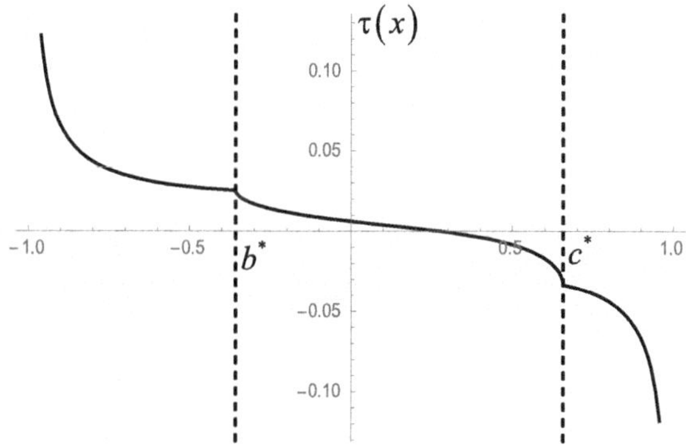

Fig. 6.3 Distribution of dimensionless tangential contact stresses ($\alpha = 0.01$, $\nu = 0.2$, $f = 0.1$)

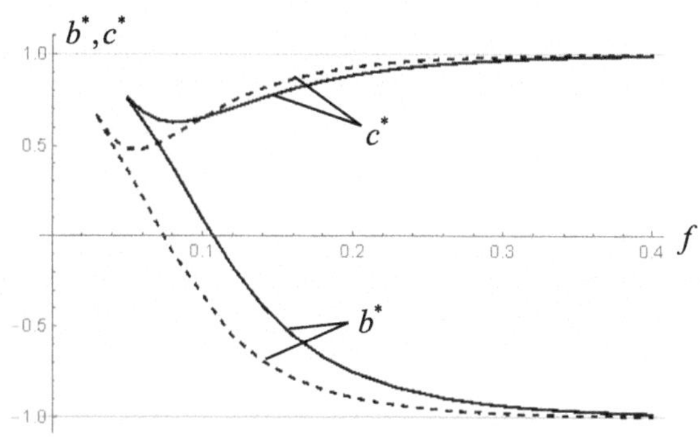

Fig. 6.4 Dependence of b^* and c^* on the coefficient of friction f.

see from the graphs in Fig. 6.4, at very low values of the friction coefficient, there is no adhesion zone. The adhesion zone appears at a certain, "critical" value of the friction coefficient, and it is the larger, the greater this coefficient. In addition, the location of pairs of curves corresponding to different values of the Poisson's ratio indicates that a larger value of the Poisson's ratio corresponds to a smaller "critical" value of the friction coefficient. A study of the last dependence showed that for possible, but different from the extreme, values of the Poisson's ratio, the "critical" value changes approximately in the interval $(0.2, 0.5)$.

Figure 6.5 shows the dependence curves of b^* and c^* on Poisson's ratio ν for $\alpha = 0.02$ and $f = 0.01$. According to Fig. 6.5, for materials with a large Poisson's ratio, the adhesion zone is larger.

The dependence of these quantities on the angle of inclination α of the applied force P is also interesting. Figure 6.6 shows the curves of these dependencies for $\nu = 0.2$ and $f = 0.1$. Obviously, for $\alpha = 0$, i.e. under normal external force, the adhesion zone is symmetrical. The same can be seen in Fig. 6.6: when $\alpha = 0$ we have $b^* = -c^* = -0.528$. The gradual deviation of the force to the right leads to a displacement of the adhesion zone to the right and to a fairly rapid decrease in its length.

6.4 Conclusion

A system of governing equations is derived for a rather complex contact problem on the indentation of a thin rigid inclusion of finite length into a crack edge of the same length located in an elastic plane. The problem is considered in the frame-

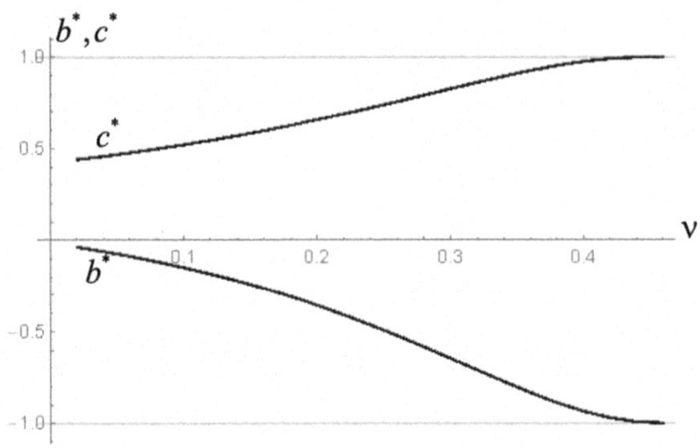

Fig. 6.5 Dependence of b^* and c^* on Poisson's ratio ν.

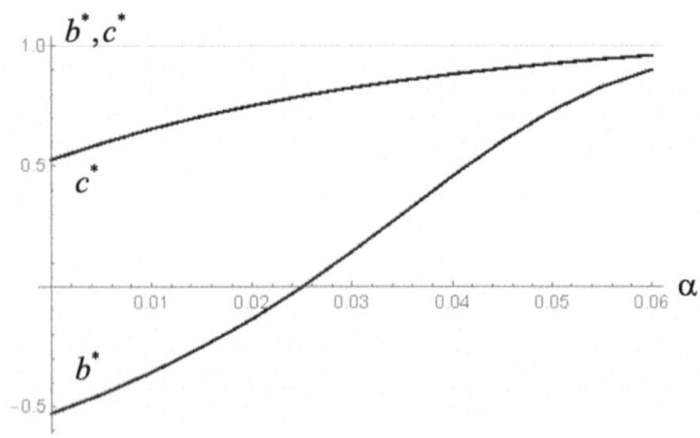

Fig. 6.6 Dependence b^* and c^* on the angle of inclination α.

work of the Galin's contact problem, i.e. in the presence of zones of adhesion and slippage. The governing system of equations consists of ten singular integral equations under four additional conditions. The features of the behavior of the unknown functions at the ends of the integration interval are revealed. Based on the mechanical quadrature method, a calculation program has been developed in the Wolfram Mathematica package environment, which allows you to find both the laws of distribution of contact stresses and construct the form of crack opening over an inclusion. With the help of this program, a detailed numerical analysis of the dependence of the coordinates of the ends of the adhesion zone on all parameters of the problem was carried out.

Acknowledgements This study was carried out within the framework of the grant 21T-2C209 (Science Committee, Ministry of Education, Science, Culture and Sports of the Republic of Armenia).

References

1. L.A. Galin. *Contact Problems in the Theory of Elasticity and Viscoelasticity (in Russ.)*. Moscow, Nauka, 1980
2. G. Ya. Popov. *Concentration of Elastic Stresses Near the Punches, Slits, Thin Inclusions and Stiffened Constructions (in Russ.)*. Moscow, Nauka, 1982.
3. V. Panasiuk, M. Savruk, A. Dacyshin. *Distribution of Stresses Near the Cracks in Plates and Shells (in Russ.)*. Kiev. Naukova Dumka, 1976.
4. N.I. Muskhelishvili. *Some Basic Problems of the Mathematical Theory of Elasticity*. Dordrecht, Springer, 1977.
5. L. Berezhnitsky, V. Panasiuk, N. Staschuk. *Interaction of Rigid Linear Inclusions and Cracks in Deformable Body (in Russ.)*. Kiev, Naukova Dumka, 1983.
6. V.N. Hakobyan. *Mixed Boundary Value Problems on Interaction of Continuum Deformable Bodies with Different Types Stress Concentrators (in Russ.)*. Gitutyun Publishing House, NAS RA, 2014.
7. V.N. Hakobyan. On One Mixed Problem for Composite Plate, Weakened by a Crack (in Russ.). *Mechanics. Proc. of NASRA*. 48(4):57-65, 1995.
8. I. Ilyina, V. Silvestrov. The problem of thin rigid interphase inclusion, detached from media along one side (in Russ.). *Proc. Russian Academy of Sciences. Mechanics of Solis* (3):153-166, 2005.
9. S.M. Mkhitaryan. On the stress-strain state of an elastic infinite plate with a crack expanding by means of smooth thin inclusion indentation (in Russ.). *Proc. of NAS RA. Mechanics.* 72(4):38-64, 2019.
10. V.N. Hakobyan, L.V. Hakobyan. Contact problem for a homogeneous plane with crack (in Russ.). *Proc. NASRA*, 73(4):3-12, 2020.
11. L.A. Galin. Indentation of a punch in the presence of friction and adhesion (in Russ.). *J. Appl. Math. Mech.* 9(5):413-424, 1945
12. V.I. Mossakovski, A.G. Biskup. Impression of a stamp with friction and adhesion present (in Russ.). *Reports of Academy of Sciences, USSR*. 206(5):1068-1070, 1972.
13. Yu.A. Antipov, N.Kh. Arutyunyan. Contact problems of the theory of elasticity with friction and adhesion. *Journal of Applied Mathematics and Mechanics*, 55(6):887-901, 1991.
14. V.I. Ostrik, A.F. Ulitko. *The Wiener-Hopf Method in Contact Problems of the Theory of Elasticity (in Russ.)*. Kiev, Naukova Dumka, 2006.

15. A.V. Sahakyan. Solution of a contact problem with contact and adhesion regions (Galin's problem) by the method of discrete singularities. In: *Development of L. A. Galin's Ideas in Mechanics. Collection of Research Papers*, Moscow - Izhevsk, pp. 103–120, 2013.
16. V.N. Hakobyan, A.V. Sahakyan. Stress state of elastic plane with partially detached from matrix absolutely rigid inclusion within Galin's model (in Russ.). *Proc. of VIII Int. Conf. The Problems of Dynamics of Interaction of Deformable Media, September 22–26, Goris - Stepanakert, Armenia, 2014*, Goris - Stepanakert, Inst. Mekh. NAN RA, pp. 43–47, 2014
17. N.I. Muskhelishvili. *Singular Integral Equations*. Dordrecht, Springer, 1958.
18. A.V. Sahakyan, H.A. Amirjanyan. Method of mechanical quadratures for solving singular integral equations of various types. *IOP Conf. Series: Journal of Physics* 991:012070, 2018.

Chapter 7
Biomechanics as a Basis for Clinical Decision Support Systems in the Surgery of the Spine-pelvic Complex

Dmitriy V. Ivanov, Irina V. Kirillova, and Leonid Yu. Kossovich

Abstract The scientific problem considered in this paper consists in the development, testing and introduction into clinical practice of biomechanical methods for preoperative evaluation of treatment options and diagnosis of pathologies of the vertebral-pelvic complex. The objectives of the study were to develop a technique for noninvasive determination of mineral density and Young's modulus of spongy bone tissue according to computer tomography, to develop quantitative criteria for evaluating the success of surgical treatment, to develop generalizing formulas for calculating the theoretical values of the parameters of sagittal balance, as well as to carry out a pilot implementation of biomechanical modeling in the process of preoperative planning.

Key words: Preoperative planning, Biomechanical modeling, Clinical decision support systems, Equivalent stress, Stress-strain state, Criteria for the success of surgical treatment, Finite-element method, Vertebral-pelvic complex

7.1 Introduction

In the structure of the Russia population general morbidity, cardiovascular and musculoskeletal system diseases have not changed their positions for more than 15 years: 13.3-15.2% and 7.5-8.2%, respectively. Despite the fact that in the last decade, according to Rosstat (Federal State Statistic Service), the level of injuries has been steadily decreasing, it remains one of the leading causes of disability and mortality of the population. Degenerative-dystrophic diseases of the musculoskeletal system also often lead to a serious deterioration in the quality of life and disability of the population. Especially noteworthy is such a combined pathology as coxover-

Dmitriy V. Ivanov · Irina V. Kirillova · Leonid Yu. Kossovich
Saratov State University, 83 Astrakhanskaya Street, Saratov, 410012, Russian Federation,
e-mail: ivanovdv@gmail.com,nano-bio@sgu.ru

H. Altenbach et al. (eds.), *Advances in Solid and Fracture Mechanics*,
Advanced Structured Materials 180,
https://doi.org/10.1007/978-3-031-18393-5_7

tebral syndrome, which consists in the presence of pathology in the patient both in the lumbosacral segment of the spine and in the hip joint. The danger of this disease is not only in its high incidence (up to 95% of patients with degenerative changes of the hip joint), but also in the fact that doctors' experience serious difficulties in diagnosis and treatment [1].

The complexity of the diagnosis and complexity of the above diseases, as well as the fact that different treatment options are possible for each of them, poses the task of developing modern quantitative methods and tools for their research in order to select and justify a successful treatment option in each case. One of the modern tools for helping surgeon and evaluating treatment options are preoperative planning systems or clinical decision support systems (CDSS).

The issues of developing preoperative planning systems, including biomechanical support for choosing an operation option for each individual patient, have been raised in the scientific literature since the end of the 20th century [2]. However, at the moment, preoperative planning systems with biomechanical support for choosing the optimal treatment option have not been developed anywhere in the world. The development and implementation of quantitative methods for assessing the severity of the disease and its treatment options based on biomechanical modeling can significantly improve the quality of treatment, as well as improve the postoperative prognosis and quality of patient's life.

The purpose of this study is to create the biomechanical foundations necessary for the development and introduction into preoperative diagnostics and planning of clinical decision support systems with biomechanical support. The research objective are:

- To develop and test a technique for noninvasive determination of mineral density and Young's modulus of spongy bone tissue according to computer tomography.
- To investigate the relationship between the sagittal balance parameters and stress-strain state of the vertebral-pelvic complex elements (VPC).
- To develop generalizing formulas for calculating the theoretical values of the sagittal balance parameters.
- To develop and test quantitative criteria for evaluating the success of surgical treatment.
- To perform a pilot implementation of biomechanical modeling into preoperative planning process and demonstrate the effectiveness of biomechanical modeling when choosing a treatment option within the framework of preoperative planning in musculoskeletal surgery.
- To develop the concept of clinical decision support systems with biomechanical support.

7.2 Noninvasive Method of Obtaining Bone Tissues Mechanical Properties by Computer Tomography

7.2.1 Results of Experiments on Scanning Samples of the Femoral Heads Spongy Bone in a Computer Tomograph

To assess the effect of convolutional kernel, the "beam hardening" effect and other characteristics of a computed tomograph (CT), samples of spongy bone tissue of 150 patients were scanned (of which 5 patients were in the phantom, the rest were "in the air"), as well as samples of an aqueous solution of potassium hydroorthophosphate imitating a certain bone mineral density (BMD). Of these, 4 cylindrical samples with BMD of 50 mg/cm^3, 100 mg/cm^3, 150 mg/cm^3 and 200 mg/cm^3 imitating spongy bone, and 4 hexahedral samples with BMD of 250 mg/cm^3, 350 mg/cm^3, 450 mg/cm^3 and 550 mg/cm^3 imitating cortical bone. Samples were placed in the center of phantom, deionized water was poured into phantom. Figure 7.1 shows samples on CT scanner table and in phantom.

The samples were scanned in various modes (Table 7.1) three times. CT table was fixed in such a position that the samples were located in the center of its aperture. Convolutional kernels FC17 and FC03 were investigated, since they are used for postprocessing of CT images on this tomograph. The thickness of the slice during scanning was 1 mm. When analyzing Hounsfield units, 2-3 mm retreated from the end edges of each sample.

It was revealed that an increase in density does not always entail an increase in Hounsfield units for calibrated samples with BMD corresponding to spongy bone

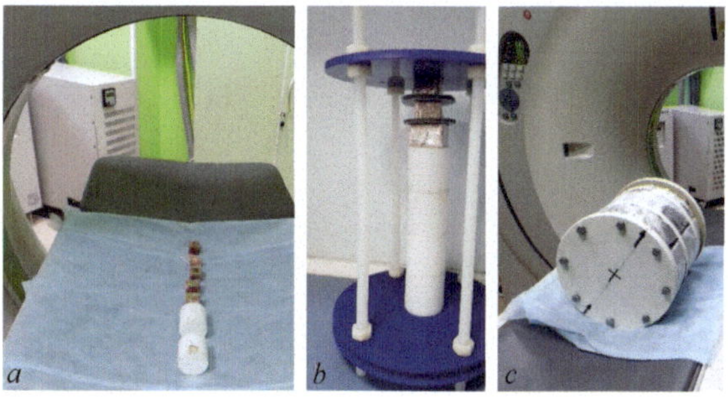

Fig. 7.1 Samples before scanning in CT: *a* - on CT table; *b* - in the holding frame of the phantom; *c* - on the CT table inside the phantom filled with water

Table 7.1 Examined CT scan modes and settings

	in air/in phantom
Voltage	120 kV
Tube current	40 mA / 300 mA
Convolutional kernel	FC17 / FC03
FOV	400 mm

(Fig. 7.2). In the phantom (with both FC03 and FC17 kernels), this effect was absent in the same way as "in the air" with FC17 kernel. To convert the Hounsfield units of K2HPO4 samples from experiments "in the air" into Hounsfield units "in phantom", regression dependencies were selected (Fig. 7.3) for FC17 kernel. It is revealed that the current has practically no effect on the Hounsfield units. Different convolutional kernels give significantly different Hounsfield units when scanning the same samples "in the air". It is shown that there are no differences in measurements "in the air" and in the phantom at a significance level of 5% for the FC03 kernel, which confirms the fact that it is intended to correct the "beam-hardening" effect. It is shown that the effect of strengthening the rigidity of X-ray radiation "beam hardening" significantly affects the Hounsfield units and should be considered when evaluating BMD. The difference in Hounsfield units for FC03 and FC17

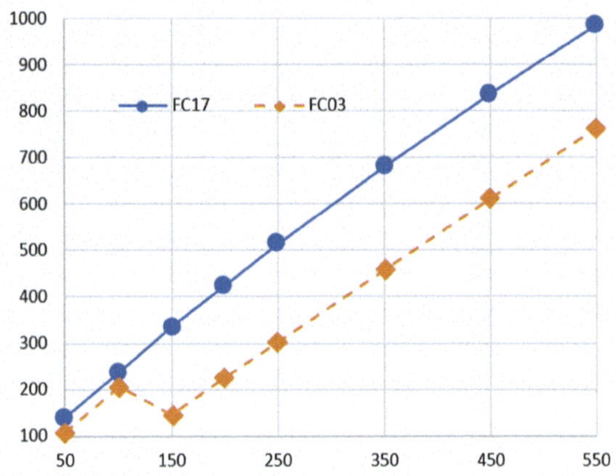

Fig. 7.2 Dependence of Hounsfield units (vertically) of K2HPO4 samples on their BMD (horizontally mg/cm^3). Scanning "in air", 300 mA current

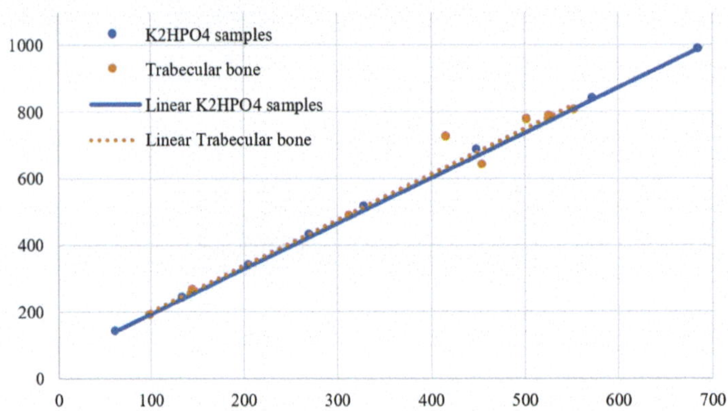

Fig. 7.3 Regression dependencies of recalculation of HU from "in the air" (horizontal axis) to "in phantom" (vertical axis) for calibrated samples K2HPO4 and trabecular bone samples for FC17 kernel (solid line $y = 0.7x - 40.8$, $R^2 = 0.99$ – linear regression for K2HPO4 samples, dotted line $y = 0.7x - 33.9$, $R^2 = 0.97$ – linear regression for trabecular bone samples)

kernels for the same samples when scanning "in the air" reached up to 130% (about 56% on average) for calibrated samples and up to 128% (about 81% on average) for spongy bone. An almost "ideal" regression dependence was obtained for the FC17 kernel when comparing scans in the "in-air" and in the phantom modes (Fig. 7.3), which allows scanning bone tissue samples in the "air" with subsequent recalculation of the data obtained into BMD values, or into Hounsfield units in a phantom simulating the patient's body.

7.2.2 Results of Uniaxial Compression Mechanical Experiments of Femoral Heads Spongy Bone Samples

The samples were prepared using a hand-held metal hacksaw with the initial fragments fixed in a vise. From one fragment of the femoral head, from 1 to 4 samples were prepared (Fig. 7.4 a, b) in the form of rectangular parallelepipeds with a rib size of at least 5 and not more than 10 mm. The prepared fragments of the spongy bone were delivered within one or two hours to the Research Institute of Traumatology, Orthopedics and Neurosurgery of the Saratov State Medical University (NIITON SSMU), where they were scanned on a Toshiba Aquilion 64 CT. FC17 kernel was selected as CT settings, tube current was 40 mA, slice thickness was 1 mm. Samples were scanned "in the air", the resulting Hounsfield units were recalculated, for this the dependence shown in Fig. 7.3 was used. Samples of femoral heads of 150 patients were examined on CT. For each head the first letter of the surname, age (year of birth), gender, sample volume and the average value of Hounsfield units were fixed.

After CT, fragments of femoral heads were subjected to uniaxial compression experiments in order to determine the Young's modulus on Instron 5944 test machine. Preload value was 10 N, loading speed was 30 mm/min. Table 7.2 shows the results of the study for three randomly selected patients.

Regression analysis was used to identify the dependence that allows calculating the Young's modulus of spongy bone tissue through Hounsfield units. Models of linear multiple regression (additive model) and linear regression in logarithms (multiplicative model) were constructed and analyzed. For all models, the effect of the patient's age was statistically insignificant. The level of statistical significance of the coefficients was less than 5%.

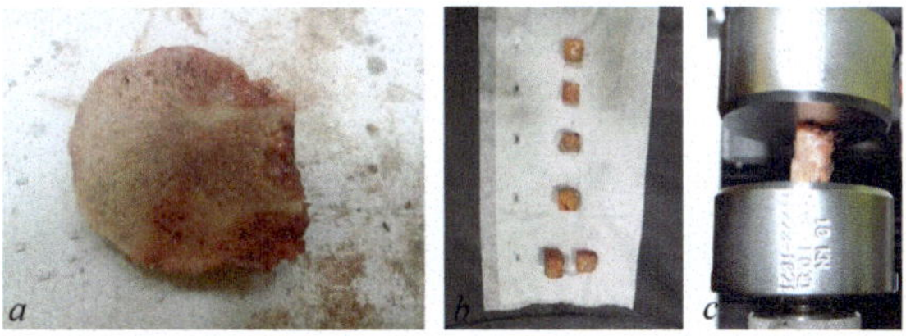

Fig. 7.4 Samples of bone tissue of the femoral heads: *a* – bone tissue sample before the experiment; *b* – bone tissue samples on CT table; *c* – bone tissue sample during uniaxial compression experiment

Table 7.2 Young's moduli of femoral head fragments determined from mechanical experiments

No.	ID	ICD-10 code	Young's module, MPa	
			Experimental value	Weighted average harmonic value
1	B., 1951, male	M 16.0	48,2	42,6
			37,7	
			40,2	
			45,7	
2	B., 1954, female	M 16.0	116,7	133,5
			150,3	
3	T., 1953, female	M 16.1	74,6	73,1

The specified model of linear multiple regression, presented in Fig. 7.5, has the form:

$$E = 0,20\text{HU} + 5,98;\ R^2 = 0,67, \tag{7.1}$$

where E is Young's modulus, HU are Hounsfield units. R^2 is the coefficient of determination. Multiplicative model was also obtained:

$$E = \mathrm{e}^{1,17\ln(\text{HU})-2,55},\ R^2 = 0,68. \tag{7.2}$$

Model (7.2) explains more than 68% of the variation of the dependent variable. The advantage of this model can be the high (at a level of less than 0.001%) significance of the coefficients.

The predicted values of the Young's modulus for the studied tissues and their errors were calculated. The difference between the values of Young's modules determined by the data from mechanical experiment and predicted using the model (7.1) averaged 19%, and when using the model (7.2) – 20%.

To improve the result of the prediction of the Young's module, an analysis of its dependence on Hounsfield units was carried out, considering the code of the International Classification of Diseases (ICD-10). In the experiment, fragments of bone tissue of patients with ICD-10 codes were studied: M 16.0, 16.1, 16.2, 16.3,

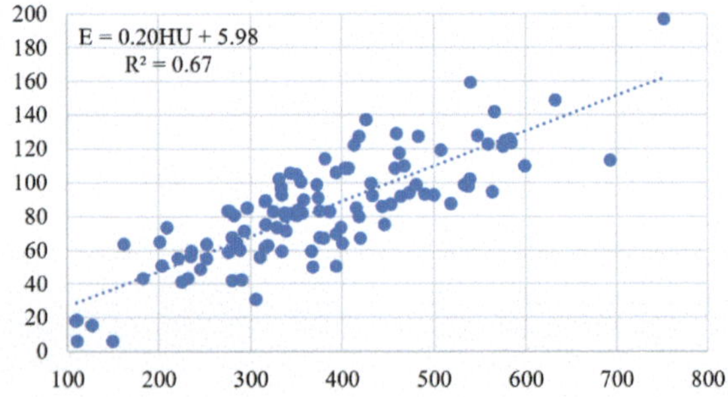

Fig. 7.5 Dot diagram: Hounsfield units (HU) – along abscissa axis, Young's modulus E – along ordinate axis

16.5, 16.6, 16.7, M17.0, M21.9, M84.1, M87.0, M87.2, M95.8 and M95.9. For these data were generalized models were obtained, predictive values of Young's modulus and their errors for the studied tissues were determined (Table 7.3).

It can be seen from Table 7.3 that when considering the data separately for each ICD-10 code, the difference in values between the Young's modulus determined from the mechanical experiment and the predicted one does not exceed 16% on average. The exception is linear regression models for the M16.0 code and the last line, which contains codes with studies of less than 5 dimensions. Multiplicative model for the M16.0 code gives results with an average error of no more than 16%.

The revealed dependencies between the Young module and Hounsfield units can be applied to the processing of CT studies conducted on the Toshiba Aquilion 64 tomograph, since different models of scanning devices can give different HU values [3]. The study of the influence of the manufacturer and the model of the tomograph on the reliability of the results obtained using the proposed formulas seems relevant. To determine the correction coefficients, calibrated samples of an aqueous solution of potassium hydroorthophosphate [4] with a known mineral density can be scanned in each tomograph, after which it will be possible to obtain similar regression dependencies and final models.

Table 7.3 Linear multiple regression models and multiplicative models considering ICD-10 code

ICD-10 code	Regression models		Average difference
M16.0	$E = 0,19\text{HU} + 14,46;\ R^2 = 0,64$	(3)	27%
	$E = e^{1,05\ln(\text{HU}) - 1,77};\ R^2 = 0,74$	(4)	16%
M16.1	$E = 0,24\text{HU} - 5,79;\ R^2 = 0,71$	(5)	14%
	$E = e^{1,08\ln(\text{HU}) - 2,02};\ R^2 = 0,75$	(6)	16%
M16.2	$E = 0,18\text{HU} + 50,10;\ R^2 = 0,67$	(7)	16%
	$E = e^{0,64\ln(\text{HU}) + 0,95};\ R^2 = 0,61$	(8)	14%
M84.1	$E = 0,18\text{HU} + 13,67;\ R^2 = 0,75$	(9)	11%
	$E = e^{0,79\ln(\text{HU}) - 0,28};\ R^2 = 0,78$	(10)	12%
M87.0	$E = 0,17\text{HU} + 12,20;\ R^2 = 0,80$	(11)	16%
	$E = e^{0,82\ln(\text{HU}) - 0,49};\ R^2 = 0,78$	(12)	16%
M16.3, M16.5, M16.6, M16.7, M17.0, M21.9, M87.2, M95.8, M95.9	$E = 0,20\text{HU} + 4,55;\ R^2 = 0,60$	(13)	20%
	$E = e^{1,56\ln(\text{HU}) - 4,96};\ R^2 = 0,66$	(14)	27%

7.3 Sagittal Balance and its Relation to Vertebral-pelvic Complex Biomechanics

For the first time, the term "vertebral-pelvic balance" (or sagittal balance) is mentioned by Duval-Beaupere [5]. The static sagittal balance of the normal spine (or optimal balance) is its physiological alignment in the most effective way with the help of muscle forces [6]. Optimal vertebral-pelvic relationships make it possible to form a biomechanically effective sagittal profile of the spine, reduce energy consumption and the risk of degeneration of adjacent vertebral motion segments (VMS) [7]. From the standpoint of biomechanics, the necessity of correction of the patient's sagittal profile during VPC operations was justified [7, 8]. It is believed [9] that maintaining sagittal balance or neutral vertical alignment of the spine in the sagittal plane is the main goal of surgical, ergonomic and physiotherapy procedures.

Sagittal balance can be described using angular vertebral-pelvic parameters: pelvic index (PI), lumbar lordosis (LL), pelvic tilt (PT) and sacral tilt (SS) [10] (Fig. 7.6). PI is the angle between the line drawn through the center of the femoral head and the middle of the closure plate of the vertebra S1. PI is anatomically fixed, does not change after adolescence and is individual for each person [11]. The angle of inclination of the sacrum SS is the compensation angle PT, characterizes the position of the end plate of the vertebra S1. PT, PI and SS are mathematically related by the following formula: $PI = PT + SS$.

To calculate the theoretical (or optimal) values of the sagittal balance parameters, the authors attempt to develop mathematical formulas [11]. The development of formulas will allow for each patient at the stage of preoperative planning to calculate the optimal values of the parameters of his sagittal balance and assess the necessary level and degree of correction of deformation. We present the results of the development of original generalizing formulas for calculating the parameters of the sagittal balance based on the analysis of scientific literature and the results of our own research presented in [12] (Table 7.4). Figure 7.7 shows, for example, the dependencies of LL on PI according to literature data and measurement results.

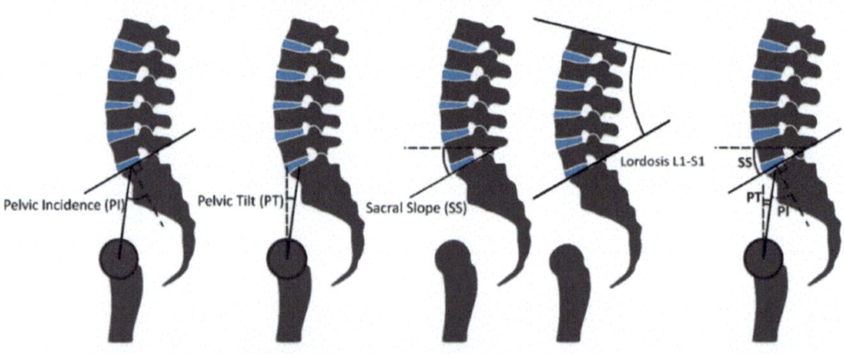

Fig. 7.6 Sagittal balance main parameters

Table 7.4 Literature sources found and used in this study, as well as data from our own research

Reference	Sample characteristics	Formulas
J.C. Le Huec, K. Hasegawa [13]	268 humans (117 male and 161 female), average age 37,2 (18-76), without symptomes	$LL = 0,54PI + 27,6$ $PT = 0,4429PI - 11.389$ $SS = 0,54PI + 11.90$
S.J. Hyun, S. Han, Y.B. Kim, Y.J. Kim, G.B. Kang, J.Y. Cheong [14]	150 males at age of 64,1 ± 6,4, without symptomes	$LL = 0,62PI + 27,61$
T.B. Sullivan, N. Marino, F.G. Reighard, P.O. Newton [11]	125 teenagers (47 boys, 78 girls), average age 13 ± 2, without symptomes	$LL = 0,66PI + 24.2$
F. Tanguay, J.M. Mac-Thiong, J.A. de Guise, H. Labelle [15]	60 children (57 girls and 3 boys) diagnosed with adolescent idiopathic scoliosis who underwent PSIF surgery	$LL = 0,56PI + 33,43$
J. Legaye, G. Duval-Beaupère [16]	49 people (28 men and 21 women), from 19 to 30 years old without symptomes	$SS = 0,5481PI + 12,7$
D.V. Ivanov, I.V. Kirillova, L.Yu. Kossovich et al. [17]	52 patients, average age 46 (32-57), 28 female and 24 mail after spine trauma and spondylolistesis	–

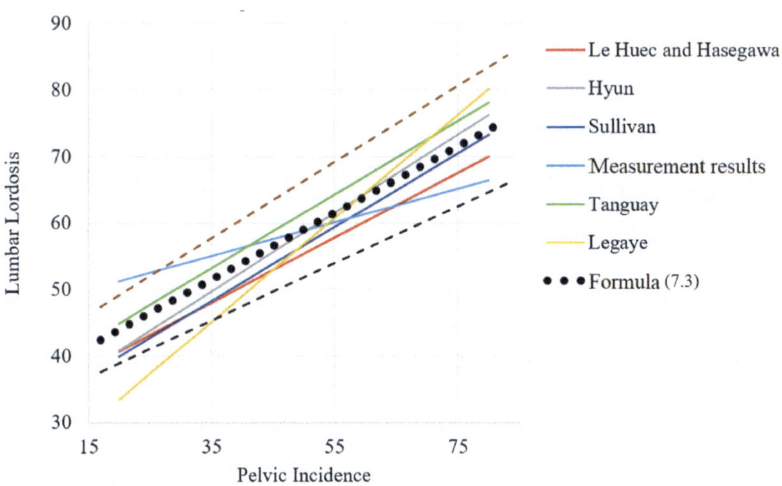

Fig. 7.7 Dependencies of Lumbar Lordosis on Pelvic Incident proposed in [11, 13, 14, 15, 16] and measurement results

Relationship Between LL and PI The relationship between LL and PI was studied on the basis of measurement results [12] supplemented with data from the works [11, 13, 14, 15, 16]. According to 638 results of pairwise measurements, including those presented in the literature [11, 13, 14, 15, 16], as well as the data of patients of NIITON SSMU, a linear regression relationship was constructed

$$\mathrm{LL} = 0,502\mathrm{PI} + 33,90. \tag{7.3}$$

Relationship Between SS and PI The dependence of SS on PI was studied on the basis of measurement results [12] supplemented by data from [13, 16]. A linear regression dependence was constructed based on 191 results of pairwise measurements

$$\mathrm{SS} = 0,476\mathrm{PI} + 15,62. \tag{7.4}$$

Relationship Between SS and PI, LL For practical application, a formula linking three parameters – SS, PI and LL may be of great interest. Since PI changes little both with age and in the case of most diseases and injuries, it can be assumed that SS compensates for the changing LL over time, considering PI value that is individual for each person and changes little during life. For statistical analysis, the results of studies of patients of NIITON SSMU were supplemented with data from League [16] and Le Huec [13]. Based on the results of SS, PI, LL measurements in 52 patients of NIITON SSMU supplemented with data [13, 16] – a total of 171 data sets – a linear regression relationship was constructed

$$\mathrm{SS} = 0,32\mathrm{PI} + 0,37\mathrm{LL} + 2,4. \tag{7.5}$$

Despite the fact that many formulas linking the main angular parameters of sagittal balance have been published in the modern scientific literature, their application in clinical practice is difficult. This is due to the fact that different authors publish different formulas obtained from their own limited samples. There is a problem of generalizing the experience accumulated in science and obtaining generalizing formulas covering all the most well-known dependencies to date. Such a generalization at this stage is formulas (7.3) and (7.4) that satisfactorily describe the data of different researchers for patients of different genders, ages and the presence or absence of pathologies [17]. The analysis showed that the dependencies between PI, LL and SS in patients with pathologies of VPC on average can be described by the same dependencies as in healthy people. The advantage of the fundamentally new formula (7.5), which establishes the relationship between SS, PI and LL, is also confirmed by the value of the average relative approximation error, which was 9.6% for this dependence, while for formulas (7.3) and (7.4) (dependencies $\mathrm{LL}(\mathrm{PI})$ and $SS(\mathrm{PI})$) it was 14.6% and 13.2%, respectively. Considering the average values of SS and their variation, a decrease in the approximation error by 3.6-5% can have a significant impact on the practical application of the developed dependence. It should be noted that the geometric relationship between the main parameters of the sagittal balance is not accidental and is confirmed by biomechanical calculations [8].

7.4 Quantitative Criteria for Assessing the Success of Surgical Treatment

In traumatology practice, the main goal of traditional preoperative planning is the selection of implants, modes and techniques of their installation, ensuring the stability of the structure and reposition of bone fragments. In orthopedics, traditional preoperative planning should ensure the selection of implants to recreate the optimal biomechanics of the operated segment.

In this regard, at the stage of treatment planning, it is necessary not only to describe qualitatively, but also quantitatively, the optimal anatomy and biomechanics of the operated segment. In particular, when planning hip replacement, the length of the limbs should be the same, the center of rotation of the femoral component of the implant should correspond to the center of rotation of a healthy joint, etc. When planning the treatment of spinal injuries, the sagittal balance (hereinafter referred to as SB) should be observed, or, in other words, the SB parameters should be optimal. Therefore, it is necessary to develop a set of numerical parameters with which it would be possible to distinguish a successful operation from an unsuccessful one from the point of view of the anatomy (or geometry) of the VPC segment and to ensure adequate geometric preoperative planning.

7.5 Geometric Criteria for Evaluating the Success of Spinal Surgery

Vertical alignment of the spine in the sagittal plane is considered the main goal of surgical, ergonomic and physiotherapeutic procedures [9] and is described by the parameters of sagittal balance. Previously, we justified [7] the need to correct the sagittal profile of the patient when performing operations on VPC. It was shown [28] that the sagittal profile of the spine is strictly correlated with the quality of life, and VPC plays a key role in compensatory mechanisms of spine imbalance. It is especially important that in surgical correction of deformity, the impact on the SB and lumbo-pelvic parameters significantly improves the quality of life of patients.

The intervals of changes in the main parameters of the SB have been studied [18], and many authors are trying to develop formulas for calculating optimal balance parameters for a particular patient. The PI parameter is individual for each person and remains unchanged throughout life. It was shown [11, 16] that there is a relationship between PI and other basic angular parameters of SB, the values of which can change with the development of degenerative-dystrophic diseases and injuries of VPC.

In connection with the above, the geometric criteria for evaluating success should include the optimal parameters (parameter intervals) of the SB, which can be calculated individually from a radiograph performed in a standing position. Here are the formulas [17] for calculating PT, SS, LL through PI, derived from the indicators of healthy patients.

$$\begin{aligned} LL &= 0,502PI + 33,90°, \\ SS &= 0,476PI + 15,62°, \\ PT &= PI - SS. \end{aligned} \tag{7.6}$$

Formulas (7.6) allow one to calculate optimal angular parameters of the SB for a particular patient through the parameter PI.

SVA (sagittal vertical alignment) parameter is measured as the distance from the plumb line passing through the center of the seventh cervical vertebra (C7PL) to the posterior edge of the surface of the upper end plate of the sacrum. It is believed [19] that a balanced spine has SVA < 5 cm. It was shown [20] that poor functional results of surgical treatment of degenerative injuries of the thoracolumbar spine correlate well with insufficient sagittal alignment (SVA > 5 cm). The authors [21] recommend restoring the SVA value < 5 cm and achieving optimal lordosis (ratio of PI and LL angles) to achieve the best quality of life after treatment. It was revealed [22] that patients with compression fractures of the vertebrae and balanced vertebral-pelvic parameters, as well as sagittal alignment (SVA < 5 cm) are more likely not to require surgical intervention than patients with suboptimal SB parameters. Thus, the values of SVA, together with the angular parameters of SB described above, will be considered criteria for evaluating the success of surgical reconstructive treatment of diseases and injuries of the spine.

7.6 Geometric Criteria for Assessing the Success of Hip Replacement

When planning treatment, the position of the endoprosthesis components should be evaluated, as well as the geometric characteristics of the operated limb based on the results of X-ray examination. Based on anatomical considerations, it is logical to attribute the values of the parameters of the position of the center of rotation of the acetabulum component, offset (distance from the center of rotation to the axis of the leg of the endoprosthesis), as well as the length of the operated lower limb to the geometric criteria for assessing the success of hip replacement.

Studies of the wear of the acetabulum component liner allow us to formulate another criterion, which is that the angle of inclination of the acetabulum component in the frontal plane should be from 40 to 50 degrees [23]. The substantiations of these and other criteria values of radiological parameters are given in the literature, and the criteria themselves are summarized in Table 7.5.

Table 7.5 Geometric criteria for evaluating the success of hip replacement

No.	Criteria description	Criteria value	
		Satisfactory	Unsatisfactory
1	Vertical rotation center of the acetabulum component	Matching a healthy hip joint[1].	Inconsistency with a healthy hip joint.
2	Horizontal rotation center of the acetabulum component	Matching a healthy hip joint[2].	Inconsistency with a healthy hip joint.
3	Angle of inclination (in the frontal plane) of the acetabulum component	40 – 50° (neutral position).	< 40°, > 50°
4	Anteversion angle (slope in the sagittal plane) of the acetabulum component	10 – 15°[3].	< 10° and > 15°.
5	Undercovering of the acetabulum component	< 30%.	> 30%.
6	Endoprosthesis leg position in medullar channel	Endoprosthesis leg location is central (central axis of endoprosthesis leg coincides with central axis of medullar canal).	Significant valgus or varus deviation of the leg from the central axis of femur – more than 5°.
7	Filling of medullar canal	On the X-ray in the straight anterior–posterior projection, the leg of the endoprosthesis fills the canal by 80% or more, in the lateral projection by 70% or more[4].	On the X-ray in the straight anterior-posterior projection, the leg of the endoprosthesis fills the channel less than 80%, in the lateral projection – less than 70%. The size of the structure is not large enough.
8	Offset	Is equal to the offset of a healthy hip joint.	More or less offset of a healthy hip joint.
9	Lower limb length	The length of the lower limbs is the same.	The length of the lower limbs is different[5].

[1] When endoprosthetics in difficult cases, it is allowed to shift the center of rotation up to 2.5 cm.
[2] When endoprosthetics in difficult cases, medialization of the acetabulum component is allowed (it's going beyond the Kohler line).
[3] When using rear access, the angle of the anteversion should be 20 – 25°.
[4] It matters for cementless legs.
[5] In the case of rigid compensated frontal deformity in senile persons, a slight (up to 1 cm) elongation of the lower limb is allowed in order to preserve the usual vertebral-pelvic relations.

7.7 Biomechanical Criteria for Evaluating the Success of Treatment: Assessment of Mechanical Strength, Fixation Stability, Implant Life

Biomechanical criteria for evaluating the success of treatment should allow assessing the stability of fixation, the strength of implantable structures, the risk of damage to bone and soft tissues, as well as the strength of the bone-implant system as a whole. The stability of fixation can be estimated by the relative movements of the fixed elements of the VPC based on the values of the amplitude of motion (range of motion, ROM) in vertebral motion segment (VMS) [24, 25]. Under bending loads (torso tilts forward and backward), ROM parameter is calculated as follows. In the sagittal plane in the spine model, the angle between the upper closure plates of adjacent vertebrae is measured before the load is applied. After applying the load and changing the position of the vertebrae relative to each other, this angle is measured again. ROM represents the difference in absolute values of two measured angles (before and after deformation).

At loads corresponding to the tilt of the torso to the side, ROM is measured similarly, but measurements are carried out in the frontal plane [26]. Under twisting loads, ROM is calculated as the angle by which the vertebrae of the studied VMS turned relative to each other during rotation in the axial plane [26]. At critical ROM values, damage to the intervertebral disc occurs. In the case of rotational loading, based on the study of VMS with healthy disks and VMS with disks subject to degenerative changes [27], it was shown that ROM in 16° for healthy and in 14.5° for diseased disks is critical, leading to damage (averaged values for the sample from the experiment are indicated). Thus, the ROM parameter can be considered as a criterion for evaluating the success of surgical treatment, the critical values of which for different types of loads are summarized in Table 7.6.

If we talk about the relative movements of the fixed elements of the VPC, then in this case, during the operation, the surgeon achieves the best stability, that is, minimal relative movements. There is evidence in the literature that when a person moves, the VMS is considered unstable if the relative movements of the vertebrae of one VMS exceed 3 mm [28]. Among several treatment options, the most successful is the one in which the displacement within the VMS [29, 30] are minimal.

The strength of the "bone-implant" system as a whole and its individual elements is determined by the mechanical stresses arising in it under typical loads simulating the state of rest and human movement. Internal stresses in implants are compared with their strength characteristics: endurance limit (under cyclic loads), yield strength and strength limits [31, 32] with a certain safety margin factor (Table 7.7). The allowable stresses for metals can be calculated based on their properties. As a rule, for steels, the yield strength with a margin factor equal to 1.5 is taken as the allowable stress. For titanium alloys, the tensile strength with a margin factor of 3 is taken.

Table 7.6 Criteria (threshold) ROM values (in degrees)

Disk condition	Loads		
	Leaning forward/backward	Leaning to the sides	Body rotations (twisting)
Healthy disk	15	15	16.0 (20)
Degenerative disk	15	15	14.5 (20)

Table 7.7 Strength characteristics of bone tissues and implants

Material	Tensile strength, MPa
Stainless steel (316LS/316LVM)	up to 1241 (yield strength up to 945)
Ti6Al4V ELI Titanium Alloy	up to 970
Cortical bone	up to 161
Trabecular bone	up to 15

From the point of view of structural strength, the operation option for which the strength conditions for the allowable stresses are met will be considered successful. If two or more variants of the operation satisfy the strength conditions for the allowable stresses, then the most successful among them will be the one for which the stresses in the implants and bone tissues will be the least.

7.8 Results of the Pilot Implementation of Biomechanical Modeling in the Process of Preoperative Planning

7.8.1 Development of the Accord Software Platform

The results of this study were used in the development of a prototype (platform) of a clinical decision support system with biomechanical support for a doctor's decision in surgery of the vertebral-pelvic complex (hereinafter referred to as the Accord platform). For the first time in the world, the Accord platform provides not only a stage of geometric planning of the operation, but also biomechanical modeling of treatment options. The Accord platform is a base and ecosystem for the development of systems to support clinical solutions in surgery of the musculoskeletal system and other areas of medicine.

The development of the Accord platform was based on the following scientific results:

- method of noninvasive determination of Young's modulus of spongy bone tissue according to CT data;
- original formulas for calculating the optimal values of the main parameters of the sagittal balance according to radiography data;
- criteria for evaluating the success of surgical treatment of the consequences of diseases and injuries of VPC.

The main functional capabilities of the developed platform (in relation to biomechanical modeling of treatment options) include the following:

- automated construction of solid-state models of vertebrae based on CT data;
- calculation of the Young's modulus of bone tissue by CT;
- setting typical loads (forces and moments) and anchors;

- calculation of strength (stability, service life);
- justification of the optimal operation option.

The central component of the platform is the "Biomechanical Modeling Control Module". The "Biomechanical modeling control module" numerically solves the spatial static problem of elasticity theory for VPC elements together with the installed implants using the finite element method. The developed biomechanical criteria for evaluating the success of treatment are implemented in the "Biomechanical Modeling Control Module".

7.8.2 Substantiation of the Effectiveness of Biomechanical Modeling in Preoperative Planning

The effectiveness of biomechanical modeling implemented in the Accord platform during preoperative planning has been confirmed in the framework of demonstration experiments (DE). DE are devoted to geometric planning and biomechanical modeling of the results of surgical reconstructive treatment of a particular patient: on planned treatment for lumbar spine spondylolisthesis (first DE); on planned treatment for degenerative-dystrophic hip joint disease (second DE); on routine treatment for lumbar spine spondylolisthesis (third DE).

The purpose of the DE was to show the possibilities and effectiveness of biomechanical modeling when choosing a successful option for surgical reconstructive treatment of spinal pathology (first and third DE) and hip joint (second DE). Within the framework of each DE, a mathematical biomechanics problem was solved to determine the stress-strain state of the simulated VPC segment with installed implant models under the action of typical external surface forces and moments [35]. Simulated VPC segments, as well as implants, were considered to be isotropic elastic bodies. To simplify the formulation of the problem, such elements of the VPC as the ligamentous apparatus were modeled by one-dimensional elastic elements of the spring type. The elastic modulus of spongy bone tissue was calculated based on the patient's CT data [33]. The initial data for modeling were the results of radiation examination (computed tomogram and radiograph) of the VPC segment.

DE were performed jointly with specialists of the Research Institute of Traumatology, Orthopedics and Neurosurgery of the Saratov State Medical University (NIITON SSMU).

First DE A group of experts from NIITON SSMU selected a patient K., born in 1985, with a diagnosis of lumbar spine spondylolisthesis. The attending physician planned the following treatment options:

a transpedicular fixation (TPF) with 4 screws, ALIF fusion, Seohan Care Adinis cage;
b TPF with 6 screws, TLIF fusion, Unilif cage;
c TPF with 4 screws, TLIF fusion, Unilift cage.

Solid models of operation options are shown in Fig. 7.8.

The loading of the simulated segment of the spine was determined by the weight of the human body and was carried out by a tracking load of 100 N for each vertebra and twisting, as well as bending moment of 7.5 Nm [34, 35]. Results of biomechanical modeling are presented in Figs. 7.9 and 7.10. Maximum total displacements and equivalent stresses for each planned treatment and loading option are presented in Tables 7.8-7.10.

The consultation of doctors of the NIITON SSMU, based on the planning and modeling carried out, patient K., born in 1985, chose the third treatment option: TPF, TLIF fusion, Stryker's Unilif cage. NIITON SSMU staff performed surgical reconstructive treatment considering the recommendations for the selected option. Treatment results are shown in Fig. 7.11.

Second DE Experts from NIITON SSMU selected a patient I.T.R. born in 1953 with a diagnosis M16.0 "Primary bilateral coxarthrosis" according to ICD-10. The following options of surgical reconstructive treatment are offered for the patient: total endoprosthetics of the right hip joint with the installation of the acetabulum component (press fit) at an angle of inclination *a* - 35 degrees, *b* - 45 degrees and *c* - 55 degrees. The femoral component was cementless fixation. 2D solid models of hip joint and pelvic bones with implants installed are shown in Fig. 7.12.

Biomechanical modeling of each treatment option was carried out. Loading of the simulated segment of the vertebral-pelvic complex (VPC) was carried out with a load of 450 N, corresponding to the patient's weight of 90 kg. Results of biomechanical modeling are presented in Figs. 7.13 and 7.14.

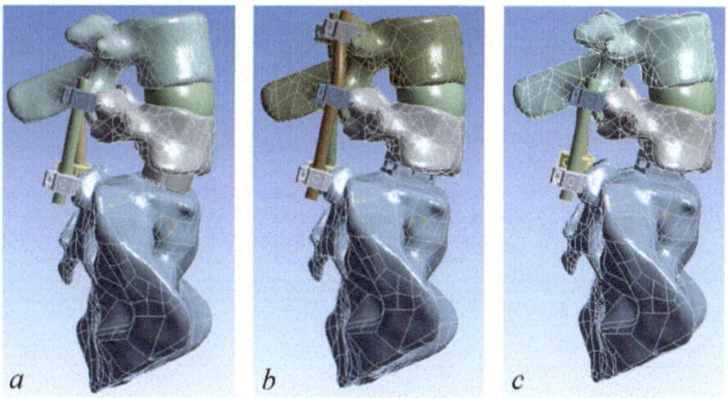

Fig. 7.8 3D solid models of the spine segment with implants installed: *a* – TPF with 4 screws and an Adonis cage, *b* – TPF with 6 screws and a Unilif cage, *c* – TPF with 4 screws and a Unilif cage

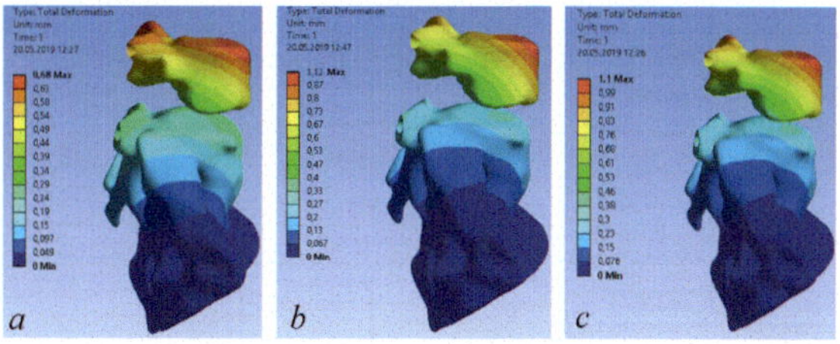

Fig. 7.9 Displacement fields (following load loading and bending moment – forward tilt): *a* – TPF with 4 screws and Adonis cage, *b* – TPF with 6 screws and Unilif cage, *c* – TPF with 4 screws and Unilif cage

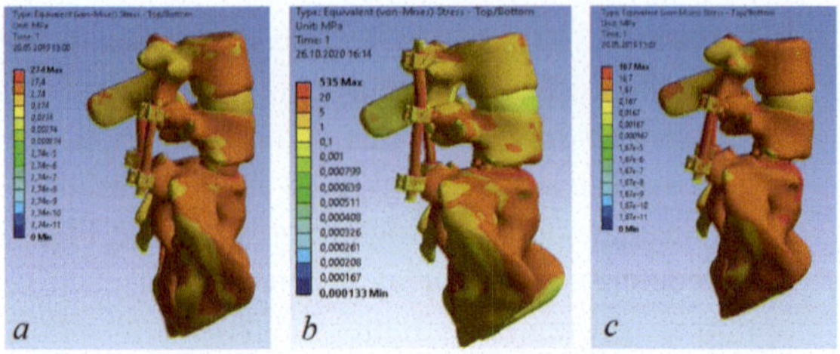

Fig. 7.10 Equivalent stress fields (following load loading and bending moment – forward tilt): *a* – TPF with 4 screws and Adonis cage, *b* – TPF with 6 screws and Unilif cage, *c* – TPF with 4 screws and Unilif cage

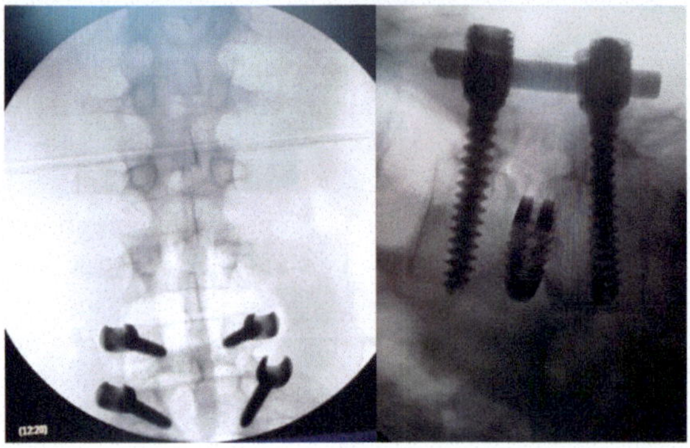

Fig. 7.11 Treatment results. Postoperative CT

Maximum total displacements and equivalent stresses for each proposed treatment and loading option are presented in Table 7.11. Based on the performed

Table 7.8 Maximum total displacement, mm

Treatment option	Tilt forward	Tilt back	Tilt to the right	Tilt to the left	Counterclockwise rotation	Clockwise rotation
TPF with 4 screws and Adonis cage	0,7	1	0,8	0,8	0,8	0,9
TPF with 6 screws and Unilif cage	1,1	0,2	0,6	0,7	0,7	0,8
TPF with 4 screws and Unilif cage	1,1	0,1	0,6	0,5	0,7	0,5

Table 7.9 Maximum equivalent stresses in implants, MPa

Treatment option	Tilt forward	Tilt back	Tilt to the right	Tilt to the left	Counterclockwise rotation	Clockwise rotation
TPF with 4 screws and Adonis cage	274	84	193,5	205,3	242	225
TPF with 6 screws and Unilif cage	535	52,2	225,2	219,4	207,6	240,3
TPF with 4 screws and Unilif cage	167	52,1	170,2	201,4	195	216,5

Table 7.10 Maximum equivalent stresses in spongy/cortical bone, MPa

Treatment option	Tilt forward	Tilt back	Tilt to the right	Tilt to the left	Counterclockwise rotation	Clockwise rotation
TPF with 4 screws and Adonis cage	15,5 / 66,1	28,2 / 20,1	8,8 / 37,6	11,5 / 45,7	21,1 / 51,9	17,8 / 38,1
TPF with 6 screws and Unilif cage	18,2 / 95,5	7,5 / 17,3	9,5 / 35,8	10,7 / 41,1	16 / 49,2	16,6 / 39,7
TPF with 4 screws and Unilif cage	10,9 / 30	4,5 / 5,2	8,3 / 34	7,2 / 32,6	13,5 / 37,7	14,2 / 36,5

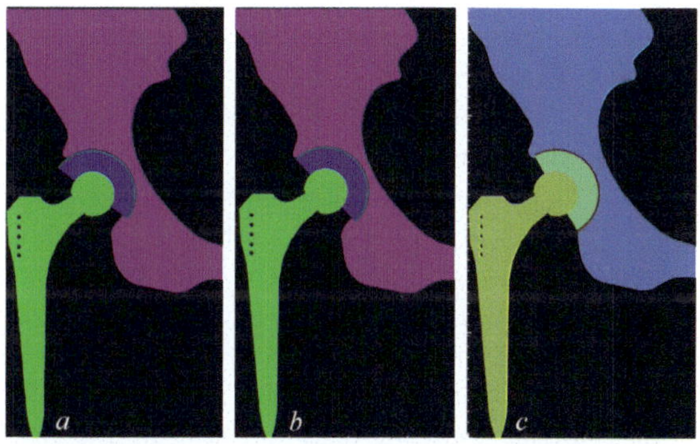

Fig. 7.12 2D solid models with installed implants: *a* – 35 degrees inclination angle, *b* – 45 degrees inclination angle, *c* – 55 degrees inclination angle

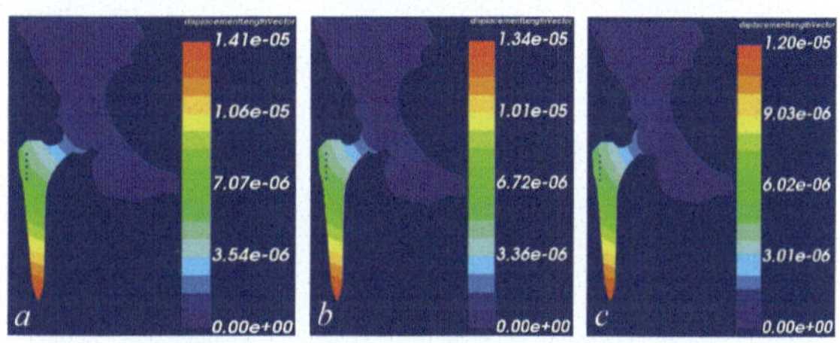

Fig. 7.13 Displacement field in pelvic bones and endoprosthesis for three inclination angles: *a* – 35 degrees inclination angle, *b* – 45 degrees inclination angle, *c* – 55 degrees inclination angle

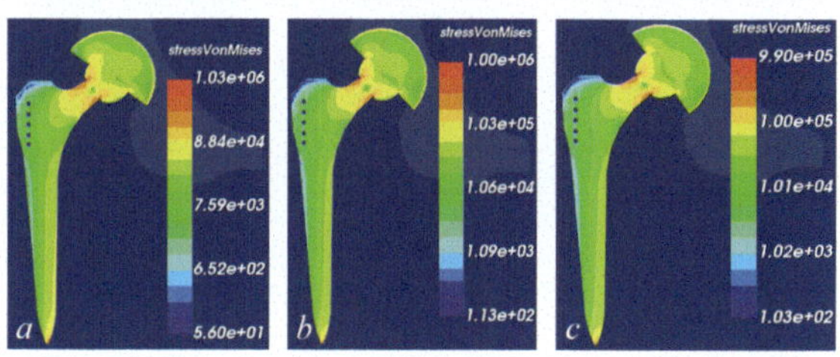

Fig. 7.14 Equivalent stresses (MPa) in endoprosthesis for three inclination angles: *a* – 35 degrees inclination angle, *b* – 45 degrees inclination angle, *c* – 55 degrees inclination angle

Table 7.11 Stress-strain state parameters of the studied treatment options

Treatment option	Maximum total displacement, m	Maximum equivalent stresses in pelvic bones, MPa	Maximum equivalent stresses in implants, MPa	Maximum equivalent stresses in polyethylene liner, MPa
35° inclination angle	$1,41 \cdot 10^{-5}$	16,9	103	40
45° inclination angle	$1,34 \cdot 10^{-5}$	19,3	100	29
55° inclination angle	$1,20 \cdot 10^{-5}$	20,1	99	60

geometric planning, biomechanical modeling, the optimal – the second treatment option was selected.

Third DE The experts of NIITON SSMU selected patient A. born in 1976 with a diagnosis of unstable spondylolisthesis of the L3 vertebra. The attending physician has planned the following treatment options: *a* transcutaneous TPF of the L3-L4 segment with reduction of the L4 vertebra, anterior interbody fusion according to the OLIF technique is delayed and is planned during the 2nd stage of the operation; *b* open TPF of the L3-L4 segment with reduction of the L4 vertebra and simultaneous interbody fusion L3-L4 according to the TLIF method; *c* transcutaneous TPF of the L3-L4 segment with reduction of the L4 vertebra and simultaneous interbody fusion using the DLIF technique. Solid-state models of the L2-L5 segment with implants are shown in Fig. 7.15. The results of biomechanical modeling are presented in Fig. 7.16 and 7.17. The maximum total displacements and equivalent stresses for each proposed treatment option are presented in Table 7.12. The analysis of the maximum values of displacements and equivalent stresses allowed us to determine that the first treatment option is successful.

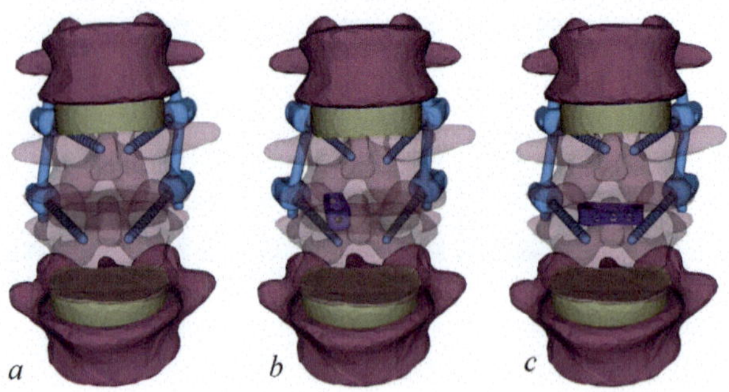

Fig. 7.15 3D solid models of the L2-L5 spine segment with installed implants: *a* – transcutaneous transpedicular fixation L3-L4 with reduction of the L4 vertebra; *b* - open transpedicular fixation L3-L4 with reduction of the L4 vertebra and simultaneous interbody fusion L3-L4 by the TLIF method; *c* – transcutaneous transpedicular fixation L3-L4 with reduction of the L4 vertebra and simultaneous interbody fusion using the DLIF technique

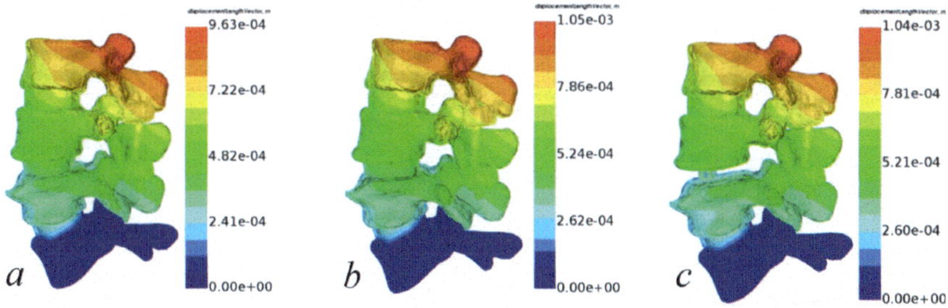

Fig. 7.16 Total displacements (in mm) in spine L2-L5 segment calculated using the Accord platform in the L2-L5 segment of the spine and implants: *a* – transcutaneous TPF L3-L4 with reduction of the L4 vertebra; *b* – open TPF L3-L4 with reduction of the L4 vertebra and simultaneous interbody fusion L3-L4 by the TLIF method; *c* – transcutaneous TPF L3-L4 with reduction of the L4 vertebra and simultaneous interbody fusion by the DLIF method

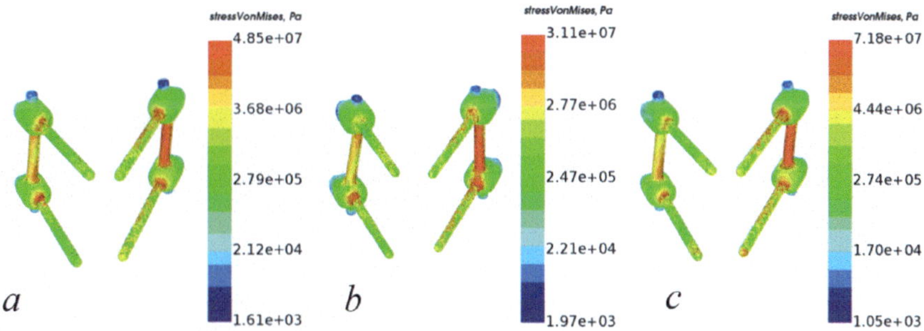

Fig. 7.17 Equivalent stresses (in Pa) in implants calculated using the Accord platform: *a* – transcutaneous TPF L3-L4 with reduction of the L4 vertebra; *b* – open TPF L3-L4 with reduction of the L4 vertebra and simultaneous interbody fusion L3-L4 by the TLIF method; *c* – transcutaneous TPF L3-L4 with reduction of the L4 vertebra and simultaneous interbody MIS fusion by the DLIF method

Based on the performed biomechanical modeling, the council of experts of NIITON SSMU decided to choose the following option of surgical reconstructive treatment and its recommendations to the surgeon: transcutaneous TPF L3-L4 with reduction of the L4 vertebra, anterior interbody fusion according to the OLIF technique is postponed and is planned during the 2nd stage of the operation. The staff of NIITON SSMU performed treatment considering the recommendations. Treatment results are shown in Fig. 7.18.

Let us analyze the results of DE from the standpoint of biomechanics. All the variants of surgical reconstructive treatment considered in the framework of DE provide the necessary stability of fixation under the studied loads (Tables 7.8, 7.11 and 7.12). Analysis of the maximum values of displacements and equivalent stresses allowed us to determine that in the first DE the third variant of surgical reconstructive treatment is the most successful (Tables 7.9 and 7.10), in the second DE successful the second treatment option is recognized (Table 7.11), and in the third DE – the

Table 7.12 Stress-strain state parameters of the studied treatment options

Treatment option	Maximum total displacement, mm	Maximum equivalent stresses in vertebrae fixed with implants, MPa	Maximum equivalent stresses in implants, MPa
transcutaneous TPF L3-L4 with reduction of the L4 vertebra	0.96	48.5	43.9
open TPF L3-L4 with reduction of the L4 vertebra and simultaneous interbody fusion L3-L4 by the TLIF method	1.05	31.1	31.0
transcutaneous TPF L3-L4 with reduction of the L4 vertebra and simultaneous interbody MIS fusion by the DLIF method	1.04	71.8	60.1

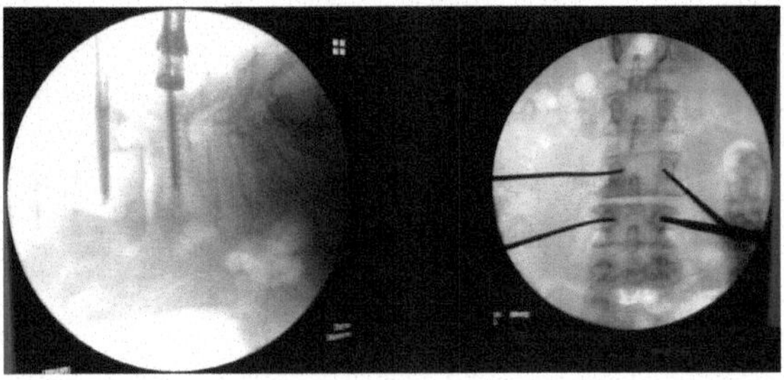

Fig. 7.18 Results of treatment in the third DE

first (Table 7.12). The results of the operation performed in accordance with the recommendations for choosing the optimal treatment option in the first TE are shown in Fig. 7.11. In the second DE, a group of experts from NIITON SSMU confirmed the clinical validity of the conclusions from biomechanical modeling and concluded that the selected treatment option corresponded to the option implemented in practice. As part of the third DE, the staff of NIITON SSMU treated the patient on April 21, 2021, considering the recommendations for the selected option (Fig. 7.19).

7.8.3 The Concept of Clinical Decision Support Systems with Biomechanical Support

High-quality personalized preoperative planning (PP) is impossible without the use of high-precision implant templates, the storage and use of which in the CDSS is carried out through databases (DB). For this purpose, the Implants database was developed, which contains high-precision templates (for geometric planning) and flat,

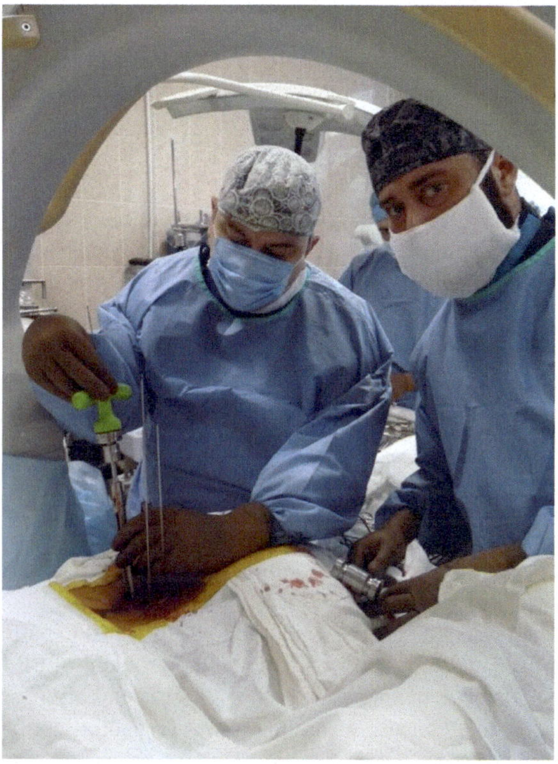

Fig. 7.19 A snapshot during the treatment of patient A., born in 1976

as well as spatial solid-state models (for biomechanical modeling) of implants registered in Russia. Within the framework of the developed platform, geometric planning (GP) can be carried out on a desktop computer, as well as using the SpinoMeter mobile application [12]. Both versions of GP programs calculate their optimal (theoretical) values for the basic geometric parameters of VPC measured by the surgeon. Thus, the doctor sees which parameters need to be adjusted during surgery, which at the PP stage allows to assess the level and degree of correction. For noninvasive measurement of the Young's modulus of spongy bone tissue, a technique has been developed and implemented in the platform [1]. The mechanical properties of bone tissues, intervertebral discs, ligaments, and implants are stored in "Mechanical" database and can be used in personalized biomechanical modeling (BM) of treatment options [37]. In order to automate the BM process and simplify working with the platform, a neural network was developed that implements automated segmentation of CT images and builds solid models of vertebral bodies based on them [38].

When working with the platform, the doctor can perform BM treatment options, solving the problem of static mechanics of a deformable solid [31, 35]: he has access to a tool for assessing the strength of the bone-implant system under the influence of external typical loads that simulate the static position of the human body, as well as its various rotations and rotations [39, 40]. To make an individual postoperative prognosis within the framework of the PP, the results of pre- and postoperative surveys of patients about their quality of life are stored and analyzed using the developed platform using "Medical" database. "Medical" database is also used for the formation of register uploads underlying medical registers of patients with pathologies of VPC.

A number of clinical cases (for example, combined pathologies, congenital deformities, and others) require high-precision and three-dimensional GP and BM necessary to justify the choice of a successful treatment option. At the same time, in their practice, doctors, as a rule, face "standard" clinical cases, in the treatment of which "standard" treatment algorithms can be dispensed with. To solve the tasks of PP and "complex" and "standard" clinical cases, the Accord platform can be used in two modes: personal virtual operating room (PVO) and regional center (RC). In the air defense mode, the doctor independently plans the treatment of "standard" clinical cases and uses a desktop computer in the clinic. In the air defense mode, the basic version of the CDSS software is used with limitations in the formulation of the biomechanics problem. In the RC mode, an extended version of the platform software is used, which implies the use of powerful computers and allows solving the biomechanics problem in the most complete formulation. In this mode, biomechanics engineers are supposed to be involved to perform BM. The operating modes of the platform allow us to solve the problem of PP for all clinical cases arising in the practice of surgeons, traumatologists and orthopedists.

In the database of the platform, the central entity is the patient with whom his clinical cases are associated (described cases of diseases and damage to VPC elements), the results of PP treatment (including GP and BM), implants planned to be installed during surgery, as well as pre- and postoperative quality of life surveys. All medical data in the database of the platform is stored in an impersonal form. The concept of CDSS in VPC surgery was formulated. The concept includes requirements for the development of CDSS, providing PP options for the treatment of the consequences of diseases and injuries of VPC and operating on the basis of the PP method "planning-modeling-forecast". Modern CDSS in VPC surgery should:

- Be based on the "planning-modeling-prognosis" method.
- Have a database with implants templates and solid models, mechanical properties of VPC elements and patient's medical data.
- Have a modular structure and operate with depersonalized patient data, as well as data in DICOM format.
- Provide a full cycle of preoperative planning, including geometric planning, biomechanical modeling and prediction of treatment results.
- Calculate the optimal (theoretical) values of the main VPC sagittal balance geometric parameters.
- Noninvasively determine the mechanical properties of VPC bone tissues.
- Have as part a register of patients with pathologies of VPC.
- Store in the database all the results of preoperative planning (including the results of geometric planning and biomechanical modeling, forecasting).

- Automatically segment CT images and create solid-state models of VPC elements based on them.
- Operate in two modes and cover all possible PP tasks for "standard" and "complex" clinical cases of diseases and injuries of VPC.

7.9 Conclusion

In the diagnosis and preoperative planning of surgical treatment in traumatology and orthopedics, doctors use the technology of traditional geometric planning. This involves using the results of computed tomography, magnetic resonance imaging or radiography and performing various kinds of measurements on them (angles, distances, areas, volumes), as well as positioning implants and/or fixation systems on them. In some clinical cases, especially with combined pathologies, only geometric planning may not be enough to choose a successful treatment option. There is a problem of developing and introducing into preoperative planning additional methods for quantifying treatment options and choosing a successful one among them.

For the first time in the world, the basics and technology for the implementation of biomechanical modeling as one of the stages of preoperative planning in surgery have been developed and tested. It is shown that for the elements of the musculoskeletal system, quantitative criteria for evaluating the success of treatment (geometric and biomechanical) can be developed, with the help of which it is possible to evaluate treatment options and choose a successful one among them. On the example of pathologies of the vertebral-pelvic complex, criteria for assessing the strength, stability of implantable structures and the service life of implants, the risk of bone damage are proposed. Developed and tested: methods for calculating the Young's modulus of spongy bone by CT; operating modes of the software platform (preoperative planning with biomechanical support can be performed by a doctor at his automated workplace or when working with a biomechanical engineer using a high-performance computer), implementing preoperative planning with biomechanical support.

There are no analogues of the completed developments underlying the Accord software platform in the world. There are known attempts to introduce biomechanical modeling into the process of teaching doctors, made by representatives of the organization AO (AO Foundation). However, colleagues offer a tool only for qualitative assessment of the biomechanics of osteosynthesis of fractures of long tubular bones by plates.

It should also be noted that the results of this work, namely biomechanical support of the doctor's decision, were introduced into routine preoperative planning on the basis of NIITON SSMU and showed their effectiveness and reliability in choosing a successful treatment option for injuries and degenerative-dystrophic diseases of the spine of NIITON SSMU patients.

References

1. A.O. Denisov, V.A. Shilnikov, S.A. Barns. Coxa-vertebral syndrome and its significance in hip arthroplasty (review). *Traumatology and orthopedics of Russia*. 63(1): 121-127, 2012.
2. R.V. O'Toole, B. Jaramaz, A.M. Digioia, C.D. Visnic, R.H. Reid. Biomechanics for preoperative planning and surgical simulations in orthopaedics. *Computers in Biology and Medicine*. 25(2):183–191, 1995.
3. J. Free, F. Eggermont, L. Derikx, R. van Leeuwen, Y. van der Linden, W. Jansen, E. Raaijmakers, E. Tanck, R. Kaatee. The effect of different CT scanners, scan parameters and scanning setup on Hounsfield units and calibrated bone density: a phantom study. *Biomed. Phys. Eng. Express*. 4(5):055013, 2018.
4. R.M. Witt, J.R. Cameronand. Improved bone standard containing dipotassium hydrogen phosphate solution for the intercomparison of different transmission bone scanning systems, Technical Report, University of Wisconsin–Madison, Department of Radiology, 1971.
5. G. Duval-Beaupere, C. Schmidt, P. Cosson. A barycentremetric study of the sagittal shape of spine and pelvis: the conditions required for an economic standing position. *Ann. Biomed. Eng*. 20(4):451–462, 1992.
6. J.C. Le Huec, R. Saddiki, J. Franke, J. Rigal, S. Aunoble. Equilibrium of the human body and the gravity line: the basics. *Eur Spine J*. (Suppl 5):558-563, 2011.
7. A.L. Kudiashev, V.V. Khominets, A.V. Teremshonok, K.E. Korostelev, E.B. Nagorny, A.V. Dol, D.V. Ivanov, I.V. Kirillova, L.Yu. Kossovich. Biomechanical prerequisites for the formation of proximal transitional kyphosis after transpedicular fixation of the lumbar spine. *Russian Journal of Biomechanics*. 2017
8. D.V. Ivanov, V.V. Hominets, I.V. Kirillova, L.Yu. Kossovich, A.L. Kudyashev, A.V. Teremshonok. Biomechanics of compensatory mechanisms in spinal-pelvic complex. *Journal of Physics: Conference Series*. 5th International Conference on Topical Problems of Continuum Mechanics with a Special Session in Honor of Alexander Manzhirov's 60th Birthday. 991:012036, 2018.
9. S.K. Makirov, A.A. Yuz, M.T. Jahaf. Method of assessing the parameters of the sagittal spinal pelvic balance (in Russ.). *Hirurgiâ pozvonočnika (Spine Surgery)*. 12(3):55-63, 2015.
10. J.C. Le Huec, A. Faundez, D. Dominguez, P. Hoffmeyer, S. Aunoble. Evidence showing the relationship between sagittal balance and clinical outcomes in surgical treatment of degenerative spinal diseases: a literature review. *Int. Orthop*. 39(1):87-95, 2015.
11. T.B. Sullivan, N. Marino, F.G. Reighard, P.O. Newton. Relationship between lumbar lordosis and pelvic incidence in the adolescent patient: normal cohort analysis and literature comparison. *Spine Deform*. 6(5):529-536, 2018.
12. D.V. Ivanov, I.V. Kirillova, L.Yu. Kossovich, S.V. Likhachev, A.V. Polienko, A.V. Kharlamov, A.E. Shulga, A.E.. Comparative analysis of the SpinoMeter mobile application and Surgimap system for measuring the sagittal balance parameters: interobserver reliability test. *Genij Ortopedii*. 27(1)74-79, 2021.
13. J.C. Le Huec, K. Hasegawa. Normative values for the spine shape parameters using 3D standing analysis from a database of 268 asymptomatic Caucasian and Japanese subjects. *Eur Spine J*. 25(11):3630-3637, 2016.
14. S.J. Hyun, S. Han, Y.B. Kim, Y.J. Kim, G.B. Kang, J.Y. Cheong. Predictive formula of ideal lumbar lordosis and lower lumbar lordosis determined by individual pelvic incidence in asymptomatic elderly population. *Eur Spine J*. 28(9):1906-1913, 2019.
15. F. Tanguay, J.M. Mac-Thiong, J.A. de Guise, H. Labelle, H.. Relation between the sagittal pelvic and lumbar spine geometries following surgical correction of adolescent idiopathic scoliosis. *Eur Spine J*. 123:299-302, 2006.
16. J. Legaye, G. Duval-Beaupere. Sagittal plane alignment of the spine and gravity: a radiological and clinical evaluation. *Acta Orthop Belg*. 71:213-220, 2005.
17. D.V. Ivanov, A.S. Falkovich, A.M. Donnik, A.V. Polienko, E.S. Olenko, A.V. Krutko. Generalization of dependencies between geometric parameters of sagittal balance (in Russ.). *Russian Journal of Biomechanics*. 6(1):8–24, 2022.

18. J.C. Le Huec, S. Charosky, C. Barrey, J. Rigal, S. Aunoble, S.. Sagittal imbalance cascade for simple degenerative spine and consequences: algorithm of decision for appropriate treatment. *Eur. Spine J.* (Suppl 5):699-703, 2011.
19. S.R. Manoharan, D. Joshi, M. Owen, S.M. Theiss, D. Deinlein. Relationship of Cervical Sagittal Vertical Alignment After Sagittal Balance Correction in Adult Spinal Deformity: A Retrospective Radiographic Study. *Int J Spine Surg.* 12(2):269-275, 2018.
20. K.T. Yeh, R.P. Lee, I.H. Chen, T.-C. Yu, K.-L. Liu, C.-H. Peng, J.-H. Wang, W.-T. Wu . Correlation of Functional Outcomes and Sagittal Alignment After Long Instrumented Fusion for Degenerative Thoracolumbar Spinal Disease. *Spine (Phila Pa 1976).* 43(19):1355-1362, 2018.
21. K.D. Than, P. Park, K.M. Fu, S. Nguyen, M.Y. Wang, D. Chou, P.D. Nunley, N. Anand, R.G. Fessler, C.I. Shaffrey, S. Bess, B.A. Akbarnia, V. Deviren, J.S. Uribe, F. La Marca, A.S. Kanter, D.O. Okonkwo, G.M. Mundis Jr, P.V. Mummaneni. Clinical and radiographic parameters associated with best versus worst clinical outcomes in minimally invasive spinal deformity surgery. *J Neurosurg Spine.* 25(1):21-5, 2016.
22. F.-C. Kao, Y.-J. Huang, P.-Y. Chiu, M.-K. Hsieh, T.-T. Tsai, T.-T.. Factors Predicting the Surgical Risk of Osteoporotic Vertebral Compression Fractures. *Journal of Clinical Medicine.* 8(4):501, 2019.
23. J.L. Tian, L. Sun, R.Y. Hu, W. Han, X.B. Tian. Correlation of Cup Inclination Angle with Liner Wear for Metal-on-polyethylene in Hip Primary Arthroplasty. *Orthop Surg.* 9(2):186-190, 2017.
24. K. Chun, I. Yang, N. Kim, D. Cho. Effect of Device Rigidity and Physiological Loading on Spinal Kinematics after Dynamic Stabilization: An In-Vitro Biomechanical Study. *J Korean Neurosurg Soc.* 58(5):412-418, 2015.
25. Y. Más, L. Gracia, S. Gabarre, D. Peña, A. Herrera. Finite element simulation and clinical follow-up of lumbar spine biomechanics with dynamic fixations. PLoS One. 12(11):e0188328, 2017.
26. A.A. White, M.M. Panjabi. *Clinical Biomechanics of the Spine*, 2nd ed. New York, J.B. LIPPINCOTT Company, 1990.
27. T.H. Hansson, T.S. Keller, M.M. Panjabi. A study of the compressive properties of lumbar vertebral trabeculae: effects of tissue characteristics. *Spine.* 12(1):56-62, 1987.
28. E.S. Baykov, A.A. Baykalov. Connection of biomechanical and biochemical parameters of spinal-motor segments with recurrence of hernias of lumbar intervertebral discs. *Hir. Pozvonochnika.* 4:61-68, 2017.
29. R.N. Natarajan, K. Watanabe, K. Hasegawa. Biomechanical Analysis of a Long-Segment Fusion in a Lumbar Spine - A Finite Element Model Study. *J Biomech Eng.* 140(9), 2018.
30. Y. Duan, H.H. Wang, A.M. Jin, L. Zhang, S.X. Min, C.L. Liu, S.J. Qiu, X.Q. Shu. Finite element analysis of posterior cervical fixation. *Orthop Traumatol. Surg. Res.* 101(1):23-29, 2015.
31. Y. Su, D. Ren, P.C. Wang. Comparison of Biomechanical Properties of Single- and Two-segment Fusion for Denis Type B Spinal Fractures. *Orthopaedic Surgery.* 5(4):266-273, 2013.
32. A.L. Kudiashev, V.V. Khominets, A.V. Teremshonok, E.B. Nagorny, S.Yu. Stadnichenko, A.V. Dol, D.V. Ivanov, I.V. Kirillova, L.Yu. Kossovich, A.L. Kovtun. Biomechanical modeling in surgical treatment of a patient with true lumbar spondylolisthesis. *Hir. Pozvonochnika.* 15(4):87-94, 2018.
33. D.V. Ivanov, I.V. Kirillova, L.Yu. Kossovich, L.V. Bessonov, A.V. Petraikin, A.V. Dol, E.S. Ahmad, S.P. Morozov, A.V. Vladzymyrskyy, K.A. Sergunova, A.V. Kharlamov. Influence of convolution kernel and beam-hardening effecton the assessment of trabecular bone mineral density using quantitative computed tomography. *Izvestya of Saratov University. New Series. Series: Mathematics. Mechanics. Informatics.* 20(2):205-219, 2020.
34. M. Dreischarf. Comparison of eight published static finite element models of the intact lumbar spine: predictive power of models improves when combined together. *J. Biomech.* 47(8):1757-6176, 2014.

35. A.V. Dol, E.S. Dol, D.V. Ivanov. Biomechanical modeling of options for surgical reconstructive treatment of spinal spondylolisthesis at the L4–L5 level. *Russian Journal of Biomechanics*. 22(1):25-36, 2018.
36. L.V. Bessonov, A.A. Golyadkina, P.O. Dmitriev, A.V. Dol, V.S. Zolotov, D.V. Ivanov, I.V., Kirillova, L.Yu. Kossovich, Yu.I. Titova, V.Yu. Ulyanov, A.V. Kharlamov. Constructing the dependence between the Young's modulus value and the Hounsfield units of spongy tissue of human femoral heads. *Izvestiya of Saratov University. Mathematics. Mechanics. Informatics.* 21(2):182-193, 2021.
37. A.S. Beskrovny, L.V. Bessonov, A.A. Golyadkina, A.V. Dol, D.V. Ivanov, I.V. Kirillova, L.Yu. Kossovich, D.A. Sidorenko. Development of a decision support system in traumatology and orthopedics. Biomechanics as a tool for preoperative planning. *Russian Journal of Biomechanics*. 25(2):99-112, 2021
38. A.S. Beskrovny, L.V. Bessonov, D.V. Ivanov, V.S. Zolotov, D.A. Sidorenko, I.V. Kirillova, L.Yu. Kossovich: Construction of 3D solid vertebral models using convolutional neural networks. *Izvestiya of Saratov University. Mathematics. Mechanics. Informatics.* 21(3):368–378, 2021.
39. Y.H. Kim, B. Khuyagbaatar, K. Kim. Recent advances in finite element modeling of the human cervical spine. *Journal of Mechanical Science and Technology*. 32(1):1-10, 2018.
40. S.N. Zahari, M.J.A. Latif, N.R.A. Rahim, M.R.A. Kadir, T. Kamarul. The Effects of Physiological Biomechanical Loading on Intradiscal Pressure and Annulus Stress in Lumbar Spine: A Finite Element Analysis. *Journal of Healthcare Engineering*. 9618940, 2017.

Chapter 8
Dispersion of the Bending Wave in a Fluid-loaded Elastic Layer

Julius Kaplunov, Ludmila Prikazchikova, and Sheeru Shamsi

Abstract A plane strain problem is considered for an elastic layer immersed into a compressible fluid. The dispersion relation for anti-symmetric waves is studied. The associated three-term long-wave low-frequency expansion for a fluid-borne bending wave is derived, along with similar expansions corresponding to Kirchhoff and Timoshenko-Reissner type fluid-loaded plates. The results of comparative asymptotic analysis are presented. The role of plate inertia and fluid compressibility are discussed.

Key words: Fluid-loaded elastic layer, Plate theories, Dispersion, Asymptotic

8.1 Introduction

Fluid-structure interaction problems for elastic plates have been investigated since long ago. However, asymptotic considerations in this area were usually restricted to the classical Kirchhoff theory, e.g. see [2, 3]. Only a few publications has approached the subject using original equations in dynamic elasticity, e.g. see [5, 6, 9]. Until now to the best of authors' knowledge there is no direct comparisons of the asymptotic (not just numerical) results, obtained from linear elasticity and approximate plate models. At the same time, nowadays there is a significant demand of more rigorous and accurate predictions inspired by advanced industrial applications, including soft robotics, e.g. see [8].

In this paper we study a plane strain problem for an elastic layer, governed by 2D equations in elasticity, in contact with a compressible non-viscous fluid. The related dispersion equation for anti-symmetric waves is analysed at the long-wave low-frequency limit. The ratio between dimensionless wavelength and frequency is

---

Julius Kaplunov · Ludmila Prikazchikova · Sheeru Shamsi
Keele University, ST5 5GB,
e-mail: j.kaplunov@keele.ac.uk,l.prikazchikova@keele.ac.uk,s.s.shamsi@keele.ac.uk

H. Altenbach et al. (eds.), *Advances in Solid and Fracture Mechanics*,
Advanced Structured Materials 180,
https://doi.org/10.1007/978-3-031-18393-5_8

not taken to be the same as for a bending wave on a free plate and corresponds to a specific fluid-borne bending wave, e.g. see [10]. A three-term asymptotic expansion of the aforementioned dispersion relation is derived and implemented for testing the approximate dispersion relations for two simplified fluid-structure interaction models based on thin plate asymptotic theories including the classical Kirchhoff theory as well as Timoshenko-Reissner type theory, e.g. see [1, 7] and references therein. The roles of fluid compressibility and plate inertia are also addressed.

It is shown in particular that the leading order term in the derived expansion of the "exact" dispersion relation also follows from the dispersion relation for a Kirchhoff plate immersed into incompressible fluid, neglecting the plate inertia. Moreover, fluid compressibility appears to be outside the range of validity of both classical and refined plate based formulations studied in the paper. It is also established that the adapted refined theory has a higher asymptotic accuracy than the classical one. This observation is far from being obvious, since consideration starts from the assumption that fluid loading may be considered as a prescribed external stress field. The latter assumption has been proved to be justified even at a higher order, although it formally supports the asymptotic scaling characteristic of a bending wave on a plate with traction free faces not incorporating accurately enough the effect of the fluid.

The paper is organised as follows. The linear equations in plane elasticity and fluid dynamics are presented in Sect. 8.2, along with the approximate formulations based on the Kirchhoff and Timoshenko-Reissner types plate theories. All associated dispersion relations are derived in Sect. 8.3. The Sect. 8.4 is concerned with a comparative analysis of asymptotic expansions.

8.2 Basic Equations

Consider free in-plane vibrations of an elastic layer of thickness $2h$ immersed in a non-viscous compressible fluid. Let the mid-line of the layer be the $x_2 = 0$ axis of the Cartesian coordinate system ($-\infty < x_1, x_2 < \infty$), see Fig. 8.1. Throughout the paper we use the following notation: ρ and ρ_0 are solid and fluid densities, respectively; E is Young's modulus, ν is the Poisson ratio, λ and μ are Lamé elastic constants, c_1 and c_2 are the longitudinal and transverse wave speeds in solid, c_0 is the wave speed in fluid.

The equations of motion in terms of the elastic potentials ϕ and ψ and the fluid potential φ can be written as

$$\Delta\phi - \frac{1}{c_1^2}\frac{\partial^2\phi}{\partial t^2} = 0, \qquad \Delta\psi - \frac{1}{c_2^2}\frac{\partial^2\psi}{\partial t^2} = 0, \tag{8.1}$$

and

$$\Delta\varphi - \frac{1}{c_0^2}\frac{\partial^2\varphi}{\partial t^2} = 0, \tag{8.2}$$

where

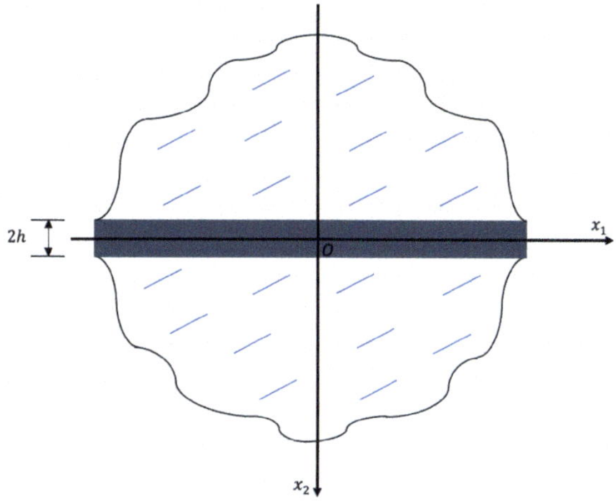

Fig. 8.1 Fluid-loaded elastic layer.

$$\Delta = \frac{\partial^2}{\partial x_1^2} + \frac{\partial^2}{\partial x_2^2}, \quad c_1 = \sqrt{\frac{\lambda + 2\mu}{\rho}} \quad \text{and} \quad c_2 = \sqrt{\frac{\mu}{\rho}},$$

with c_0 staying for the fluid compressibility, e.g. see [4] for more detail.

The contact conditions along the interfaces $x_2 = \pm h$ are given by

$$\sigma_{21} = 0, \qquad \sigma_{22} = \rho_0 \frac{\partial^2 \varphi}{\partial t^2}, \qquad v_2 = \frac{\partial \varphi}{\partial x_2}, \tag{8.3}$$

where the stresses σ_{21} and σ_{22} and the vertical displacement v_2 are expressed as

$$\sigma_{21} = \mu \left(2\frac{\partial^2 \phi}{\partial x_1 \partial x_2} - \frac{\partial^2 \psi}{\partial x_1^2} + \frac{\partial^2 \psi}{\partial x_2^2} \right), \quad \sigma_{22} = \lambda \frac{\partial^2 \phi}{\partial x_1^2} + (\lambda + 2\mu)\frac{\partial^2 \phi}{\partial x_2^2} - 2\mu \frac{\partial^2 \psi}{\partial x_1 \partial x_2},$$
$$v_2 = \frac{\partial \phi}{\partial x_2} - \frac{\partial \psi}{\partial x_1}. \tag{8.4}$$

The set of 2D equations above may be considered as a benchmark formulation. Its refined long-wave low-frequency approximation, e.g. see [1, 7], is governed by the 1D equation

$$\frac{2Eh^3}{3(1-\nu^2)} \frac{\partial^4 u}{\partial x_1^4} + 2\rho h \left[1 + h^2 \frac{7\nu - 17}{15(1-\nu)} \frac{\partial^2}{\partial x_1^2} \right] \frac{\partial^2 u}{\partial t^2}$$
$$- 2\rho_0 \left[1 - h^2 \frac{8 - 3\nu}{10(1-\nu)} \frac{\partial^2}{\partial x_1^2} \right] \frac{\partial^2 \varphi}{\partial t^2} = 0, \tag{8.5}$$

and the impenetrability conditions at $x_2 = \pm h$

$$u + \frac{\nu h^2}{2(1-\nu)} \frac{\partial^2 u}{\partial x_1^2} = \frac{\partial \varphi}{\partial x_2}. \tag{8.6}$$

Here u is an approximate mid-line vertical displacement, i.e. $v_2(x_1,0,t) \approx u(x_1,t)$; the fluid potential φ satisfies equation (8.2).

Neglecting the terms with the factor h^2 inside the brackets in (8.5) as well as in (8.6) we arrive at the most popular approximation of the original model above given by

$$\frac{2Eh^3}{3(1-\nu^2)} \frac{\partial^4 u}{\partial x_1^4} + 2\rho h \frac{\partial^2 u}{\partial t^2} - 2\rho_0 \frac{\partial^2 \varphi}{\partial t^2} = 0, \tag{8.7}$$

and

$$u = \frac{\partial \varphi}{\partial x_2} \quad \text{at} \quad x_2 = \pm h. \tag{8.8}$$

The latter corresponds to the classical Kirchhoff theory for plate bending, while the refined equations (8.5)-(8.6) originate from Timoshenko-Reissner type asymptotic version of plate bending theory. The goal of the paper is to evaluate the accuracy of the dispersion relations associated with both long-wave low-frequency approximations above for a fluid-borne bending wave using the "exact" solution, in which the motion of the layer is governed by the 2D equations in plane elasticity.

8.3 Dispersion Relations

Begin with travelling wave solutions of 2D hyperbolic equations (8.1)-(8.2). For the anti-symmetric in the vertical coordinate x_2 harmonic motion we obtain at $|x_2| \leqslant h$

$$\begin{aligned} \phi &= C_1 \sinh(k\alpha x_2) e^{i(kx_1 - \omega t)}, \\ \psi &= C_2 \cosh(k\beta x_2) e^{i(kx_1 - \omega t)}, \end{aligned} \tag{8.9}$$

and at $|x_2| \geqslant h$

$$\varphi = C_3 e^{-k\eta(|x_2|-h)} e^{i(kx_1 - \omega t)}, \tag{8.10}$$

where k is wave number, ω is angular frequency, $i = \sqrt{-1}$, C_j $(j = 1,2,3)$ are arbitrary constants and

$$\alpha = \sqrt{1 - \frac{\omega^2}{k^2 c_1^2}}, \qquad \beta = \sqrt{1 - \frac{\omega^2}{k^2 c_2^2}}, \qquad \eta = \sqrt{1 - \frac{\omega^2}{k^2 c_0^2}}. \tag{8.11}$$

Then, on substituting the formulae above into contact conditions (8.3), we arrive at the set of three linear algebraic equations

$$
\begin{aligned}
&2i\alpha\cosh(kh\alpha)C_1 + (1+\beta^2)\cosh(kh\beta)C_2 = 0,\\
&\mu(1+\beta^2)\sinh(kh\alpha)C_1 - 2i\beta\mu\sinh(kh\beta)C_2 + \rho_0\omega^2k^{-2}C_3 = 0,\\
&\alpha\cosh(kh\alpha)C_1 - i\cosh(kh\beta)C_2 + \eta C_3 = 0.
\end{aligned} \tag{8.12}
$$

Now, equating to zero the determinant of the associated 3x3 matrix we obtain the sought for dispersion relation

$$
(2K^2-\Omega^2)^2\tanh A - 4K^2AB\tanh B + \frac{\Omega^4Ar}{H} = 0, \tag{8.13}
$$

where $r = \frac{\rho_0}{\rho}$ denotes a relative density, $K = kh$ and $\Omega = \frac{\omega h}{c_2}$ stand for dimensionless wavenumber and frequency, whereas

$$
A = \sqrt{K^2-\Omega^2\kappa^2}, \qquad B = \sqrt{K^2-\Omega^2}, \qquad H = \sqrt{K^2-\Omega^2\delta^2}, \tag{8.14}
$$

with

$$
\delta = \frac{c_2}{c_0} \quad \text{and} \quad \kappa = \frac{c_2}{c_1} = \sqrt{\frac{1-2\nu}{2-2\nu}}
$$

denoting wave speed ratios.

Next, consider the 1D refined equation (8.5) together with the impenetrability condition (8.6). Setting $u = C_4e^{i(kx_1-\omega t)}$, where C_4 is another arbitrary constant and using the same solution (8.10) for the fluid potential we derive a dispersion relation for the bending wave on a fluid-loaded Timoshenko-Reissner type plate. It is given by

$$
\begin{aligned}
\Omega^2\Big(&3r\left(\nu(8-3\nu)K^4 - 16(1-\nu)^2K^2 - 20(1-\nu)^2\right)\\
&+4H(1-\nu)\left((7\nu-17)K^2 - 15(1-\nu)\right)\Big) + 40(1-\nu)HK^4 = 0.
\end{aligned} \tag{8.15}
$$

Finally, we write down the dispersion equation for fluid-loaded Kirchhoff plate, starting from the formula (8.7) along with the solution (8.10). The result is

$$
3(1-\nu)(r+H)\Omega^2 - 2HK^4 = 0. \tag{8.16}
$$

As might be expected the last dispersion relation follows from the previous one by neglecting the terms $O(K^2)$ inside the coefficient at Ω^2.

8.4 Asymptotic Expansions

Consider the exact dispersion relation (8.13) over the long-wave low-frequency domain

$$\Omega \ll 1, \qquad K \ll 1, \tag{8.17}$$

assuming $\Omega \sim K^{5/2}$ instead of the canonical scaling $\Omega \sim K^2$ underlying the bending wave on an elastic plate in absence of fluid loading [7]. Let us substitute the polynomial

$$\Omega^2 = \Omega_0 K^5 + \Omega_1 K^6 + \Omega_2 K^7 + \ldots \tag{8.18}$$

into (8.13) and expand there all the square roots and hyperbolic functions into asymptotic series. After straightforward but lengthy algebraic manipulations we derive

$$\begin{aligned} \Omega_0 &= \frac{2}{3r(1-\nu)}, \\ \Omega_1 &= -\frac{2}{3r^2(1-\nu)}, \\ \Omega_2 &= -\frac{8}{15r(1-\nu)} + \frac{2}{3r^3(1-\nu)}. \end{aligned} \tag{8.19}$$

The substitution of the formula (8.18) into the shortened dispersion relation (8.15) gives the same expressions (8.19) for all three coefficients. As might be expected, for the dispersion relation (8.16) the first two coefficients in expansion (8.18) coincide with those in (8.19), while the third one is given by

$$\Omega_2 = \frac{2}{3r^3(1-\nu)}. \tag{8.20}$$

It is worth noting that the leading order term in (8.18) corresponds even to a simpler fluid-structure interaction model than that based on the full Kirchhoff dynamic plate theory. In this case one may start from the equations

$$\frac{2Eh^3}{3(1-\nu^2)} \frac{\partial^4 u}{\partial x_1^4} - 2\rho_0 \frac{\partial^2 \varphi}{\partial t^2} = 0, \tag{8.21}$$

and

$$\Delta \varphi = 0. \tag{8.22}$$

Thus, plate inertia is neglected in the first of these equations, while the second one is for incompressible fluid. It can be also easily verified that the two term expansion, neglecting the term with Ω_2 in (8.13), follows from the full dynamic equation (8.7). A weak compressibility of the fluid appears outside the range of validity of the three-term expansion (8.18), where both of the studied approximate fluid-structure interaction formulations are also not applicable. Indeed, the formulae for the coefficients (8.19) do not involve the fluid wave speed c_0. It is worth noting that the classical and refined models predict the correct values for the first two and three coefficients, respectively, in the asymptotic expansion (8.18), in spite of originating from the asymptotic analysis of thin-walled structures not in contact with the fluid assuming $\Omega \sim K^2$ rather than $\Omega \sim K^{5/2}$, see [1, 7] for more detail.

Numerical results for a steel layer immersed in water are presented in Figs. 8.2 and 8.3. The problem parameters are $c_0 = 1480\,\mathrm{ms}^{-1}, \nu = 0.2, c_2 = 3189\,\mathrm{ms}^{-1}, \rho = 7800\,\mathrm{kgm}^{-3}, \rho_0 = 1000\,\mathrm{kgm}^{-3}$. The dispersion curves for fluid-borne bending wave calculated from the exact dispersion relation (8.13) and its one, two and three-term asymptotic expansions, associated with formula (8.18) with the coefficients given by (8.19), are displayed in Figure 8.2.

Figure 8.3 shows comparison of three-term expansion (8.18) with the coefficient Ω_2 from (8.19) and (8.20). The proximity of the dispersion curves is due to relatively small value of the density ratio ($r \approx 0.13$). In this case the same term with r^{-3} dominates in the expressions for Ω_2 in formulae (8.19) and (8.20). In fact, the range $r \ll 1$ corresponds to a light fluid loading, assuming a special asymptotic analysis and is not tackled in the paper.

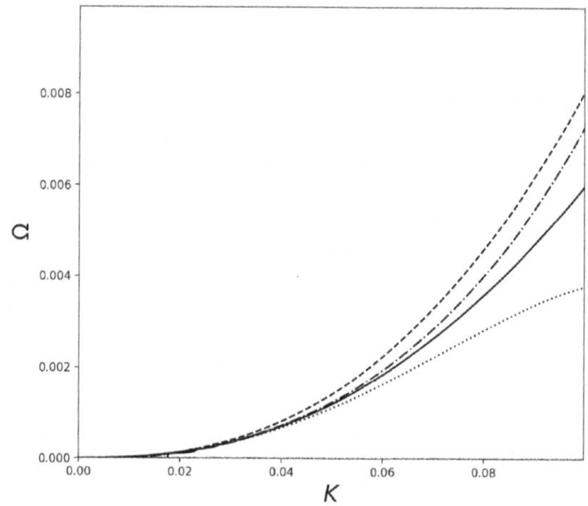

Fig. 8.2 Comparison of the solution of the dispersion relation (8.13) (solid line) with its asymptotic expansion (8.18) with the coefficients (8.19) at leading (dashed line), first (dotted line) and second (dashdot line) order.

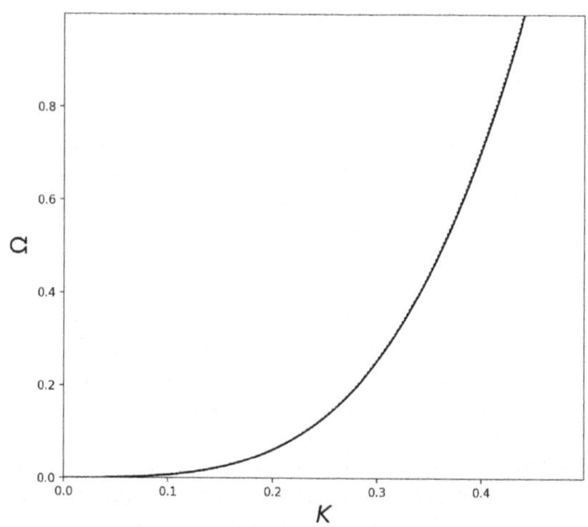

Fig. 8.3 Comparison of three-term expansions (8.18) with the coefficient Ω_2 from (8.19) (solid line) and (8.20) (dotted line).

8.5 Concluding Remarks

The three-term asymptotic expansion (8.18) with (8.19) of the exact dispersion relation (8.13) corresponds to a fluid-borne bending wave governed by the scaling $\Omega \sim K^{5/2}$. At leading order, this wave is affected by the plate stiffness and fluid inertia, see also equations (8.21)-(8.22). The plate inertia arises only at the next order. The effect of fluid compressibility does not enter the coefficients of the expansion (8.18) through the fluid wave speed c_0, i.e. it is negligible in comparison with the transverse shear deformation and other phenomena of the second order characteristic of a Timoshenko-Reissner type plate, see [7] for more detail. It is remarkable that the adapted formulation (8.5)-(8.6), based on the asymptotic Timoshenko-Reissner type theory (see [1, 7]) results in a correct coefficient Ω_2, although this is originally deduced for a plate under prescribed surface loading for the scaling $\Omega \sim K^2$. This observation also motivates establishment of a refined asymptotic model starting from the initial coupled fluid-structure interaction problem, e.g. see (8.1)-(8.4) within a plane strain setup. In the paper we do not make further assumptions on the problem parameters such as that on a light fluid loading. In the latter case some of the obtained results need to be amended. The developed asymptotic framework also may be adapted for a practically important setup of an elastic layer lying on a fluid half-space, e.g. see [6].

Acknowledgements Sponsorship by Keele University for S.Shamsi's PhD is gratefully acknowledged.

References

1. A. Belov, J. Kaplunov, and E. Nolde. A refined asymptotic model of fluid-structure interaction in scattering by elastic shells. *Flow, Turbulence and Combustion*, 61(1):255–267, 1998.
2. D. G. Crighton. The free and forced waves on a fluid-loaded elastic plate. *Journal of Sound and Vibration*, 63(2):225–235, 1979.
3. D. G. Crighton, A. P. Dowling, J. Ffowcs-Williams, M. Heckl, F. Leppington, and J. F. Bartram. *Modern Methods in Analytical Acoustics*. Acoustical Society of America, 1992.
4. R. Feynman. *The Feynman Lectures on Physics*. Addison Publishing Company, 1969.
5. M. Johansson, P. D. Folkow, A. Hägglund, and P. Olsson. Approximate boundary conditions for a fluid-loaded elastic plate. *The Journal of the Acoustical Society of America*, 118(6):3436–3446, 2005.
6. J. Kaplunov and D. Markushevich. Plane vibrations and radiation of an elastic layer lying on a liquid half-space. *Wave Motion*, 17(3):199–211, 1993.
7. J. D. Kaplunov, L. Y. Kossovitch, and E. Nolde. *Dynamics of thin walled elastic bodies*. Academic Press, 1998.
8. Z. Lin, A. Hess, Z. Yu, S. Cai, and T. Gao. A fluid–structure interaction study of soft robotic swimmer using a fictitious domain/active-strain method. *J Comp Phys*, 376:1138–1155, 2019.
9. S. Sorokin. Analysis of time harmonic wave propagation in an elastic layer under heavy fluid loading. *Journal of Sound and Vibration*, 305(4-5):689–702, 2007.
10. M. Talmant, H. Überall, R. D. Miller, M. F. Werby, and J. Dickey. Lamb waves and fluid-borne waves on water-loaded, air-filled thin spherical shells. *The Journal of the Acoustical Society of America*, 86(1):278–289, 1989.

Chapter 9
Mathematical Models of Local Ice Strength and Problems Elastic-plastic Bending of Hydraulic Structures Supports

Sergei M. Kovalev and Galina V. Pavilaynen

Abstract The modern science of strength and plasticity of building provides ample opportunities for the design and constructions of cost-effective structures while ensuring their high reliability under extreme operating conditions. Such structures include drilling platforms for offshore hydrocarbon production and helipads. Studies by glaciologists document the anisotropy and plasticity of ice and the huge loads that ice fields can exert on the vertical supports of drilling platforms.The method for determining the compression ice strength in the boreholes without extraction of the ice samples using a borehole jack is being developed at the Laboratory of Ice Physics of the Arctic and Antarctic Research Institute and considered in this article.

Key words: Compression ice strength, Plasticity, Vertical beam, Bending, Borehole jack

9.1 Introduction

Many studies, for example [1, 2], confirm the anisotropy and plasticity of ice and the huge loads that ice fields can exert on the vertical supports of drilling platforms [14, 16]. Physical-mechanical ice properties are traditionally studied on the basis of cores and samples drilled from level ice floes, ice ridges, stamukhas and icebergs. What is the main difficulty in determination of the strength of ice samples at uniaxial compression? For calculation of ice loads on engineering structures one uses the

Sergei M. Kovalev
Arctic and Antarctic Research Institute, 28 Bering Street, St. Petersburg, 199397, Russian Federation,
e-mail: skovalev@aari.ru
Galina V. Pavilaynen
St. Petersburg State University, 7/9 University Embankment, St. Petersburg, 199034, Russian Federation,
e-mail: g_v_pavilaynen@mail.ru

H. Altenbach et al. (eds.), *Advances in Solid and Fracture Mechanics*,
Advanced Structured Materials 180,
https://doi.org/10.1007/978-3-031-18393-5_9

strength of samples drilled parallel to the ice freezing surface. In order to prepare such samples first, an ice block is cut in the ice cover. Then this block is extracted to the surface. The required horizons are identified on the block, specific places are marked and then ice cores are drilled out. However, studies show a significant anisotropy of ice and the fibrous structure [3, 4]. Production of samples in the field requires a lot of time, as a result, the properties of ice change significantly. Therefore one should develop new methods for determination of mechanical ice properties without extraction of ice samples.

As part of the Research and Technological Works (NITR) of Roskomhydromet in 2020-24 the Arctic and Antarctic Research Institute develops the topic "Study of large-scale dynamics, physical processes. Mechanics of deformation and destruction of sea ice with the aim of improvement of short-range prediction of compression and ridging". For successful implementation of this topic it is necessary to know the true compression and flexural ice strength values [5, 6, 7]. It is supposed to determine the compression ice strength under full-scale conditions by means of the borehole jack.

The method for determining the compression ice strength in the boreholes without extraction of the ice samples using a borehole jack is being developed at the Laboratory of Ice Physics of the Arctic and Antarctic Research Institute from the beginning of the XXI century. At this time already the third "borehole jack" generation was developed. The complex was used for compression ice strength determination during numerous expeditions. A large amount of experimental data was obtained proving the possibilities of the methodology to determine the local ice strength for different ice-covered seas of Russia.

The methodology for determining the compression ice strength by means of the borehole jack is included into the international and national normative documents [8]. The technique and equipment for determining the strength of ice in natural conditions has long been known. The downhole probe-indenter is used in numerous studies of ice strength. Currently, this technique and equipment are used at the research station "Ice base Cape Baranov" (Severnaya Zemlya, Russian Federation).

9.2 Complex System "Borehole Jack"

A downhole probe-indenter (borehole jack) is a complex technical device of a hydraulic type that provides force measurement when an indenter (stamp) of a certain area is introduced into the borehole wall with ice destruction [9]. The development of this device has been carried out since 2000 and several modifications have already been carried out. In 2012, together with CJSC AVA Hydrosystems and NK Rosneft, a third-generation borehole probe-indenter was developed and manufactured.

The modern layout of the device is schematically shown in Fig. 9.1, where presents a diagram of the complex system for determination of ice strength characteristics under the full-scale conditions (in boreholes). On the ice floe *1* there is a hydraulic station *2* with the flow regulator *3* for setting the prescribed pullout rate

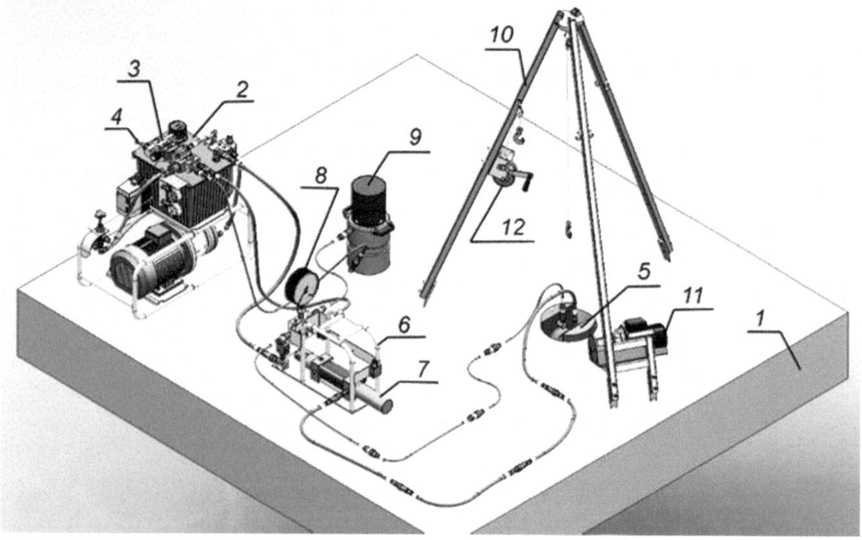

Fig. 9.1 Diagram of the complex system for determination of ice strength.

of the jack rod *5* and hydraulic distributor *4*. The hydraulic station *2* by hoses via the measurement block *6*, including the measurement cylinder *7* and manometer *8* (pressure sensor and hydraulic valves of the initial position and filling with oil of the measurement cylinder are not indicated in Fig. 9.1 and multiplier *9*, is connected with the jack *5*. Above the jack a tripod is installed *10*) with an electrical winch *11* and a hand winch *12*, which provide for jack lowering and rise *5* to the borehole in the mechanical or manual modes. The measurement block *6* is connected to the registration device and the hydraulic station *2* to electrical generator.

Now we will list the main parameters of the measuring complex for performing ice strength trials under the full-scale conditions and on ice samples. Hydraulic jack has piston diameter 95 mm and piston run 50 mm. The maximum working pressure in the piston cavity is 50 MPa. The maximum rate of piston motion is 4.5 mm/s. Removable indenter has diameter 65, 90 and 120 mm. Mass without indenter is 33.0 kg. Hydraulic station has maximum working pressure equal 27.0 MPa, nominal capacity of the pump (consumption) is 5.7 l/min, power and revolution frequency of electrical engine is 3/1500 kW/(rev/min). Mass of hydraulic station (without oil) is 62.5 kg. Measurement block has nominal working pressure equal 50.0 MPa. Supply voltage of the sensor of position of the measurement hydraulic cylinder is 24 V DC, input signal of the sensor of position of the measurement hydraulic cylinder equal 4-20 mA, mass is 28 kg.

The diagram in Fig. 9.2 shows a record of test results. It is shown the result of the work on loading ice with indenter, the process of destruction begins at 25 seconds. At this point, the local strength is fixed.

Figure 9.3 presents a block-diagram of jack operation in the borehole [9]. Here

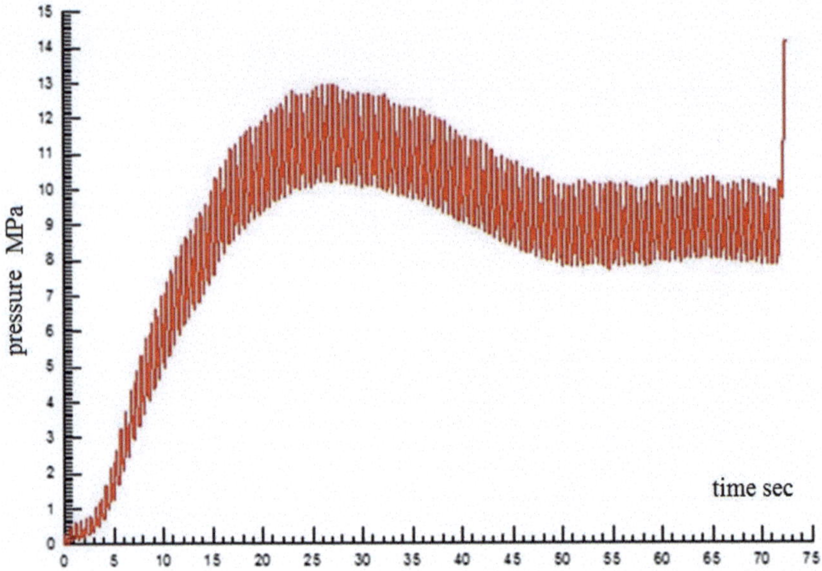

Fig. 9.2 The test results.

Fig. 9.3 Block-diagram for ice strength determination in the borehole.

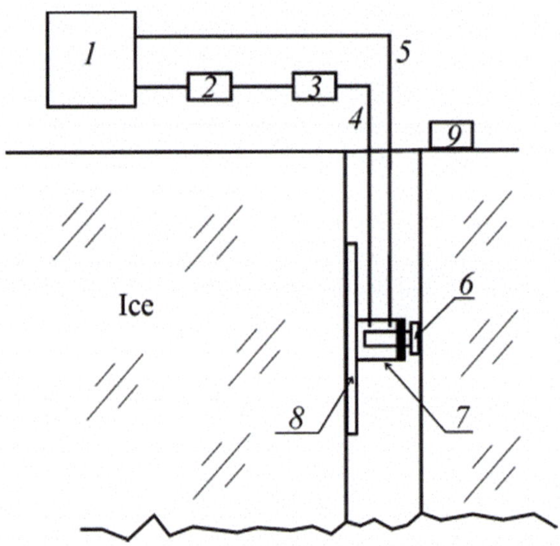

are:

1 - hydraulic station,
2 - reference manometer,
3 - pressure sensor,
4 - hose for oil feed to the working chamber of the jack hydraulic cylinder,
5 - hose for oil feed of the return piston run of the jack hydraulic cylinder,
6 - indenter,
7 - hydraulic cylinder,
8 - base plate,
9 - accelerometer.

Figure 9.4 shows photos of the probe-indenter and the press for testing ice samples. The integrated system makes it possible to determine the strength of both flat ice and of the ice in hummocks and stamukhas.

9.3 Methodology for Determination of the Local Ice Strength at Compression in Boreholes by a Borehole Jack

To determine the local ice strength in boreholes under the full-scale conditions (in situ) one applies a complex "Hydraulic jack" with electrical drives with a hand pump. Figure 9.5 presents a complex system for the determination of ice strength characteristics under the full-scale conditions in boreholes, deployed in the field. The work of devices is performed as follows. In the ice cover (level, rafted or in

a) b)

Fig. 9.4 Ice strength indenter probe and sample press. a) Indenter with a working diameter of 9 cm; b) press for the study of uniaxial compression of ice samples.

Fig. 9.5 Complex system for determination of ice strength characteristics in boreholes under the full-scale conditions.

ridged features) using a motor drill with an auger 25 cm in diameter, a borehole is drilled through the entire ice thickness if possible. The drilling chips are extracted by a special device for cleaning the boreholes. A tripod is set above the borehole on which by means of a pulley block a jack is hoisted with the indenter on the pulling out rod. Depending on the physical ice state, indenters with a diameter of 6.5 - 9.0 or 12.0 cm are used.

At the indenter's diameter of 9 cm the first test is made at a depth of 30-40 cm from the ice surface to the indenter's middle (Fig. 9.4). Such deepening is necessary to avoid ice chipping in the surface direction. The other tests are made with a step of 30 cm over the entire borehole depth. Loading is made by means of a hydraulic cylinder at its bursting impact on the borehole side and restriction of the hydraulic cylinder motion from the opposite to indenter side due to the base plate. The area of the base plate exceeds the indenter's area by more than ten times. At the pressure feed to the hydraulic cylinder this provides indenter's penetration without penetration of the base plate.

The penetration stress in ice σ_u can be calculated from ratio

$$\sigma_u = \frac{F}{S_u} = \frac{PS_n}{S_u} \tag{9.1}$$

where: F - force of indenter's penetration (N), S_u - indenter's cross-section area (m^2), P - pressure in the hydraulic system (Pa), S_n - piston working area in the jack (m^2). After carrying out the experiment and processing the results, it is possible to build a stress diagram and determine the local ice strength (Fig. 9.6).

The process of jack interaction with ice occurs as follows. Loading of the borehole occurs in three stages.

- The first stage ends with appearance of the first crack and characterizes the initial moment of indenter penetration.
- The second stage of loading ends with destruction of some ice volume (local strength), leading to formation of the zone of crushing and radial cracks (Fig. 9.7). This process has an avalanche-like character and its duration is not greater than several seconds.
- At the third stage of loading one mainly observes three types of indenter/ice interaction. The first type is characterized by indenter penetration through the zone of crushing at the practically constant stress, which is less than the destruction by 20-40 % (post-destruction stress).

The second type of interaction is characterized by the fact that the post-destruction stress is not less than the local strength. At the third type of interaction one of the radial cracks transforms into the main one immediately after achieving the destruction stresses. Figure is a photo of the ice section and its structure in polarized light [10].

The ice strength is determined in a three-dimensional stress state of the local volume, therefore the term "local ice strength" is used [7, 8, 9]. When breaking stresses are reached, a certain volume of ice is chipped in the direction of the upper or lower surface of the ice with insufficient penetration of the probe in the well (Fig. 9.8). Based on the measurements of breaking stresses along the ice thickness,

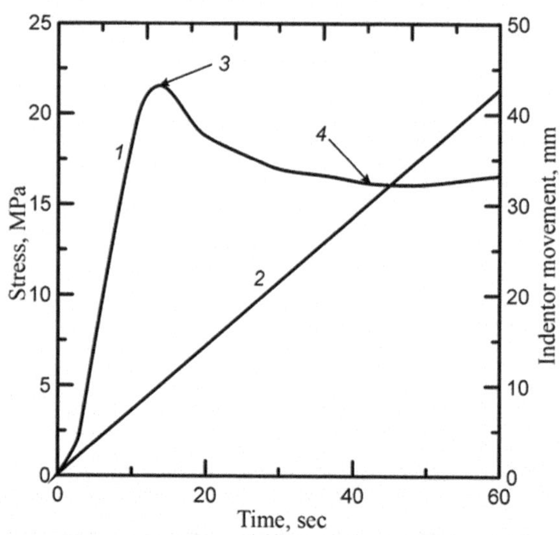

Fig. 9.6 Records of stress and indenter's motion at penetration into the ice to the borehole side. *1* - stress in the ice, *2* - indenter's motion, *3* - destruction stress, *4* - post-destruction stress.

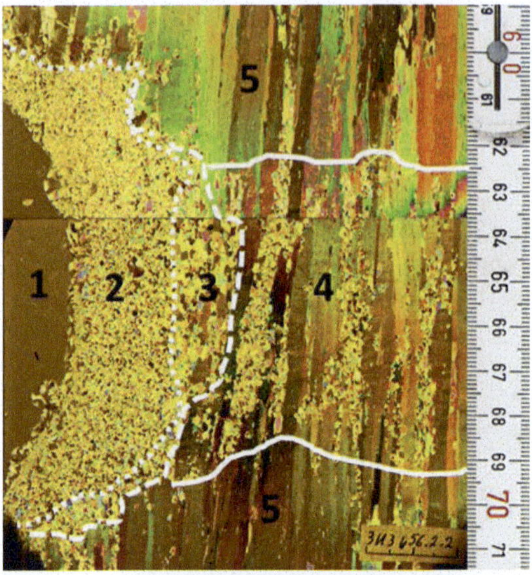

Fig. 9.7 Zone of crushing of ice of fibrous structure after indenter penetration.
1 - recess from the introduction of the indenter, 2 - zone of complete destruction of primary crystals, 3 - zone of predominant destruction of primary crystals, 4 - zone of partial destruction of primary crystals, 5 - primary fibrous crystals.

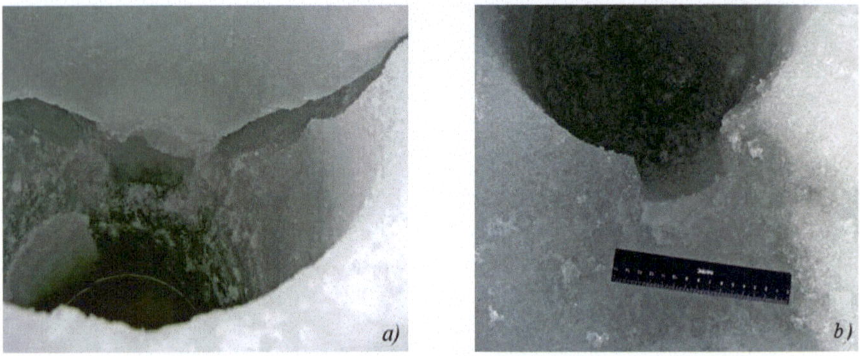

Fig. 9.8 Example of indenter penetration into the borehole side with ice chipping towards its surface: *a)* - imprint from indenter in the borehole side and crack of ice chipping (view from aside); *b)* - imprint from indenter and chipping of the surface ice part (view from above).

a diagram of the local ice strength along the vertical is plotted (Fig. 9.9).

9.4 Anisotropy of Ice

As an example, Fig. 9.10 [13] presents the results of the trials in situ by means of the borehole jack on the polygon 20×45 m in size in the northeastern part of the Kara Sea at the May 2014. The measurement step was 5 m. As can be seen from Fig. 9.10, there are two pronounced sites with the high local strength and several sites with a low strength in the area of 20×45 m. The work was carried out on the floe of

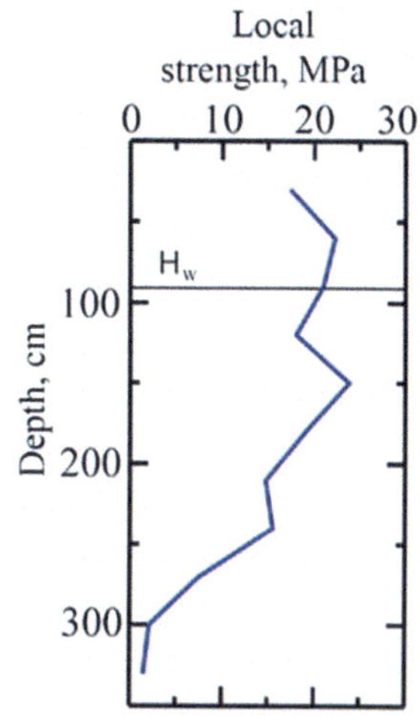

Fig. 9.9 Vertical distribution of the local ice strength in an ice ridge (H_w - water level in the borehole).

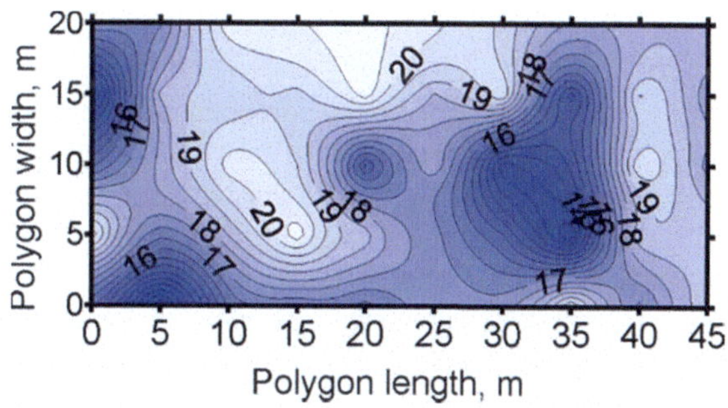

Fig. 9.10 Spatial non-uniformity of average by thickness local ice strength values (MPa) on the polygon of 20×45 m. Northeastern part of the Kara Sea, 2.-4.05.2014.

deformed level ice near ice ridges. During determination of the ice local strength, its main physical characteristics (temperature, salinity and density) are derived, the texture is described and if technical possibilities are available, the ice structure is determined.

The mechanical ice characteristics in addition to ice temperatures also depend on the other physical parameters such as salinity, density, porosity, etc. The liquid phase volume contained in the ice combines the temperature and salinity. The liquid phase volume is calculated by formula

$$V_b = S_i 10^{-3} \left(0,532 - \frac{49,185}{T_i} \right) \tag{9.2}$$

where: V_b - liquid phase volume in relative units (dimensionless), S_i - ice salinity in per mil.

Figure 9.11 uses numerous experimental results and their functional approximation:

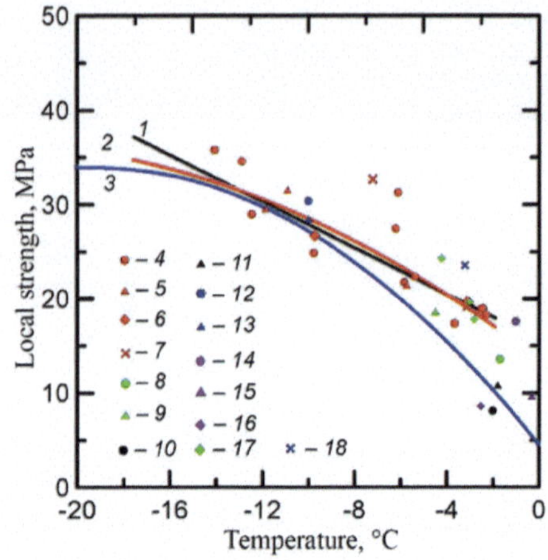

Fig. 9.11 Dependence of the local ice strength on the ice temperature.

1 - linear approximation based on data at the station "Ice Base Cape Baranov" (2017-19);
2 - approximation by the polynomial to the power of 2 based on data at station "Ice Base Cape Baranov" (2017-19);
3 - approximation by the Johnston's polynomial to the power of 2 [4];
4-7 - average local ice strength value at the station "Ice Base Cape Baranov" in 2019, 2018, 2017 and 2016, respectively;
8 - average local ice strength value in Anadyr' estuary in 2008;
9 - average local ice strength value of Nevelskoy Strait ("Sakhalin-2010" Expedition);
10-11 - average local ice strength value in Baidaratskaya Bay in 2007 and 2010, respectively;
12 - from [4];
13 - from (Sinha) [4];
14-16 - average local ice strength value in the Caspian Sea in 2001-03, 2004-08 and 2013, respectively;
17 - average local ice strength value in the "Transarctic - 2019" Expedition;
18 - average local ice strength value in the MOSAIC Expedition.

The analysis of the destructions of fibrous ice structures under the indenter action showed significant differences in ice destruction at the indenter impact parallel and perpendicular to the prevailing direction of ice growth. Figure 9.12 presents photos of the ice structure after the indenter penetration in horizon of 65 cm. The ice temperature is -6.3°C, salinity - 5.25%, indenter diameter of 9 cm. The photos of the vertical thin ice cuts (sections) in the polarized light are given. The thickness of sections is about 1 mm [12].

The size of the zones and tension at their boundaries are determined by formulas:

$$\sigma_{r_1} = \sigma_u + 4C \ln \frac{r_0}{r_1}; \tag{9.3}$$

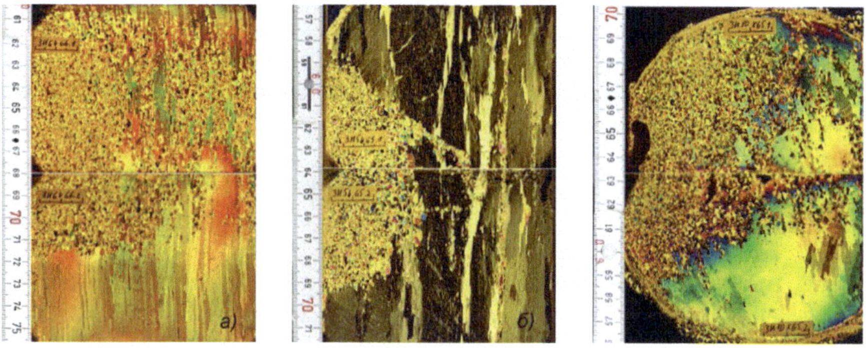

a) b) c)

Fig. 9.12 Ice destruction after the indenter impact parallel - a) and perpendicular - b) to the prevailing direction of ice growth; c) - structure in the horizontal ice sections after indenter penetration at an angle of 45° to the prevailing direction of ice growth.

$$r_1 = r_0 e^{\frac{\sigma_c - \sigma_u}{4C}}; \tag{9.4}$$

$$\sigma_{r_e} = \frac{2\sigma_c}{n} = \sigma_c \left(\frac{r_1}{r_e}\right)^2; \tag{9.5}$$

$$r_e = r_1 \sqrt{\frac{n}{2}}; \tag{9.6}$$

where: r_0, r_1 and r_e - distances to the zones of crushing, radial cracks and elastic deformation, respectively; σ_u - tension created by the jack; C - ice cover strength at shear under the conditions of comprehensive pressure or cohesion at the internal friction angle, equal to 0; σ_c - ice strength under the conditions of uniaxial pressure; σ_t - tensile ice strength. Parameter n characterizes the plastic anisotropy of ice and is equal to

$$n = \frac{\sigma_c}{\sigma_t}. \tag{9.7}$$

9.5 Estimation of Ice Pressure on the Vertical Supports of Hydraulic Structures, Taking into Account the Local Ice Strength

Work on the scale effect of ice strength is given special attention in connection with the increased tasks of designing and building engineering structures on the Arctic shelf. Traditionally, the physical and mechanical properties of ice are studied on the basis of cores drilled from even ice fields, hummocks, stamukhas and icebergs.

Determination of ice strength using a downhole probe-indenter is recommended in [8] and the international standard ISO 19906. The probe-indenter can be used to identify the profile of heterogeneity in the strength of ice roads and the consolidation of hummocks. The capabilities of the downhole probe-indenter make it possible to adequately interpret the measurement results and significantly improve the accuracy of calculating ice loads on a structure, as well as more correctly compare ice properties in different regions.

Assessment of the local strength of ice is directly related to the calculation of the pressure of ice fields, hummocks, broken ice on vertical supports. The range of changes in the local strength of ice is quite wide and ranges from 12 to 40 MPa, which is associated with the influence of such parameters as temperature, salinity, ice age and season. To assess the magnitude of the ice load on the supports of hydraulic structures, we will use the results of the work [14] given in Table 9.1.

There is also a known case of a catastrophe with the leading signs of the port of Ust-Luga, which occurred on March 6, 2013 due to an incorrect assessment of the ice pressure of hummocks 2.5 m high, formed around the supports due to the movement of an ice field 0.6 m thick under the action of a surge wind from speed 20 m/s. The results of the disaster are shown in Fig. 9.13. Underestimation of the strength of the supports led to their elastic-plastic bending with subsequent destruction [14]. To calculate the ice pressure during the bending of the supports, we will consider the possibility of the transition of the structure to the area of nonlinear deformation with an estimate of the ultimate loads of elastic bending. Two models are proposed for research - pure bending, which can simulate the impact of an ice field, and bending by a transverse concentrated force, which can simulate the pressure of an iceberg

Table 9.1 Estimation of ice load values

Estimation of ice load values	Support diameter	Support displacement	Ice load
Lighthouse Tainio, 1967	3.5 m	14 m (without tilt)	2.3–4.7 MN
Lighthouse Bjornlaken, 1969	2.9 m	17 m, tilt 12%	10.9 MN

a) b)

Fig. 9.13 Ust-Luga's catastrophe. a) Displacement of supports in the direction of maximum ice pressure; b) formation of a "bunch" due to all-round compression by ice.

or hummock. Schemes of elastic-plastic bending of vertical supports are shown in Fig. 9.14.

It was shown in [14, 15] that the solution of the problems under consideration can be reduced to the differentiation of ODEs of different orders, followed by the solution of boundary value problems. The bending moment at which a vertical beam of rectangular section $2h$ by b with length L passes to a nonlinear deformation is determined by the formula

$$M_t = \frac{2}{3} b h^2 \sigma_t. \tag{9.8}$$

The critical moment when the entire beam goes into a plastic state or a "plastic hinge" is formed near the lower clamped end of the support

$$M_k = \frac{2bh^2 n\sigma_t}{n+1}, \tag{9.9}$$

where the plastic anisotropy parameter n was defined in (9.3) - (9.6).

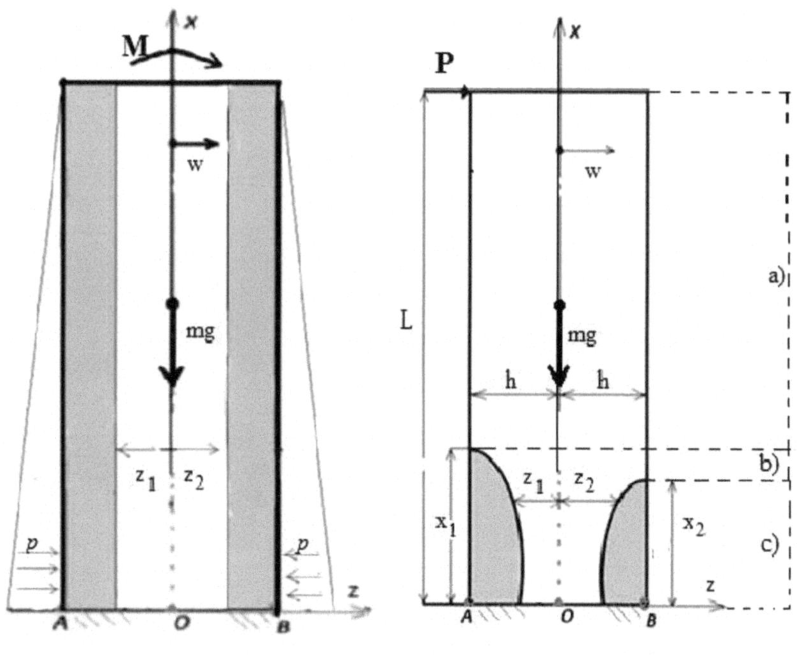

a) b)

Fig. 9.14 Mathematical models of the bending of vertical supports, taking into account the weight of the support and hydrostatic compression. a) Model of pure bending, b) model of bending with a concentrated constant force, the plastic regions are shaded.

The lateral pressure load P at which plastic deformations appear in the support is equal to

$$P_t = \frac{M_t}{L}.$$

In the case of taking into account the own weight of the beam,

$$M = P(L-x) + 4wbh(L-x)\gamma_1,$$

where γ_1 is the specific weight of the beam material, w - is the deflection of vertical beam at its free end. That problem has an analytical solution in the form of special Airy functions, but only in the elastic stage of bending [14]. In Fig. 9.15 shows the results of the analytical solution [16] for the case of pure moment (M = const) without own wight and the case of bending by load P.

Let us compare the values of the moments in Fig. 9.14 and the values of loads from ice pressure in Table 9.1. The comparison shows that the ice pressure was so strong that the lighthouse supports could not remain in an elastic state and went into a nonlinear stage of deformation.

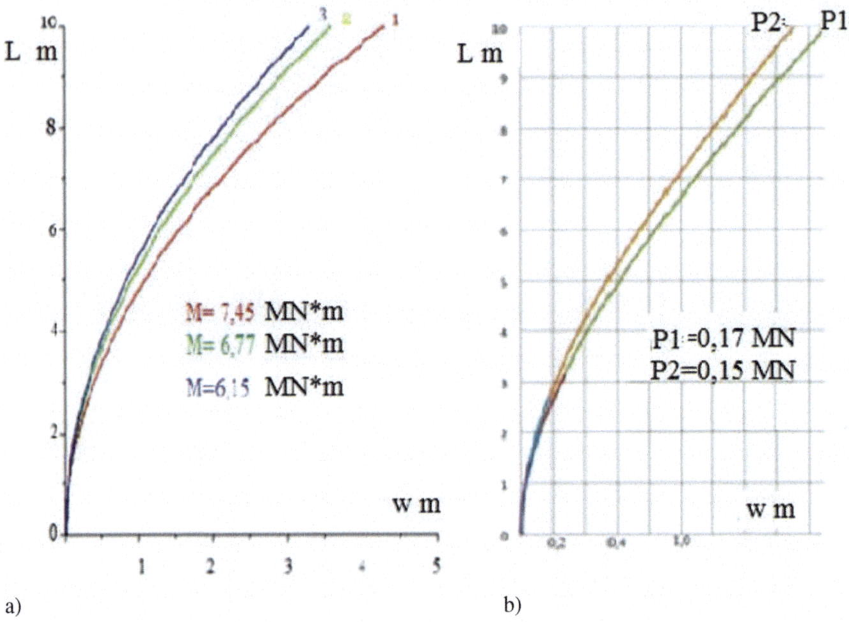

Fig. 9.15 The functions "moment - deflection" (a) and "load-deflection" (b) when solving the problem of bending vertical supports in the elastic-plastic state of deformation.

9.6 Conclusions

To calculate the ice pressure during the bending of the supports, we will consider the possibility of the transition of the structure to the area of nonlinear deformation with an estimate of the ultimate loads of elastic bending. Two models are proposed for research - pure bending, which can simulate the impact of an ice field, and bending by a transverse concentrated force, which can simulate the pressure of an iceberg or hummock. The ice pressure was so strong that the lighthouse supports could not remain in an elastic state and went into a nonlinear stage of deformation. We consider that the problem of elastic-plastic vertical beams bending may be loused analytically. It is possible to define the critical moment or critical load when the beam is crash. Analytical results can be useful as approximate when solving problems by numerical methods, for example, FEM.

Acknowledgements The work was carried out at the Federal Budgetary State Institution "Arctic and Antarctic Research Institute" and at St. Petersburg State University with the support of RFBR grant 19-01-00208.

References

1. S.M. Kovalev, G.A. Lebedev, O.A. Nedoshivin, K.K. Sukhorukov. *Mechanical Properties of Sea Ice*. St. Petersburg: Gidrometeoizdat, 2001.
2. V.V. Bogorodsky, V.P. Gavrilo. *Ice. Physical Properties. Modern Methods of Glaciology*. Leningrad: Gidrometeoizdat, 1980.
3. V.A. Borodkin, V.P. Gavrilo, S.M. Kovalev. *Ice Structure and Physical Properties in the Northwestern Part of the Weddell Sea in the Autumn-winter Period. Formation of the Database on Sea Ice and Hydrometeorology*. St. Petersburg: Gidrometeoizdat, 1995.
4. T.J.O. Sanderson. *Ice Mechanics: Risks to Offshore Structures*. London, Graham and Trotman, 1988.
5. V.N. Smirnov, S.M. Kovalev, V.A. Borodkin, A.A. Nyubom, A.I. Shushlebin. *Instrumental Monitoring and Short-term Forecast of Compression and Hummocking Phenomena in Sea Ice*. St. Petersburg, AARI, 2017.
6. V.N. Smirnov, A.I. Shushlebin, S.M. Kovalev, A.A. Nyubom. A method for determining in natural conditions the deformation and strength characteristics of an even ice cover in bending. Patent for invention No. 2614922, 2017.
7. V.N. Smirnov, A.I. Shushlebin, S.M. Kovalev, A.A. Yatzkevich, S.N. Schepanyuk, Ya.O. Efimov, K.A. Kornishin. *Complex System for Determination of Ice Strength Characteristics in Full-scale Conditions and on Ice Specimen*. Patent RF, no. 2682835, 2019.
8. ISO/FIDS 19906: 2010 (E). *Petroleum and Natural Gas Industries - Arctic Offshore Structures*, 2010.
9. V.N. Smirnov, A.I. Shushlebin, A.A. Yatzkevich. *Universal Module-block System of Monitoring of the Ice Cover State*. Patent RF, no. 65224, 2007.
10. S.M. Kovalev, V.N. Smirnov, V.A. Borodkin, A.I. Shushlebin, N.V. Kolabutin, K.A. Kornishin, Y.O. Efimov, P.A. Tarasov, D.A. Volodin. Physical and mechanical characteristics of sea ice in the Kara and Laptev seas. *International Journal of Offshore and Polar Engineering*. 29(4):369-374, 2019.
11. S.M. Kovalev, G.A. Lebedev, O.A. Nedoshivin, K.K. Sukhorukov. *Mechanical Properties of Sea Ice*. St. Petersburg, Gidrometeoizdat, 2001.

12. Complex studies of the environment of Severnaya Zemlya Archipelago and adjoining regions of the Northern Sea Route area at the research stationary base "Ice Base Cape Baranov". Report on NIR. St. Petersburg: AARI. AARI Archives, Reg. No. R-6598, 2019.
13. K.A. Kornishin, V.A. Pavlov, A.I. Shushlebin, S.M. Kovalev, Ya.O. Efimov: Determination of the local ice strength by means of the borehole jack in the Kara and Laptev Seas. *Scientific-technical herald of the JSC NK "Rosneft"* 1(42): 47-51, 2016.
14. Pavilaynen, G.V.: Mathematical models of hydraulic supports from construction materials with the effect of plastic anisotropy. *Transactions of seminar "Computer Methods in Continuum Mechanics"*, 2020-2021. pp. 5-28.
15. Pavilaynen G.V.: Elastic-plastic deformations of ribbed plates. *CRM Proceeding and Lecture Notes*. Amer. Math. Soc. Providence, R.I., pp. 227-233, 1993.
16. D.V. Lachugin, G.V. Pavilaynen. On the bending of structural materials with plastic anisotropic effect. *AIP Conference Proceeding (eds. E. Kustova, G. Leonov, N. Morosov, M. Yushkov and M. Mekhonoshina)*, p. 070019, 2018.

Chapter 10
Stress Distribution at the Wavy Surface of a Solid Incorporating Surface Stresses and Surface Tension

Sergey Kostyrko, Mikhail Grekov, and Holm Altenbach

Abstract Employing the original Gurtin-Murdoch model of surface elasticity, we investigate the stress field near the curved surface of isotropic elastic solid jointly induced by surface stresses and external tensile loading. Due to the plane strain conditions, the two-dimensional boundary value problem for half-plane with a curved boundary is formulated in terms of the complex variables. Based on the Goursat-Kolosov complex potentials and boundary perturbation method whereby the unknown functions are sought in the form of a power series in the small parameter represented by an amplitude-to-wavelength ratio of the surface undulation, the formulated boundary value problem is reduced to the recurrent sequence of the integral equations for any-order approximation. Considering the cosine-shaped surface, the first-order approximation of the stress tensor components is derived in the closed-form. The effect of the surface elasticity and surface tension on the stress field at the surface is numerically investigated.

Key words: Surface undulation, Surface elasticity, Surface tension, Complete Gurtin-Murdoch model, Complex potentials, Boundary perturbation method

10.1 Introduction

Since the failure and fracture commonly appears at the surface of a material, it is well-established that the surface properties have a tremendous effect on the material

Sergey Kostyrko and Mikhail Grekov
St. Petersburg State University, 7/9 University Embankment, St. Petersburg, 199034, Russian Federation,
e-mail: s.kostyrko@spbu.ru,m.grekov@spbu.ru

Holm Altenbach
Otto-von-Guericke-Universität Magdeburg, Universitätsplatz 2, 39106 Magdeburg, Germany,
e-mail: holm.altenbach@ovgu.de

H. Altenbach et al. (eds.), *Advances in Solid and Fracture Mechanics*,
Advanced Structured Materials 180,
https://doi.org/10.1007/978-3-031-18393-5_10

strength [8, 13, 26, 28, 29, 38, 55, 57]. The failure mechanism is traditionally associated with the local stress concentrations resulting from geometrical features of surface roughness [14, 15, 18, 19, 21, 22, 25, 43, 65]. However, surface energy come to play an increasingly important role when the size of surface asperities reduces to the nanoscale [24, 30, 44, 45, 60, 64, 68]. The interest in new theoretical models, which can accurately describe the behaviour of nanostructured materials from a perspective of surface elasticity, has considerably increased in recent decades with the rapid advances in nanoscience and -technology [11, 12, 27, 36, 37, 67]. Perhaps the most used model in the field of nanomechanics was formulated by Gurtin and Murdoch (GM) [33, 34]. They proposed to consider the surface domain as a negligibly thin layer ideally adhering to the bulk material and differing from it by the elastic moduli. The stress resultants acting in a surface layer were considered as surface stresses. As a consequence, the constitutive relations and equilibrium equations of the surface have been derived in a general form taking into account the effect of surface elasticity and surface tension. The existence and uniqueness of linear elasticity solutions considering Gurtin–Murdoch model was carefully studied in [2, 3].

It should be pointed out that the GM model has been widely applied in the analysis of the size-dependent behaviour of nanosized particles, inclusions, wires, beams, plates, and film coatings [1, 5, 7, 9, 10, 48, 49, 51, 66]. However, the majority of the studies have been concentrated on the nanostructures having perfect surfaces without any topological defects, although it is almost impossible to avoid them during production. Another simplification, that could be quite often found in the literature, is related to neglecting the effects of the surface tension on the mechanical response of nanostructures. In some papers, the surface tension was totally ignored [4, 16, 41, 42, 61]. In others, only the normal component of the surface gradient of the displacement field has been omitted [23, 31, 40]. In [35, 52, 62, 66], the gradient of the surface displacement was completely avoided. The discussion of the simplified GM models can be found in [20, 47, 56]. As a result of such simplifications, the effect of bulk traction induced by surface tension was not fully taken into consideration. Nevertheless, the importance of surface tension in elastic deformation of nanomaterials is increasingly recognized [20, 47, 53, 54, 59, 69, 70]. Therefore, in this study, we extend our previous model [23] and carefully analyse the influence of the surface tension on the stress field near surface asperities. For this purpose, we developed a general approach based on the solution of the constitutive and equilibrium equations of the GM model which were derived in the Lagrangian description for an arbitrary surface shape of solid under plane strain conditions. In order to determine the semi-analytical expressions for the components of the stress and strain tensors, we employ the complex variable method and boundary perturbation technique whereby the unknown functions are sought in the form of a power series in the small parameter represented by an amplitude-to-wavelength ratio of the surface undulation. Subsequent numerical analysis based on the first-order approximation investigates the distribution of elastic stresses along the cosine-shaped surface induced by surface stresses and remote tensile loading.

10.2 Problem Formulation

In this study, we consider a semi-infinite elastic solid with a roughened surface slightly deviated from the planar one as shown in Fig. 10.1. The surface has elastic properties differing from the same properties of the volume and, according to the theory of surface elasticity [33, 34], is represented as a very thin film which adheres to the bulk material without slipping. The plane strain conditions are assumed to be satisfied and the solid is subjected to the remote tensile loading T and the surface traction q^s induced by surface stress and surface tension.

So, we come to the two-dimensional boundary value problem for the elastic half-plane $\Omega = \{z : x_2 < \varepsilon f(x_1), x_1 \in (-\infty, +\infty)\}$ of the complex variable $z = x_1 + ix_2$ (i is the imagine unit) with the curved boundary Γ defined by the equation

$$z \equiv \zeta = x_1 + i\varepsilon f(x_1). \tag{10.1}$$

Function f can describe the periodic undulation of surface profile as in [22, 23, 42, 66], i.e. $f(x_1) = f(x_1 + a)$, or the local surface defect as in [19, 25], i.e. $f(x_1) = 0$ if $|x_1| \geq a$. In both cases, $\max|f(x_1)| = a$, $\varepsilon|f'(x_1)| < 1$, $0 < \varepsilon \ll 1$, and $A = \varepsilon a$ is the maximum deviation of the surface from a flat configuration, i.e plane $x_2 = 0$.

In the presence of surface tension and elasticity, the traction boundary condition on free from an external loading surface Γ takes the form:

$$\sigma_{nn}(\zeta) + i\sigma_{nt}(\zeta) = q^s(\zeta), \quad \zeta \in \Gamma, \tag{10.2}$$

where σ_{nn} and σ_{nt} are the components of the bulk stress tensor defined in the local Cartesian coordinates (n, t).

The additional equation determines the continuity of displacements across the surface region:

$$u^s(\zeta) = u(\zeta), \tag{10.3}$$

where $u^s = u_1^s + iu_2^s$ and $u = u_1 + iu_2$, (u_1^s, u_2^s) and (u_1, u_2) are the displacements of the surface and bulk phases along the corresponding coordinate axes x_1 and x_2.

According to [20], function q^s corresponding to the original GM model is written as

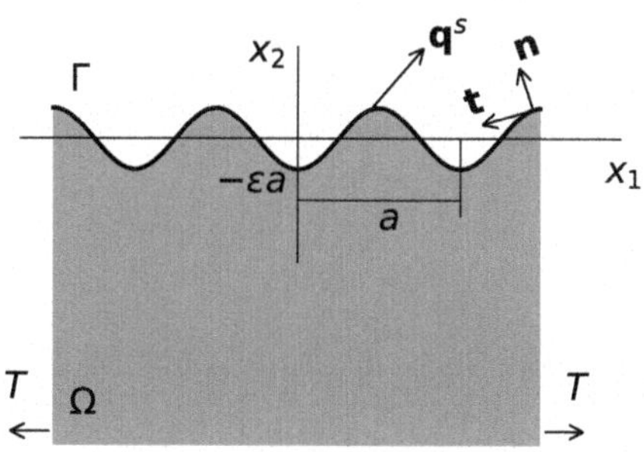

Fig. 10.1 The model of a semi-infinite elastic solid with a nanosized surface asperities.

$$
\begin{aligned}
q^s(\zeta) = \frac{\sigma_0^s}{R} + &\left[\frac{M_s}{R}\mathrm{Re}\frac{\partial u}{\partial \zeta} + \sigma_0^s \mathrm{Im}\left(\frac{\partial^2 u}{\partial \zeta^2} e^{i\alpha_0}\right)\right] \\
- i &\left[M_s \mathrm{Re}\left(\frac{\partial^2 u}{\partial \zeta^2} e^{i\alpha_0}\right) - \frac{\sigma_0^s}{R}\mathrm{Im}\frac{\partial u}{\partial \zeta}\right],
\end{aligned}
\tag{10.4}
$$

where R is the curvature radius of the boundary Γ; α_0 is the angle between the tangent to Γ and x_1-axis at the point ζ; σ_0^s is the surface tension (residual surface stress); $M_s = \lambda_s + 2\mu_s$, λ^s and μ^s are the surface elastic constants similar to the Láme constants λ and μ of the bulk material.

Taking into account that

$$
e^{i\alpha_0} = \frac{1 + i\varepsilon f'(x_1)}{h(x_1)}, \quad h(x_1) = \sqrt{1 + \varepsilon^2 f'^2(x_1)},
\tag{10.5}
$$

Eq. (10.4) can be reduced to the following one

$$
\begin{aligned}
q^s(\zeta) = \kappa\sigma_0^s + &\frac{\kappa}{2}\left[p_s \frac{\partial u}{\partial \zeta} + m_s \overline{\frac{\partial u}{\partial \zeta}}\right] \\
- &\frac{1}{2h}\left[p_s(i - \varepsilon f')\frac{\partial^2 u}{\partial \zeta^2} + m_s(i + \varepsilon f')\overline{\frac{\partial^2 u}{\partial \zeta^2}}\right],
\end{aligned}
\tag{10.6}
$$

where $p_s = M_s + \sigma_0^s$, $m_s = M_s - \sigma_0^s$, and the local principal curvature κ of the boundary Γ is defined by the equation

$$
\kappa(x_1) = \frac{\varepsilon f''(x_1)}{h^3(x_1)}.
\tag{10.7}
$$

In Eqs. (10.5)–(10.7) and hereafter, the prime denotes the derivative with respect to the argument, and the bar over a symbol denotes the complex conjugation.

The condition (10.2) with the expression (10.4) describes the generalized Young–Laplace law [11, 53, 54] in terms of the complex variables for the general case of the plane strain problem. It is worth noting that Eq. (10.4) is valid for any value of the parameter ε, not only for a small one. Assuming $\sigma_0^s = 0$ in the square brackets of Eq. (10.4) that corresponds to the equality $m_s = p_s = M_s$ in Eq. (10.6), one can come to the simplified form of Eq. (10.4) considered in [23, 31, 40] when the normal component of the surface gradient of the displacement field is omitted [47]. If the surface is flat, i.e. $\varepsilon = 0$ and $f'(x_1) \equiv 0$, Eq. (10.4) is transformed into the simplest one used in [28, 29, 66].

At infinity, the stresses σ_{ij} $(i, j = \{1,2\})$ in coordinates (x_1, x_2) and the rotation angle ω are specified as

$$
\lim_{x_2 \to -\infty} \sigma_{22} = \lim_{x_2 \to -\infty} \sigma_{12} = \lim_{x_2 \to -\infty} \omega = 0, \qquad \lim_{x_2 \to -\infty} \sigma_{11} = T.
\tag{10.8}
$$

10.3 Boundary Equation for Complex Potentials

According to [17, 50], the traction $\sigma = \sigma_{nn} + i\sigma_{nt}$ at the area with the normal **n** and complex displacement u in the point $z \in \Omega$ are related to the Goursat–Kolosov functions Φ and Υ holomorphic in the regions Ω and $\widetilde{\Omega} = \{z : \overline{z} \in \Omega\}$, respectively:

$$\sigma(z) = \Phi(z) + \overline{\Phi(z)} - \left(\Upsilon(\overline{z}) + \overline{\Phi(z)} - (z - \overline{z})\overline{\Phi'(z)}\right)e^{-2i\alpha}, \quad z \in \Omega, \tag{10.9}$$

$$2\mu\frac{du}{dz} = \varkappa\Phi(z) - \overline{\Phi(z)} + \left(\Upsilon(\overline{z}) + \overline{\Phi(z)} - (z - \overline{z})\overline{\Phi'(z)}\right)e^{-2i\alpha}, \quad z \in \Omega, \tag{10.10}$$

where α is the angle between axes t and x_1.

Assuming $\alpha = 0$ and $\alpha = \pi/2$ in Eq. (10.9) when x_2 tends to $-\infty$, one can obtain

$$\lim_{x_2 \to -\infty} \Phi(z) = \lim_{x_2 \to +\infty} \Upsilon(z) = T/4. \tag{10.11}$$

In Eqs. (10.9) and (10.10), we pass to the limit when $z \to \zeta \in \Gamma$ and $\alpha = \alpha_0$. Taking into account the boundary equation (10.2) and inseparability condition (10.3), we get the following boundary equations for complex potentials Φ and Υ and derivative of the complex displacement u':

$$\Phi(\zeta) + \overline{\Phi(\zeta)} - \left(\Upsilon(\overline{\zeta}) + \overline{\Phi(\zeta)} - \left(\zeta - \overline{\zeta}\right)\overline{\Phi'(\zeta)}\right)e^{-2i\alpha_0} = q^s(\zeta), \tag{10.12}$$

$$2\mu u'(\zeta) = (\varkappa + 1)\Phi(\zeta) - q^s(\zeta), \quad \zeta \in \Gamma, \tag{10.13}$$

where $\Phi(\zeta) = \lim\limits_{z \to \zeta - i0} \Phi(z)$, $\Upsilon(\overline{\zeta}) = \lim\limits_{z \to \zeta - i0} \Upsilon(\overline{z})$.

10.4 Boundary Perturbation Method

In the case of a flat solid surface, i.e. $\varepsilon = 0$, Eq. (10.12) is reduced to Riemann-Hilbert problem on piecewise holomorphic function, similar to that derived in [28]. Otherwise, when the surface profile is undulated, i.e $\varepsilon \neq 0$ and $f' \neq 0$, it is impossible to find the closed-form expressions for complex potentials $\Phi(z)$ and $\Upsilon(z)$ from Eq. (10.12). Assuming that the surface is slightly undulated, i.e. $\varepsilon \ll 1$ and $|f'| < 1/\varepsilon$), we obtain the asymptotic expressions for $\Phi(z)$ and $\Upsilon(z)$ using the boundary perturbation method which has been applied recently to a number of problems considering deformation of solids at the macro- [6, 18, 19, 21, 22, 25, 26, 65] and nanoscale [23, 24, 30, 40, 41, 42, 63].

According to this method, we seek the unknown functions Φ, Υ and u' as power series in the small parameter ε:

$$\Psi(z) = \sum_{n=0}^{\infty} \frac{\varepsilon^n}{n!}\Psi_{(n)}(z), \; \Psi(z) \equiv \{\Phi(z), \Upsilon(z), u'(z)\} \tag{10.14}$$

The boundary values of expansion coefficients $\Phi_{(n)}$, $\Upsilon_{(n)}$ and $u'_{(n)}$ are represented in the form of Taylor series in the vicinity of the line $x_2 = 0$ treating x_1 as a parameter:

$$\Phi_{(n)}(\zeta) = \sum_{m=0}^{\infty} \frac{[i\varepsilon f(x_1)]^m}{m!} \Phi_{(n)}^{(m)}(x_1), \tag{10.15}$$

$$\Upsilon_{(n)}(\overline{\zeta}) = \sum_{m=0}^{\infty} \frac{[-i\varepsilon f(x_1)]^m}{m!} \Upsilon_{(n)}^{(m)}(x_1),$$

$$u'_{(n)}(\zeta) = \sum_{m=0}^{\infty} \frac{[i\varepsilon f(x_1)]^m}{(m)!} u_{(n)}^{(m+1)}(x_1).$$

In view of the condition $\varepsilon |f'(x_1)| < 1$, one can write the following power series for known functions κ and h^{-1} describing local principal curvature and metric coefficient, respectively:

$$\kappa(x_1) = \varepsilon f''(x_1) \left[1 + \sum_{m=1}^{\infty} \frac{(-1)^m}{m!} \frac{(2m+1)!!}{2^m} (\varepsilon f'(x_1))^{2m} \right], \tag{10.16}$$

$$h^{-1}(x_1) = \sum_{m=0}^{\infty} \frac{(-1)^m}{m!} \frac{(2m-1)!!}{2^m} (\varepsilon f'(x_1))^{2m}.$$

Substituting series (10.14)–(10.16) into Eqs. (10.12) and equating the coefficients at ε^n $(n = 0, 1, \ldots)$, we come to the recurrent sequence of Riemann-Hilbert problems:

$$\Xi^+_{(n)}(x_1) - \Xi^-_{(n)}(x_1) = -q^s_{(n)}(x_1) + F_{(n)}(x_1), \tag{10.17}$$

where

$$\Xi^{\pm}_{(n)}(x_1) = \lim_{z \to x_1 \pm i0} \Xi_{(n)}(z)$$

and $q^s_{(n)}$ are the coefficients of the power series expansion of right-hand side of Eq. (10.12), $q^s_{(0)} = 0$; $F_{(n)}$ are the functions coming from expansion of left-hand side of Eq. (10.12) and depending on the complex potentials of the previous approximations, i.e. $\Phi_{(m)}$ and $\Upsilon_{(m)}$ $(m = 0, 1, \ldots, n-1)$ and their derivatives, $F_{(0)} = 0$; piecewise functions $\Xi_{(n)}$ are defined as

$$\Xi_{(n)}(z) = \begin{cases} \Upsilon_{(n)}(z), \ \mathrm{Im}\, z > 0 \\ \Phi_{(n)}(z), \ \mathrm{Im}\, z < 0 \end{cases} \tag{10.18}$$

The solution of the problem (10.17) can be written in terms of Cauchy type integrals [50]:

$$\Xi_{(n)}(z) = I_n(z) + J_n(z) + C_n, \tag{10.19}$$

where

$$I_n(z) = -\frac{1}{2\pi i}\int_{-\infty}^{+\infty}\frac{q^s_{(n)}(t)}{z-t}\mathrm{d}t, \quad J_n(z) = \frac{1}{2\pi i}\int_{-\infty}^{+\infty}\frac{F_{(n)}(t)}{z-t}\mathrm{d}t \tag{10.20}$$

and $C_0 = T/4$, $C_n = 0$ $(n = 1, 2, \ldots)$. It should be noted that integral J_n is the known function at the current n-th step of the approximation.

As a result, the boundary equation (10.13) can be presented as the recurrent sequence of the integral equations:

$$2\mu u'_{(n)}(x_1) - (\varkappa+1)I_n^-(x_1) + q^s_{(n)}(x_1) = (\varkappa+1)J_n^-(x_1) + Q_n, \tag{10.21}$$

where $Q_0 = (\varkappa+1)T/4$, $Q_n = 0$ $(n = 1, 2, \ldots)$.

Following to the Sokhotski–Plemelj formulas [50], the boundary values I_n^- and J_n^- of the Cauchy type integrals I_n and J_n can be written as

$$I_n^-(x_1) = \frac{1}{2}q^s_{(n)}(x_1) - \frac{1}{2\pi i}\int_{-\infty}^{\infty}\frac{q^s_{(n)}(t)}{x_1-t}\mathrm{d}t, \tag{10.22}$$

$$J_n^-(x_1) = -\frac{1}{2}F_{(n)}(x_1) + \frac{1}{2\pi i}\int_{-\infty}^{\infty}\frac{F_{(n)}(t)}{x_1-t}dt.$$

For the zero-order approximation, we arrive at the following relation

$$u'_{(0)} = \frac{(\varkappa+1)T}{8\mu}, \tag{10.23}$$

and the corresponding complex potentials can be found from Eq. (10.19):

$$\Phi_{(0)} = \Upsilon_{(0)} = T/4. \tag{10.24}$$

After that, we can obtain functions $q^s_{(1)}$ and $F_{(1)}$:

$$q^s_{(1)}(x_1) = \sigma_0^s f''(x_1) - \frac{i}{2}[p_s + m_s]u''_{(1)}(x_1), \quad F_{(1)}(x_1) = iTf'(x_1), \tag{10.25}$$

When the surface profile is described by the periodic function f, the solution of the integral equation (10.21) can be found in terms of a trigonometric series:

$$u'_{(1)}(x_1) = \sum_{k=1}^{\infty}\left[A_k^1\sin(b_kx_1) + B_k^1\cos(b_kx_1)\right], \quad b_k = 2\pi k/a. \tag{10.26}$$

When the surface undulation is presented by the cosine function, i.e.

$$f(x_1) = -a\cos(2\pi x_1/a),$$

the unknown coefficients in Eq. (10.26) can be easily derived from Eq. (10.21) as follows

$$A_1^1 = 0, \quad B_1^1 = -\frac{ab_1^2}{32\mu^2}\left[T(\kappa+1)(p_s+m_s)+16\sigma_0^s\mu\right]. \tag{10.27}$$

As a consequence, the corresponding complex potentials are obtained from Eq. (10.19) taking into account Eq. (10.25):

$$\begin{aligned}\Upsilon_{(1)}(z) &= \left[\frac{ab_1}{2}T - \frac{ab_1^2}{2}\sigma_0^s + \left\{-\frac{ab_1^2}{32\mu}(\kappa+1) - \frac{B_1 b_1}{4}\right\}(p_s+m_s)\right] e^{ib_1 z},\\ \Phi_{(1)}(z) &= \left[-\frac{ab_1}{2}T - \frac{ab_1^2}{2}\sigma_0^s + \left\{-\frac{ab_1^2}{32\mu}(\kappa+1) + \frac{B_1 b_1}{4}\right\}(p_s+m_s)\right] e^{-ib_1 z}.\end{aligned} \tag{10.28}$$

According to Eq. (10.12), the components of the stress tensor in the local coordinates (n,t) can be written for the first-order approximation as it follows

$$\begin{aligned}\sigma_{nn}(z)+i\sigma_{nt}(z) &= 2\varepsilon \mathrm{Re}\,\Phi_{(1)}(z) - \varepsilon\left[\Upsilon_{(1)}(\bar{z}) + \overline{\Phi_{(1)}(z)} - (z-\bar{z})\overline{\Phi'_{(1)}(z)}\right],\\ \sigma_{tt}(z)+\sigma_{nn}(z) &= T + 4\varepsilon \mathrm{Re}\,\Phi_{(1)}(z), \quad z \in \Omega.\end{aligned} \tag{10.29}$$

Thus, the stress field near nanosized surface asperities with geometrical features described by cosine function can be evaluated for the first-order approximation using Eqs. (10.27)–(10.29). As one can see from these equations, the original GM model leads to the same results as the simplified one taking into account the surface tension only in the first term of Eq. (10.6). This follows since $p_s + m_s = M_s + \sigma_0^s + M_s - \sigma_0^s = 2M_s$ for the original GM model, and $p_s + m_s = M_s + M_s = 2M_s$ for the simplified one. In the next section, we will analyse the effect of surface elastic parameters and surface tension on the stress distributions along the cosine surface patterns considering several numerical examples. The analysis of more complex structures of the surface relief based on the high-order approximations of the boundary perturbation method will be conducted in our further studies.

10.5 Numerical Results

In the following numerical analysis, we assume that the surface properties are described by the surface parameters derived in [46, 58] for aluminium by molecular dynamics simulations, which are $\lambda_s = 6.851$ N/m and $\mu_s = -0.376$ N/m, as a consequence, $M_s = 6.099$ N/m. The bulk elastic constants for aluminium are taken $\lambda = 58.17$ GPa and $\mu = 26.13$ GPa. The surface relief is described by the cosine function, i.e. $f(x_1) = -a\cos(2\pi x_1/a)$, and its wavelength is set as $a = 10$ nm. The small parameter is taken equal to $\varepsilon = 0.1$ that corresponds to the relief amplitude equal to $A = 1$ nm.

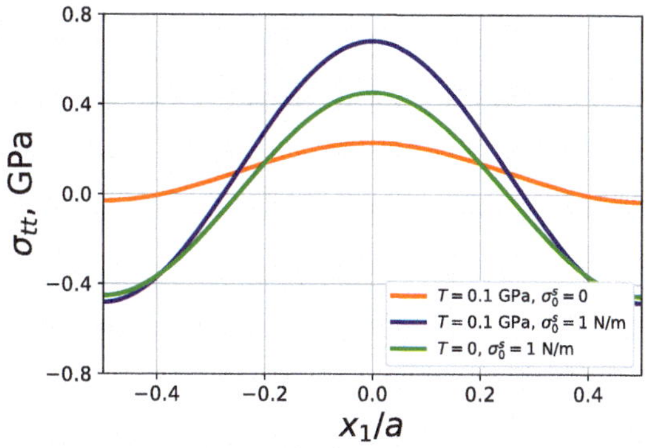

Fig. 10.2 The distribution of the hoop stress σ_{tt} along the slightly undulated surface Γ with $M_s = 6.099$ N/m.

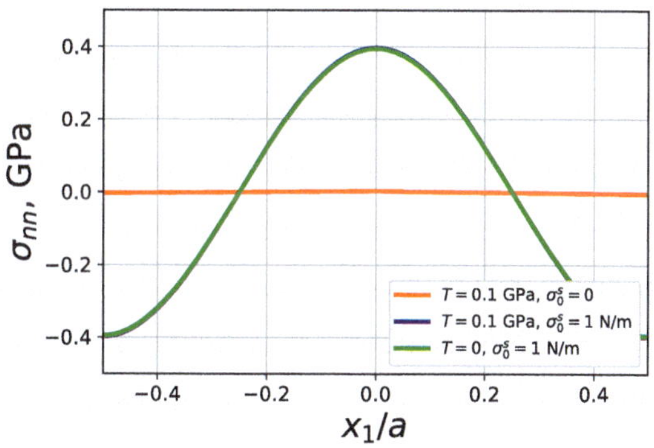

Fig. 10.3 The distribution of the normal stress σ_{nn} along the slightly undulated surface Γ with $M_s = 6.099$ N/m.

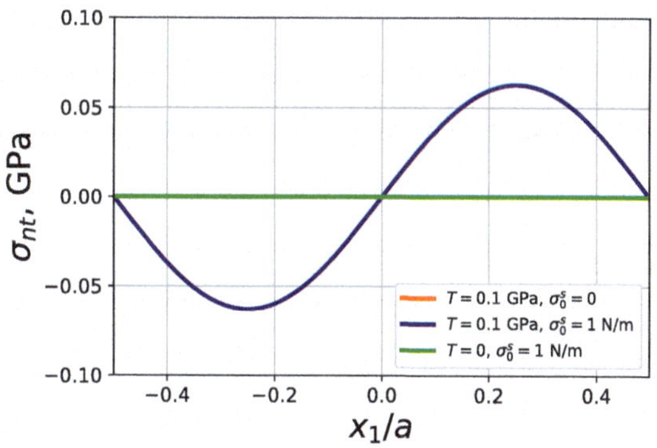

Fig. 10.4 The distribution of the tangential stress σ_{nt} along the slightly undulated surface Γ with $M_s = 6.099$ N/m.

Figures 10.2, 10.3 and 10.4 illustrate the influence of the remote loading T and surface tension σ_0^s on the distribution of hoop σ_{tt}, normal σ_{nn} and tangential σ_{nt} stresses, respectively, along the surface Γ with cosine undulated profile, and stiffness $M_s = 6.099$ N/m. From Fig. 10.2, it is seen that the maximum hoop stresses appears under the joint influence of surface tension $\sigma_0^s = 1$ N/m and remote tensile loading $T = 0.1$ GPa (blue line). Also, we can conclude that surface tension alone without remote loading, i.e. $\sigma_0^s = 1$ N/m and $T = 0$ (green line), has a greater impact on the hoop stress field than remote loading without surface tension, i.e. $\sigma_0^s = 0$ and $T = 0.1$ GPa (red line). Figure 10.3 shows that surface tension is responsible for appearance of normal stresses: their maximum level increases to 0.4 GPa if we take $\sigma_0^s = 1$ N/m, and an increase in tensile loading does not affect it (green and blue lines overlap). In contrast, the tangential stresses depend on tensile loading and does not depend on the surface tension (see Fig. 10.4 – red and blue lines corresponding to the cases $\sigma_0^s = \{0; 1\}$ N/m, respectively, overlap). However, their level induced by tensile loading is much lower than the level of normal stresses induced by surface tension.

Figures 10.5, 10.6 and 10.7 show the same dependencies as in the previous examples but for the increased stiffness of the surface layer $M_s = 60.99$ N/m. As one can see, an increase in parameter M_s leads to increase in the hoop stresses σ_{tt}. This is particularly the case for the hoop stresses induced by surface tension alone (Fig. 10.5, green line) or jointly with external loading T (Fig. 10.5, blue line). For normal σ_{nn} and tangential σ_{nt} stresses, the increase in surface stiffness M_s yields no visible results.

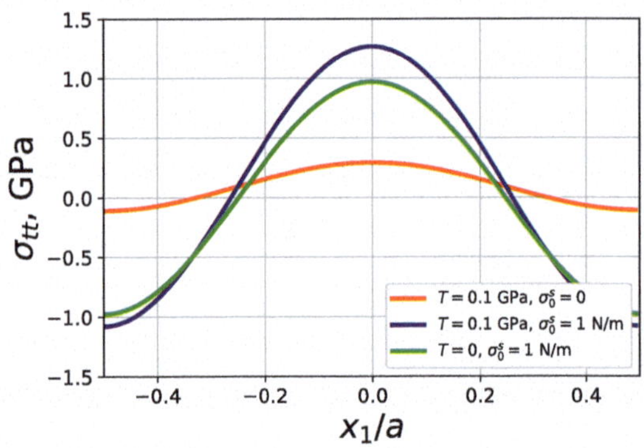

Fig. 10.5 The distribution of the hoop stress σ_{tt} along the slightly undulated surface Γ with $M_s = 60.99$ N/m.

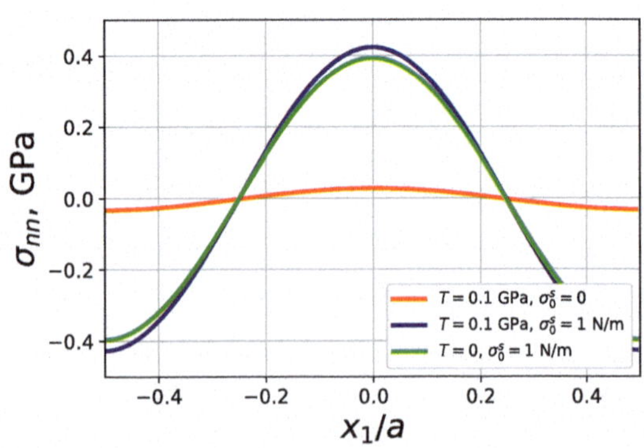

Fig. 10.6 The distribution of the normal stress σ_{nn} along the slightly undulated surface Γ with $M_s = 60.99$ N/m.

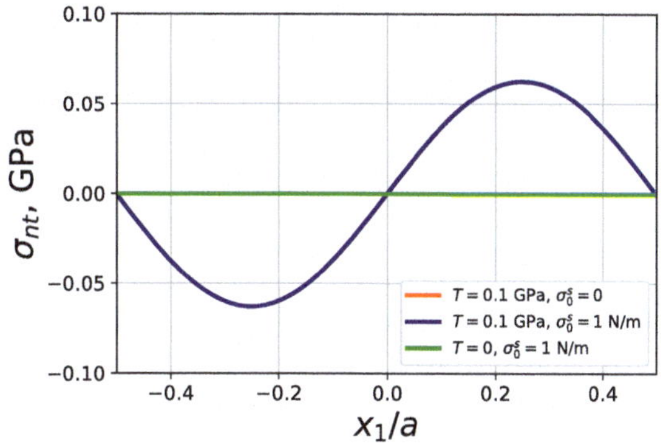

Fig. 10.7 The distribution of the tangential stress σ_{nt} along the slightly undulated surface Γ with $M_s = 60.99$ N/m.

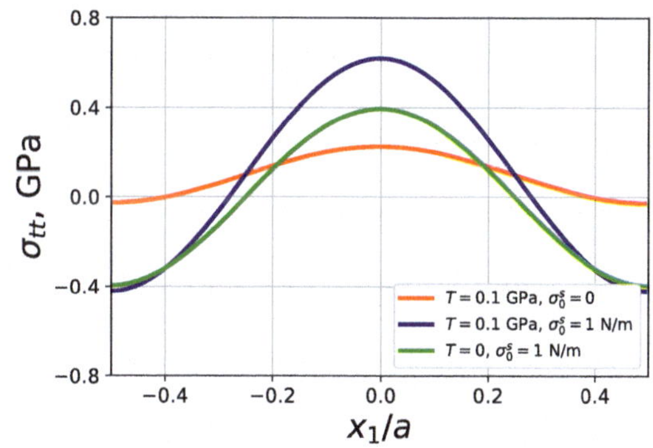

Fig. 10.8 The distribution of the hoop stress σ_{tt} along the slightly undulated surface Γ with vanishing surface stiffness, i.e. $M_s = 0$.

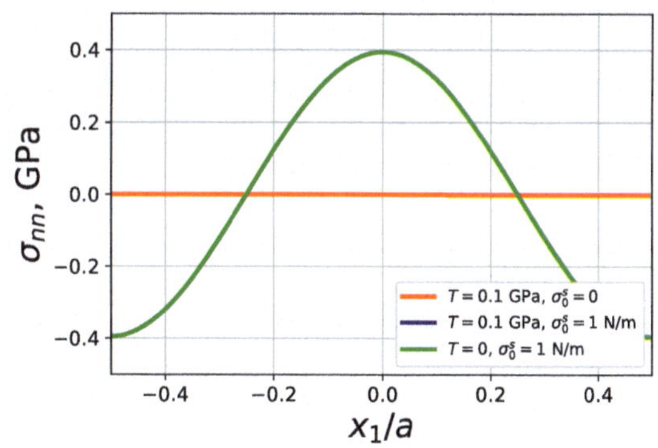

Fig. 10.9 The distribution of the normal stress σ_{nn} along the slightly undulated surface Γ with vanishing surface stiffness, i.e. $M_s = 0$.

In some studies [32, 39], authors take into account only surface tension σ_0^s neglecting the effect of surface elasticity, i.e. $M_s = 0$. Our solution allows investigating this case as well in the context of a curved solid surface. Figures 10.8, 10.9 and 10.10 give the distribution of stresses along the slightly undulated surface Γ with vanishing surface stiffness, i.e. $M_s = 0$ in Eq. (10.4) or $p_s = -m_s = \sigma_0^s$ in Eq. (10.6). Comparing the presented results, we notice that the level of the hoop stresses σ_{tt} decreases

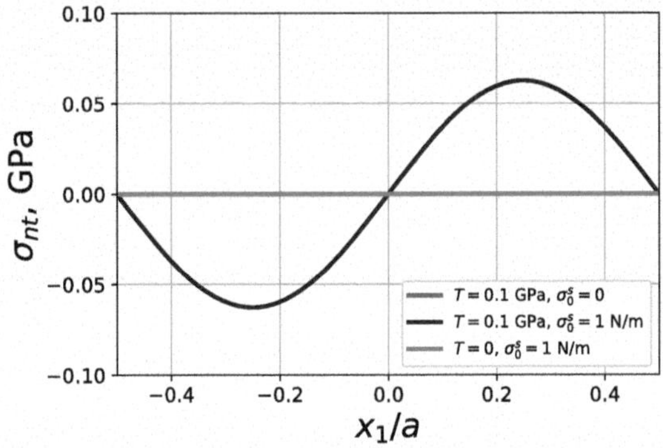

Fig. 10.10 The distribution of the tangential stress σ_{nt} along the slightly undulated surface Γ with vanishing surface stiffness, i.e. $M_s = 0$.

in contrast to the cases when $M_s \neq 0$. The level of other stresses doesn't change. It should be noted that the red lines correspond to the classical solution obtained in [14] when $M_s = 0$ and $\sigma_0^s = 0$.

10.6 Conclusions

Employing the complete Gurtin-Murdoch model of surface elasticity, the stress distribution along the perturbed solid surface has been investigated taking into account the coupled effect of surface elasticity and surface tension. Due to the plane strain conditions, the two-dimensional boundary value problem for half-plane with a curved boundary was formulated in terms of the complex variables. Based on the Goursat-Kolosov complex potentials and boundary perturbation method whereby the unknown functions are sought in the form of a power series in the small parameter represented by an amplitude-to-wavelength ratio of the surface undulation, the formulated boundary value problem was reduced to the recurrent sequence of the integral equations for any-order approximation. For the cosine-shaped surface, the first-order approximation of the stress tensor components was derived in the closed-form. Analysing the corresponding numerical results, we come to the following conclusions:

- The original GM model gives the same results for the slightly undulated surface analysed within the first-order approximation of the boundary perturbation method as the simplified one, which corresponds to omitting the normal component of the surface gradient of the displacement vector in GM constitutive equation.
- The maximum level of the hoop stresses appears under the joint influence of surface tension and remote tensile loading.
- The surface tension alone without remote loading has a greater impact on the hoop stress field than remote loading without surface tension.

- The increase in the surface stiffness leads to the increase in the hoop stresses. This effect is more significant for the surface tension-induced hoop stresses, and less significant for hoop stresses induced by the remote tensile loading.
- The surface tension induces the normal stresses the level of which is little affected by the remote tensile loading and surface stiffness.
- The surface elasticity and surface tension have a little effect on the level of tangential stresses.

Acknowledgements S. Kostyrko and M. Grekov acknowledge the support of the Russian Science Foundation under grant number 22-11-00087, https://rscf.ru/en/project/22-11-00087/

References

1. H. Altenbach, V. A. Eremeyev. On the shell theory on the nanoscale with surface stresses. *International Journal of Engineering Sciences*, 49:1294–1301, 2011.
2. H. Altenbach, V.A. Eremeyev, L.P. Lebedev. On the existence of solution in the linear elasticity with surface stresses. *Zeitschrift für Angewandte Mathematik und Mechanik*, 90:231–240, 2010.
3. H. Altenbach, V.A. Eremeyev, L.P. Lebedev. On the spectrum and stiffness of an elastic body with surface stresses. *Zeitschrift für Angewandte Mathematik und Mechanik*, 91:699–710, 2011.
4. H. Altenbach, V.A. Eremeyev, N.F. Morozov. Surface viscoelasticity and effective properties of thin-walled structures at the nanoscale. *International Journal of Engineering Science*, 59:83–89, 2012.
5. S. Baranova, S.G. Mogilevskaya, V. Mantiĉ, S. Jiménez-Alfaro. Analysis of the antiplane problem with an embedded zero thickness layer described by the Gurtin–Murdoch model. *Journal of Elasticity*, 140:171–195, 2020.
6. E.A. Bashkankova, A.B. Vakaeva, M.A. Grekov. Perturbation method in the problem on a nearly circular hole in an elastic plane. *Mechanics of Solids*, 50:198–207, 2015.
7. A.O. Bochkarev, M.A. Grekov. Influence of surface stresses on the nanoplate stiffness and stability in the Kirsch problem. *Physical Mesomechanics*, 22:209–223, 2019.
8. R.C. Cammarata. Surface and interface stress effects in thin films. *Progress in Surface Science*, 46:1–38, 1994.
9. M. Dai, M. Li, P. Schiavone. Plane deformations of an inhomogeneity-matrix system incorporating a compressible liquid inhomogeneity and complete Gurtin–Murdoch interface model. *Journal of Applied Mechanics*, 85:121010, 2018.
10. M. Dai, H.B. Yang, P. Schiavone. Stress concentration around an elliptical hole with surface tension based on the original Gurtin–Murdoch model. *Mechanics of Materials*, 135:144–148, 2019.
11. H.L. Duan, J. Wang, B.L. Karihaloo. Theory of elasticity at the nanoscale. *Advances in Applied Mechanics*, 42:1–68, 2009.
12. V.A. Eremeyev. On effective properties of materials at the nano- and microscales considering surface effects. *Acta Mechanica*, 227:29–42, 2016.
13. L.B. Freund, S. Suresh. *Thin Film Materials: Stress, Defect Formation and Surface Evolution.* University Press, Cambridge, 2003.
14. H. Gao. A boundary perturbation analysis for elastic inclusions and interfaces. *International Journal of Solids and Structures*, 28:703–725, 1991.
15. H. Gao. Some general properties of stress-driven surface evolution in a heteroepitaxial thin film structure. *Journal of the Mechanics and Physics of Solids*, 42:741–772, 1994.

16. N. Gorbushin, V.A. Eremeyev, G. Mishuris. On the stress singularity near the tip of a crack with surface stresses. *International Journal of Engineering Science*, 146:103183, 2020.
17. M.A. Grekov. *Singular Plane Problems in Elasticicty (in Russ.)*. St. Petersburg State University, St. Petersburg, 2001.
18. M.A. Grekov. The perturbation approach for a two-component composite with a slightly curved interface. *Vestnik Sankt-Petersburgskogo Universiteta. Ser 1. Mathematika Mechanika Astronomiya*, 1:81–88, 2004.
19. M.A. Grekov. Two types of interface defects. *Journal of Applied Mathematics and Mechanics*, 75:76–88, 2011.
20. M.A. Grekov. General approach to the modified Kirsch problem incorporating surface energy effects. *Continuum Mechanics and Thermodynamics*, 33:1675–1689, 2021.
21. M.A. Grekov, S.A. Kostyrko. A film coating on a rough surface of an elastic body. *Journal of Applied Mathematics and Mechanics*, 77:79–90, 2013.
22. M.A. Grekov, S.A. Kostyrko. A multilayer film coating with slightly curved boundary. *International Journal of Engineering Science*, 89:61–74, 2015.
23. M.A. Grekov, S.A. Kostyrko. Surface effects in an elastic solid with nanosized surface asperities. *International Journal of Solids and Structures*, 96:153–161, 2016.
24. M.A. Grekov, S.A. Kostyrko, A.B. Vakaeva. The model of surface nanorelief within continuum mechanics. *AIP Conference Proceedings*, 1909:020062, 2017.
25. M.A. Grekov, S.N. Makarov. Stress concentration near a slightly curved part of an elastic body surface. *Mechanics of Solids*, 39:40–46, 2004.
26. M.A. Grekov, N.F. Morozov. Some modern methods in mechanics of cracks. *In: V.M. Adamyan, I. Gohberg, A. Kochubei, G. Popov, Y. Berezansky, M. Gorbachuk, V. Gorbachuk, H. Langer (Eds.). Modern Analysis and Applications - The Mark Krein Centenary Conference - Volume 2: Differential Operators and Mechanics (pp. 127–142)*. Birkhäuser, Basel, 2009.
27. M.A. Grekov, N. Morozov. Surface effects and problems of nanomechanics. *Journal of Ningbo University*, 25:60–63, 2012.
28. M.A. Grekov, T.S. Sergeeva. Interaction of edge dislocation array with bimaterial interface incorporating interface elasticity. *International Journal of Engineering Science*, 149:103233, 2020.
29. M.A. Grekov, T.S. Sergeeva, Y.G. Pronina, O.S. Sedova. A periodic set of edge dislocations in an elastic semi-infinite solid with a planar boundary incorporating surface effects. *Engineering Fracture Mechanics*, 186:423–435, 2017.
30. M.A. Grekov, A.B. Vakaeva. Effect of nanosized asperities at the surface of a nanohole. *In: Papadrakakis, M., Onãte, E., Schrefler, B. (eds.) VII European Congress on Computational Methods in Applied Sciences and Engineering (pp. 7875–7885)*. CIMNE, Barcelona, 2016.
31. M.A. Grekov, A.A. Yazovskaya. Effect of surface elasticity and residual surface stress in an elastic body weakened by an elliptic hole of a nanometer size. *Journal of Applied Mathematics and Mechanics*, 78:172–180, 2014.
32. J. Grilhe. Study of roughness formation induced by homogeneous stress at the free surfaces of solids. *Acta metallurgica et materialia*, 41:909–913, 1993.
33. M.E. Gurtin, A.I. Murdoch. A continuum theory of elastic material surfaces. *Archive of Rational Mechanics Analysis*, 57:291–323, 2014.
34. M.E. Gurtin, A.I. Murdoch. Surface stress in solids. *International Journal of Solids Structures*, 14:431–440, 1978.
35. M.Y. Gutkin, C. Enzevaee, H.M. Shodja. Interface effects on elastic behavior of an edge dislocation in a core–shell nanowire embedded to an infinite matrix. *International Journal of Solids and Structures*, 50: 1177–1186, 2013.
36. A. Javili, A. McBride, P. Steinmann. Thermomechanics of solids with lowerdimensional energetics: On the importance of surface, interface, and curve structures at the nanoscale. A Unifying Review. *Applied Mechanics Reviews*, 65:010802, 2013.
37. A. Javili, N.S. Ottosen, M. Ristinmaa, J. Mosler. Aspects of interface elasticity theory. *Mathematics and Mechanics of Solids*, 23:1004–1024, 2018.
38. T. Kitamura, H. Hirakata, T. Sumigawa, T. Shimada. *Fracture Nanomachanics*. Pan Stanford, New York, 2011.

39. S. A. Kostyrko Effect of the undulation shape on the flat film surface stability under diffusion processes. *Ser 1. Mathematika Mechanika Astronomiya*, 3:101–111, 2011.
40. S.A. Kostyrko, M.A. Grekov. Elastic field at a rugous interface of a bimaterial with surface effects. *Engineering Fracture Mechanics*, 216:106507, 2019.
41. S. Kostyrko, M. Grekov, H. Altenbach. Stress concentration analysis of nanosized thin-film coating with rough interface. *Continuum Mechanics and Thermodynamics*, 31:1863–1871, 2019.
42. S. Kostyrko, M. Grekov, H. Altenbach. Coupled effect of curved surface and interface on stress state of wrinkled thin film coating at the nanoscale. *Zeitschrift für Angewandte Mathematik und Mechanik* , 101:e202000202, 2021.
43. S. Kostyrko, M. Grekov, T. Kitamura. Interaction of misfit dislocations with perturbated surface in epitaxial thin film. *In: Smirnov, N., Golovkina, A. (eds.) Stability and Control Processes. SCP 2020. Lecture Notes in Control and Information Sciences - Proceedings (pp. 871-879).* Springer, Cham, 2022.
44. S. Kostyrko, G. Shuvalov. Surface elasticity effect on diffusional growth of surface defects in strained solids. *Continuum Mechanics and Thermodynamics*, 31:1795–1803, 2019.
45. G.I. Mikhasev, M.G. Botogova, V.A. Eremeyev. On the influence of a surface roughness on propagation of anti-plane short-length localized waves in a medium with surface coating. *International Journal of Engineering Science*, 158:103428, 2021.
46. R.E. Miller, V.B. Shenoy. Size-dependent elastic properties of nanosized structural elements. *Nanotechnology*, 11:139–147, 2000.
47. S.G. Mogilevskaya, S.I. Crouch, H.K. Stolarski. Multiple interacting circular nano-inhomogeneities with surface/interface effects. *Journal of the Mechanics and Physics of Solids*, 56:2298–2327, 2008.
48. S.G. Mogilevskaya, A.Y. Zemlyanova, V.I. Kushch. Fiber-and particle-reinforced composite materials with the Gurtin–Murdoch and Steigmann–Ogden surface energy endowed interfaces. *Applied Mechanics Reviews*, 73:050801, 2021.
49. S.G. Mogilevskaya, A.Y. Zemlyanova, V. Mantiĉ. The use of the Gurtin–Murdoch theory for modeling mechanical processes in composites with two-dimensional reinforcements. *Composites Science and Technology*, 210:108751, 2021.
50. N.I. Muskhelishvili. *Some Basic Problems of the Mathematical Theory of Elasticity.* Springer, Netherlands, 1977.
51. L. Nazarenko, S. Bargmann, H. Stolarski. Closed-form formulas for the effective properties of random particulate nanocomposites with complete Gurtin–Murdoch model of material surfaces. *Continuum Mechanics and Thermodynamics*, 29:77–96, 2017.
52. L. Nazarenko, H. Stolarski, H. Altenbach. Effective properties of short-fiber composites with Gurtin–Murdoch model of interphase. *International Journal of Solids and Structures*, 97:75–88, 2016.
53. Y.S. Podstrigach, Y.Z. Povstenko. *An Introduction to the Mechanics of Surface Phenomena in Deformable Solids.* Naukova Dumka, Kiev, 1985.
54. Yu.Z. Povstenko. Theoretical investigation of phenomena caused by heterogeneous surface tension in solids. *Journal of the Mechanics and Physics of Solids*, 41:1499–1514, 1993.
55. Y. Pronina, A. Maksimov, M. Kachanov. Crack approaching a domain having the same elastic properties but different fracture toughness: Crack deflection vs penetration. *International Journal of Engineering Science*, 156:103374, 2020.
56. C.Q. Ru. Simple geometrical explanation of Gurtin–Murdoch model of surface elasticity with clarification of its related versions. *Science China Physics, Mechanics and Astronomy*, 53:536–544, 2010.
57. O. Sedova, Y. Pronina. The thermoelasticity problem for pressure vessels with protective coatings, operating under conditions of mechanochemical corrosion. *International Journal of Engineering Science*, 170: 103589, 2022.
58. V.B. Shenoy. Atomistic calculations of elastic properties of metallic fcc crystal surfaces. *Physical Review B*, 71:094104, 2005.
59. R. Shuttleworth. The surface tension of solids. *Proceedings of the Physical Society. Section A*, 63(5):444–457, 1950.

60. G. Shuvalov, S. Kostyrko. On the role of interfacial elasticity in morphological instability of a heteroepitaxial interface. *Continuum Mechanics and Thermodynamics*, 33:2095–2107, 2021.
61. G.M. Shuvalov, A.B. Vakaeva, D.A. Shamsutdinov, S.A. Kostyrko. The effect of nonlinear terms in boundary perturbation method on stress concentration near the nanopatterned bi-material interface. *Vestnik of Saint Petersburg University. Applied Mathematics. Computer Sciences. Control Processes*, 16:165–176, 2020.
62. L. Tian, R.K.N.D. Rajapakse. Analytical solution for size-dependent elastic field of a nanoscale circular inhomogeneity. *Journal of Applied Mechanics*, 74:568–574, 2007.
63. A.B. Vakaeva, M.A. Grekov. Effect of interfacial stresses in an elastic body with a nanoinclusion. *AIP Conference Proceedings*, 1959:070036, 2018.
64. A.B. Vakaeva, G.M. Shuvalov, S.A. Kostyrko. Evolution of the cylindrical nanopore morphology under diffusion processes. *Materials Physics and Mechanics*, 47:423–430, 2021.
65. Yu.I. Vikulina, M.A. Grekov, S.A. Kostyrko. Model of film coating with weakly curved surface. *Mechanics of Solids*, 45:778–788, 2010.
66. Yu.I. Vikulina, M.A. Grekov. The stress state of planar surface of a nanometer-sized elastic body under periodic loading. *Vestnik St. Petersburg University: Mathematics*, 45:174–180, 2012.
67. J. Wang, Z. Huang, H. Duan, S. Yu, X. Feng, G. Wang, W. Zhang, T. Wang. Surface stress effect in mechanics of nanostructured materials. *Acta Mechanica Solida Sinica*, 24:52–82, 2011.
68. H.B. Yang, M. Dai. Influence of surface roughness on the stress field around a nanosized hole with surface elasticity. *Zeitschrift für Angewandte Mathematik und Physik*, 69:1–9, 2018.
69. H.B. Yang, M. Dai, C.F. Gao. Stress field in a porous material containing periodic arbitrarily-shaped holes with surface tension. *Mathematics and Mechanics of Solids*, 23:120–130, 2018.
70. G. Yang, C.F. Gao, C.Q. Ru. A study on the Gurtin–Murdoch model for spherical solids with surface tension. *Zeitschrift für Angewandte Mathematik und Mechanik*, 72:1–15, 2021.

Chapter 11
Analytical and Numerical Methods for Analysis of Stress Singularity in Three-Dimensional Problems of Elasticity Theory

Valerii P. Matveenko, Andrey Yu. Fedorov, Tatiana O. Korepanova, Natalja V. Sevodina, and Igor N. Shardakov

Abstract Different variants of stress singularity analysis in three-dimensional problems of elasticity theory are considered. A complete system of eigensolutions is developed for different variants of circular conical bodies: solid cone, hollow cone, a composite cone under different variants of boundary conditions on the lateral surfaces. The applicability of the constructed eigensolutions for estimating the character of stress singularity at the vertices of conical bodies is considered. The numerical results presented in the study provide insight into the character of stress singularity at the vertices of solid and hollow cones under different variants of boundary conditions on the lateral surfaces. A method for constructing singular solutions for conical bodies is suggested and variants of its numerical realization based on the finite element method are considered. The results of conducted numerical experiments demonstrate the efficiency and reliability of the proposed method. The computation of eigenvalues allows us to determine the character of stress singularity in homogeneous and composite, circular and non-circular cones under different boundary conditions. The work presents an algorithm for the finite-element analysis of singular solutions to three-dimensional problems of elasticity theory for elastic bodies of isotropic, anisotropic, and functionally graded materials. The algorithm is based on determination of a power law relationship for stresses in the vicinity of singular points. The algorithm was verified by solving two- and three-dimensional problems and comparing the obtained results with those available in the literature.

Key words: Singular points, 2D and 3D problems of elasticity theory, Stress singularity, Closed-form solution, Numerical solutions, Finite element method

---

Valerii P. Matveenko · Andrey Yu. Fedorov · Tatiana O. Korepanova · Natalja V. Sevodina · Igor N. Shardakov
Institute of Continuous Media Mechanics of the Ural Branch of RAS, 614018, Academician Korolev Street, 1, Perm, Russian Federation,
e-mail: mvp@icmm.ru,fedorov@icmm.ru,ton@icmm.ru,natsev@icmm.ru,shardakov@icmm.ru

H. Altenbach et al. (eds.), *Advances in Solid and Fracture Mechanics*,
Advanced Structured Materials 180,
https://doi.org/10.1007/978-3-031-18393-5_11

11.1 Introduction

One of the important results of classical elasticity theory is that it provides the existence of singular solutions associated with the occurrence of infinite stresses at points (called singular) where smoothness of the body surface is violated, the type of boundary conditions is changed, or contact of different materials takes place, as well as inside the body, at points where the condition for smoothness of the interface between different materials is violated. An example of theoretical justification of the concept that the existence of singular solutions is possible under certain conditions can be found in work [12], where it is shown that in the vicinity of angular points the equations of linear elasticity theory have a solution in the following form

$$\sigma \sim \sum_{n=1} K_n f_n r^{\lambda_n - 1}, \quad r \to 0, \quad c < \mathrm{Re}\lambda_1 < \mathrm{Re}\lambda_2 < \ldots < \mathrm{Re}\lambda_n < \ldots, \qquad (11.1)$$

or a more complex solution with logarithmic components in the case of multiple points of the spectrum λ_n. Here, r is the distance to the angular point, K_n are constants (called the stress intensity coefficients); f_n are the functions of angular distribution of the stress field σ in the vicinity of the angular point, which in the planar case depend on a single polar angular variable φ at $c = 0$, whereas in the spatial case — on two spherical coordinates φ, θ at $c = -0.5$. The form of solution (11.1) suggests that if there are λ_n, satisfying the condition $\mathrm{Re}\,\lambda_\mathrm{n} < 1$, the stresses tend to infinity at r tending to zero.

Singular points of different types are often found in computational models constructed for solving various applied problems of the theory of elasticity. The existence of singular solutions suggests that in general the vicinities of singular points are the zones of strong stress concentration that triggers the fracture process in a body. The stress behavior in the vicinity of singular points has long been the focus of many studies. For two- and three-dimensional problems of linear elasticity theory, different variants of singular points have been considered. The results obtained in this field are presented in sufficient detail in review papers [5, 25, 28, 31, 32]. Among the variety of problems with singular points, one of the first and most studied is the problem for the crack tip, which is one of the main objects of study in fracture mechanics. The distinguishing features of problems in fracture mechanics for bodies with acute-angle notches are specified in works by N. F. Morozov [21, 22]: the stress field in the vicinity of a angular notch consists of regular and singular components, and the singularity exponent depends on the opening angle of the notch.

One of the approaches to the construction of solutions of the form (11.1) is based on studying singular regions. In two-dimensional problems, the objects of investigation are the neighborhoods of vertices of wedge-shaped regions: homogeneous or composite plane wedges with boundary conditions specified on their faces (in terms of stresses or displacements). Over a more than half-century history of studies on this topic almost all possible variants of wedge-shaped bodies have been considered: homogeneous and composite, isotropic and anisotropic, functionally gradient [7, 8], etc. For three-dimensional problems, two classes of regions can be distinguished: vicinities of points on the edge of a spatial wedge and vicinities of vertices of homogeneous and composite conical regions, such as vertices of circular and non-circular cones, triangular and polyhedral wedges. Here it should be noted that mechanical characteristics of such regions may correspond to those of isotropic, anisotropic, and even functionally graded materials. Interest in three-dimensional problems of the first class has considerably diminished due to the results of some works, including [9, 20], where it is shown that solutions to the plane and antiplane problems for wedges located in the planes perpendicular to the edge of a spatial wedge determine the type of stress singularity at the points of the edge through which the corresponding plane passes.

In the last few decades, the number of works devoted to the study of stress singularity at the vertex of a polyhedral wedge and a cone has considerably increased. Most of these problems were solved using different variants of numerical methods, mainly finite and boundary element methods. Among the works using the ideas of various numerical methods worthy of note are the studies, which are based on the finite element method [1, 6, 13, 16, 19, 23, 24], on the boundary element method [11, 30], and on the application of the Mellin transformation to initial two-dimensional boundary integral equations [2]. In [16], a numerical method was developed to estimate the nature of the stress singularity at the vertex of a cone with elliptic base and homogeneous boundary conditions. In continuation to these studies, [19] presents a series of numerical methods, which makes it possible to obtain new results for different variants of cones, in particular, for homogeneous and composite, circular and non-circular cones under homogeneous and mixed boundary conditions.

As in other sections of the theory of elasticity, the analytical methods play an important role in the construction of singular solutions, and are still considered as an effective instrument both for obtaining specific numerical results and testing numerical methods. In three-dimensional problems, analytical methods are mainly applied to circular cones (axisymmetric conical regions: homogeneous [3, 14, 15, 33] and composite [14, 26, 27]). One of the first examples of analytical treatment of these problems is [3], which considers a solid cone under axisymmetric deformation and rotation with boundary conditions specified in terms of displacements and stresses. In further studies, the analytical solutions of some particular problems were obtained. For example, works [26, 27] present the results for a composite cone under axisymmetric deformation. In this case, a composite cone is a structure consisting of two nested cones, which have a common contact area. The solutions were obtained for ideal contact and ideal sliding conditions. In [33], an axisymmetric problem for a circular cone of transversally isotropic material is considered. A fairly complete review of works dealing with the study of circular cones by analytical methods is given in [32]. Among the cited works, [15] is the most comprehensive study on the subject. Here, an analytical solution for a solid circular cone was constructed and numerical results, disclosing the nature of the stress singularity at the vertex of a solid circular cone with the stress and displacement boundary conditions on the lateral surface, were obtained. In [14], a full spectrum of analytical eigenvalues for different variants of cones (solid, hollow, composite) is specified and evaluation of stress singularity exponents for solid and hollow cones under different boundary conditions on the lateral surfaces is illustrated by some numerical simulations.

11.2 Analysis of Stress Singularity Based on the Constructed Analytical Eigensolutions for Semi-infinite Circular Conical Bodies

Let us consider a homogeneous circular cone (Fig. 11.1a) whose vertex coincides with the center of spherical coordinates r, θ, φ and its base is perpendicular to the axis $\theta = 0$. The cone occupies a volume $0 \leq r < \infty$, $\theta_1 \leq \theta \leq \theta_0$, $0 \leq \varphi \leq 2\pi$, and its boundary is defined by coordinate surfaces $\theta = \theta_1$, $\theta = \theta_0$. The variant corresponds to a solid cone.

We need to construct eigensolutions satisfying the homogeneous equilibrium equations

$$(1+S)\,\mathrm{grad}\,\mathrm{div}\mathbf{U} - \mathrm{rot}\,\mathrm{rot}\mathbf{U} = 0 \tag{11.2}$$

and one of the homogeneous boundary conditions on the surfaces $\theta = \theta_1$, $\theta = \theta_0$ for displacements

$$u_r = 0, \;\; u_\theta = 0, \;\; u_\varphi = 0, \tag{11.3}$$

and stresses

$$\sigma_{r\theta} = 0, \;\; \sigma_{\theta\theta} = 0, \;\; \sigma_{\theta\varphi} = 0, \tag{11.4}$$

or mixed boundary conditions, which in terms of mechanics, correspond to ideal sliding on the lateral surface

$$u_\theta = 0, \;\; \sigma_{r\theta} = 0, \;\; \sigma_{\theta\varphi} = 0. \tag{11.5}$$

For the examined body of rotation and boundary conditions (11.3)–(11.5), the eigen solutions can be represented as a Fourier series in the circular coordinate φ

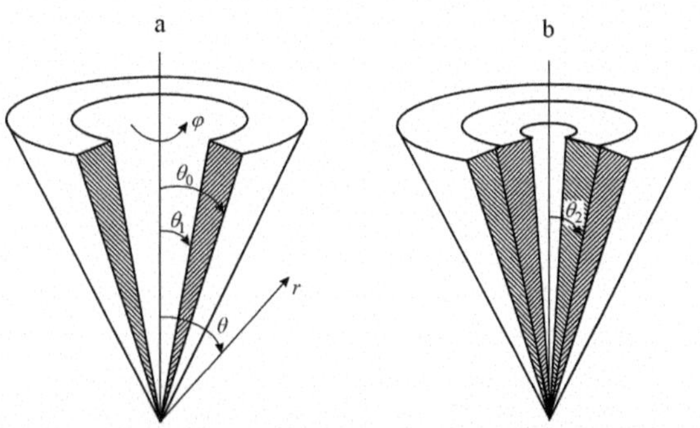

Fig. 11.1 Variants of conical bodies: hollow cone (a); hollow composite cone (b)

$$
\begin{aligned}
u_r(r,\theta,\varphi) &= u_0(\theta)\, r^\alpha + \sum_{k=1}^{\infty} [u_k(\theta)\, r^\alpha \sin(k\varphi)],\\
u_\theta(r,\theta,\varphi) &= \mathrm{v}_0(\theta)\, r^\alpha + \sum_{k=1}^{\infty} [\mathrm{v}_k(\theta)\, r^\alpha \sin(k\varphi)],\\
u_\varphi(r,\theta,\varphi) &= w_0(\theta)\, r^\alpha + \sum_{k=1}^{\infty} [w_k(\theta)\, r^\alpha \cos(k\varphi)].
\end{aligned}
\tag{11.6}
$$

Here, the dependence on the radius is represented according to (11.1) $S = 1/(1-2\nu)$; ν is Poisson's ratio; $\mathbf{U}$ is the displacement vector, u_r, u_θ, u_φ are the components of the vector of displacements along the axes r, θ, φ; $\sigma_{r\theta}$, $\sigma_{\theta\theta}$, $\sigma_{\theta\varphi}$ are the components of the stress tensor, α is the characteristic exponent.

If $\theta_1 = 0$, then the examined region is bounded by only one coordinate surface $\theta = \theta_0$, and at $\theta = 0$ the regularity conditions must be satisfied

$$
\partial u_r/\partial\theta = 0, \quad u_\theta = 0, \quad u_\varphi = 0. \tag{11.7}
$$

Within the framework of the suggested problem formulation we can also consider a composite cone occupying the domain $V = V^{(1)} + V^{(2)}$, where the subdomain $V^{(1)}$ (subdomain $V^{(2)}$) represents the cone segment made of the material with shear modulus $\mu^{(1)}$ ($\mu^{(2)}$) and Poisson's ratio $\nu^{(1)}$ ($\nu^{(2)}$) and its geometry is determined by the relations $0 \le r \le \infty$, $0 \le \varphi \le 2\pi$, $\theta_2 \le \theta \le \theta_0$ ($\theta_1 \le \theta \le \theta_2$). In particular cases, θ_1 and θ_0 can be equal to 0 and π, respectively.

For a composite cone (Fig. 11.1b), the eigensolutions (11.6) are constructed for each of the subdomains, and at the contact boundary $\theta = \theta_2$ one can set ideal bonding conditions

$$
\begin{aligned}
u_r^{(1)} &= u_r^{(2)}, \quad & u_\varphi^{(1)} &= u_\varphi^{(2)}, \quad & u_\theta^{(1)} &= u_\theta^{(2)},\\
\sigma_\theta^{(1)} &= \sigma_\theta^{(2)}, \quad & \tau_{r\theta}^{(1)} &= \tau_{r\theta}^{(2)}, \quad & \tau_{\varphi\theta}^{(1)} &= \tau_{\varphi\theta}^{(2)},
\end{aligned}
\tag{11.8}
$$

or ideal sliding conditions

$$
u_\theta^{(1)} = u_\theta^{(2)}, \quad \sigma_\theta^{(1)} = \sigma_\theta^{(2)}, \quad \tau_{r\theta}^{(1)} = \tau_{r\theta}^{(2)} = \tau_{\varphi\theta}^{(1)} = \tau_{\varphi\theta}^{(2)} = 0. \tag{11.9}
$$

After substituting equations (11.6) into equilibrium equations (11.2) and changing to a new independent variable $x = (1-\cos\theta)/2$, we obtain for each of the harmonics of the Fourier series the following equations:

$$
\begin{aligned}
&x(1-x)\frac{d^2u_k(x)}{dx^2} + (1-2x)\frac{du_k(x)}{dx} + \frac{\left[4xR_1(x-1)+k^2\right]}{4x(x-1)}u_k(x) +\\
&+\frac{x(1-x)R_2}{\sqrt{x(1-x)}}\frac{d\mathrm{v}_k(x)}{dx} + \frac{R_2}{\sqrt{x(1-x)}}\left[\left(\frac{1}{2}-x\right)\mathrm{v}_k(x) - \frac{kw_k(x)}{2}\right] = 0,
\end{aligned}
\tag{11.10a}
$$

$$G_1 x(1-x)\frac{d^2 \upsilon_k(x)}{dx^2} + G_1(1-2x)\frac{d\upsilon_k(x)}{dx} + \frac{\left[4xG_2(x-1)+k^2+G_1\right]}{4x(x-1)}\upsilon_k(x) + \\ +G_3\sqrt{x(1-x)}\frac{d}{dx}u_k(x) + \left[\frac{k(1-G_1)}{2}\frac{dw_k(x)}{dx} + \frac{(G_1+1)k(2x-1)}{4x(x-1)}w_k(x)\right] = 0, \tag{11.10b}$$

$$x(1-x)\frac{d^2 w_k(x)}{dx^2} + (1-2x)\frac{dw_k(x)}{dx} + \frac{\left[4xG_2(x-1)+G_1k^2+1\right]}{4x(x-1)}w_k(x) + \\ +\frac{kG_3}{2\sqrt{x(1-x)}}u_k(x) + \left[\frac{(G_1-1)k}{2}\frac{d\upsilon_k(x)}{dx} + \frac{(G_1+1)k(2x-1)}{4x(x-1)}\cdot\upsilon_k(x)\right] = 0. \tag{11.10c}$$

Here, the following representations are used

$$R_1 = \frac{2(1-\nu)(1-\alpha)(\alpha+2)}{(2\nu-1)}; \quad R_2 = \frac{(3-\alpha-4\nu)}{(-1+2\nu)};$$

$$G_1 = \frac{2(1-\nu)}{(1-2\nu)}; \quad G_2 = \alpha(1+\alpha); \quad G_3 = \frac{2(\alpha+4-4\nu)}{(1-2\nu)}.$$

In view of equation (11.6), the boundary conditions (11.3)–(11.5) and the regularity condition (11.7) are transformed exactly in the same way:

$$u_k(x) = 0; \qquad \upsilon_k(x) = 0; \qquad \mathrm{w}_k(x) = 0; \tag{11.11}$$

$$\mu\left[\sqrt{x(1-x)}\frac{du_k(x)}{dx} + (\alpha-1)\upsilon_k(x)\right] = 0; \tag{11.12a}$$

$$\mu\left[(2S-\alpha+\alpha S)u_k(x) + (1+S)\sqrt{x(1-x)}\frac{d\upsilon_k(x)}{dx} + \right. \\ \left. + \left(\frac{1}{2}-x\right)\frac{(S-1)}{\sqrt{x(1-x)}}\upsilon_k(x) + \frac{k(1-S)}{2\sqrt{x(1-x)}}w_k(x)\right] = 0; \tag{11.12b}$$

$$\mu\left[\sqrt{x(1-x)}\frac{dw_k(x)}{dx} - \frac{(1-2x)}{2\sqrt{x(1-x)}}w_k(x) + \frac{k}{2\sqrt{x(1-x)}}\upsilon_k(x)\right] = 0; \tag{11.12c}$$

$$\sqrt{x(1-x)}\frac{du_k(x)}{dx} = 0; \qquad \upsilon_k(x) = 0; \qquad \mathrm{w}_k(x) = 0. \tag{11.13}$$

The variant for the zero harmonic of the Fourier series is considered separately, since it does not explicitly follow from the algorithm for constructing partial solutions of the system of differential equations (11.10) for any value of $k \neq 0$. At $k = 0$ there are two problems: axisymmetric rotation and axisymmetric deformation. In the first problem, the component of the displacement vector w_0 is determined by equation (11.10c). In the axisymmetric deformation problem, the displacement vector components u_0, υ_0 are defined by equations (11.10a), (11.10b).

Solutions for the function w_0 are derived in the form of a generalized power series

$$w_0(x) = \sum_{m=0}^{\infty} \left[A_m x^{(m+\beta)} \right], \tag{11.14}$$

where A_m are the coefficients of the power series; β is the characteristic exponent.

The possibility of constructing a solution in the form (11.14) is substantiated in [18]. The point $x = 0$ for equation (11.10c) is a regular singular point. In this case, one of the partial solutions is written in the form of series (11.14), for which the region of convergence is the range of the variable $0 \le x \le 1$, since the value $x = 1$ is a zero of the function nearest to the point $x = 0$ for a higher derivative.

To find the coefficients of the series A_m and the characteristic exponent β, equation (11.14) is substituted into (11.10c). By equating the expressions with similar powers of x to zero, we obtain the recurrence relation for A_m:

$$\begin{aligned} &(2\beta+2m+1)(2\beta+2m-1)A_m+ \\ &+4\left[\alpha(\alpha+1)-(2\beta+2m-1)\ (\beta+m-1)\right]A_{m-1}- \\ &-4(\alpha+2-m-\beta)(\alpha-1+m+\beta)A_{m-2}=0, \quad (m=0,1,2,\ldots) \end{aligned} \tag{11.15}$$

From the condition for the existence of a nonzero solution with respect to A_0 we get the characteristic equation

$$(2\beta+1)(2\beta-1)=0, \tag{11.16}$$

where $\beta_1 = 0.5$ and $\beta_1 = -0.5$ are its roots.

According to the theory of ordinary differential equations [18], there is always a solution in the form of a generalized power series (11.14) that corresponds to the largest root β_1. Substituting the value of root β_1 into (11.15), we obtain a recurrence relation for $A_m^{(1)}$:

$$\begin{aligned} A_m^{(1)} = \frac{(2m^2-\alpha-\alpha^2-m)}{m(1+m)}A_{m-1}^{(1)} + \frac{(2\alpha-1+2m)(2\alpha+3-2m)}{4m(1+m)}A_{m-2}^{(1)}, \\ (m>0,\ A_0^{(1)}=1). \end{aligned} \tag{11.17}$$

Here and hereafter, the upper index defines the number of the partial solution.

The transformations performed allow us to obtain the first partial solution, which has the form of a generalized power series for equation (11.10c):

$$w_0^{(1)}(x) = \sum_{m=0}^{\infty} \left[A_m^{(1)} x^{\left(m+\frac{1}{2}\right)} \right]. \tag{11.18}$$

The difference in roots of the characteristic equation [10], i.e. $\gamma = \beta_1 - \beta_2$, is crucial for constructing a second linearly independent partial solution in the form of a generalized power series. If γ is not a positive integer, there exists a second linearly independent solution in the form of a generalized power series (11.14). If

γ is a positive integer, then in the general case the existence of a second partial solution in the form of generalized power series (11.14) is not guaranteed.

To exclude this uncertainty, we applied an approach, which is based on a sequential reduction of the original differential equation by making use of the first partial solution and keeping a fixed number of terms in the series. A series segment for the second partial solution of the original differential equation is obtained as follows. After reduction, the resulting series segment is integrated and the result of the integration is multiplied by the generalized power series corresponding to the first partial solution. The form of the obtained series segment for the second partial solution determines the characteristic exponent of the generalized power series and the terms including the logarithmic functions. It should be noted that partial solutions subsequent to the second partial solution [10] include the logarithmic functions of higher degree (compared to the first function).

Thus, the proposed method makes it possible to successively determine the types of generalized power series of all partial solutions of the original differential equation and to single out from all partial solutions the regular and irregular ones, in our case, at value $x = 0$. These capabilities of the method hold much promise for constructing solutions to particular problems, for example, that of a hollow cone.

Using the proposed method, a second partial solution $\omega_0^{(2)}$ is obtained :

$$\omega_0^{(2)}(x) = \sum_{m=0}^{\infty} \left\{ \left[A_m^{(2)} + B_m^{(2)} \cdot \ln(x) \right] x^{(m-1/2)} \right\}, \tag{11.19}$$

where the coefficients $A_m^{(2)}, B_m^{(2)}$ are determined from the recurrence relations

$$B_m^{(2)} = \frac{\left[(m-1)(2m-3) - \alpha^2 - \alpha\right]}{m(m-1)} B_{m-1}^{(2)} - \frac{(2m-3+2\alpha)(2m-5-2\alpha)}{4m(m-1)} B_{m-2}^{(2)},$$

$$A_m^{(2)} = \frac{(1-2m)}{m(m-1)} B_m^2 + \frac{\left[(m-1)(2m-3) - \alpha^2 - \alpha\right]}{m(m-1)} A_{m-1}^{(2)} + \frac{(4m-5)}{m(m-1)} B_{m-1}^{(2)} -$$
$$- \frac{(2m-3+2\alpha)(2m-5-2\alpha)}{4m(m-1)} A_{m-2}^{(2)} - \frac{2(m-2)}{m(m-1)} B_{m-2}^{(2)}. \tag{11.20}$$

From the form of the obtained solutions $w_0^{(1)}, w_0^{(2)}$ it follows that $w_0^{(1)}$ is a regular solution, and $w_0^{(2)}$ is an irregular solution at $x = 0$.

The general solution of the differential equation (11.10c) can be written as

$$w_0(x) = C_1 \cdot w_0{}^{(1)}(x) + C_2 \cdot w_0{}^{(2)}(x), \tag{11.21}$$

where C_1, C_2 are the constants determined from a preset combination of boundary conditions (11.3)–(11.5). To construct partial solutions to equations (11.10a) and (11.10b) corresponding to the axisymmetric deformation variant, we solve this system for v_0 [18]:

$$v_0(x) = \frac{\sqrt{x(1-x)}}{(1-\alpha)S+2} \times \\ \times \left\{ \frac{(S+1)}{(\alpha+\alpha^2)} \left[\left(x^2 - x\right) \frac{d^3 u_0(x)}{dx^3} + (4x-2) \frac{d^2 u_0(x)}{dx^2} \right] - (2S+1) \frac{du_0(x)}{dx} \right\} \tag{11.22}$$

and obtain for the function u_0 the fourth-order differential equation.

$$x^2(x-1)^2 \frac{d^4 u_0(x)}{dx^4} - x(x-1)(4-8x) \frac{d^3 u_0(x)}{dx^3} + \\ + [2 - 2x(\alpha+3)(\alpha-2)(-1+x)] \frac{d^2 u_0(x)}{dx^2} - \\ -\alpha(2+2\alpha)(2x-1) \frac{du_0(x)}{dx} - \left(\alpha+\alpha^2\right)(1-\alpha)(2+\alpha) u_0(x) = 0. \tag{11.23}$$

This equation is a differential equation with a regular singular point, so that linearly independent partial solutions can be represented in the form of convergent generalized power series. Using the above approach for constructing such series, we obtain four partial solutions $u_0^{(1)}, u_0^{(2)}, u_0^{(3)}, u_0^{(4)}$ in the following form:

$$u_0^{(1)}(x) = \sum_{m=0}^{\infty} \left[A_m^{(1)} x^{(m+1)} \right]; \\ u_0^{(2)}(x) = \sum_{m=0}^{\infty} \left[A_m^{(2)} x^m \right]; \\ u_0^{(3)}(x) = \sum_{m=0}^{\infty} \left\{ \left[A_m^{(3)} + B_m^{(3)} \ln(x) \right] x^{(m+1)} \right\} \\ u_0^{(4)}(x) = \sum_{m=0}^{\infty} \left\{ \left[A_m^{(4)} + B_m^{(4)} \ln(x) \right] x^m \right\}, \tag{11.24}$$

where the coefficients $A_m^{(1)}, A_m^{(2)}, A_m^{(3)}, A_m^{(4)}, B_m^{(3)}, B_m^{(4)}$, are determined from the recurrence relations available on https://www.icmm.ru/compcoeff/.

Substituting (11.24) into expression (11.22), we obtain partial solutions $v_0^{(1)}, v_0^{(2)}, v_0^{(3)}, v_0^{(4)}$ for the function v_0:

$$v_0^{(1)}(x) = \frac{\sqrt{x(1-x)}}{[(\alpha-1)S-2](\alpha+\alpha^2)} \sum_{m=0}^{\infty} \left[P_m^{(1)} x^m \right], \\ v_0^{(2)}(x) = \frac{\sqrt{x(1-x)}}{[(\alpha-1)S-2](\alpha+\alpha^2)} \sum_{m=0}^{\infty} \left[P_m^{(2)} x^m \right], \\ v_0^{(3)}(x) = \frac{\sqrt{x(1-x)}}{[(\alpha-1)S-2](\alpha+\alpha^2)} \left\{ \frac{(1+S)}{x} + \sum_{m=0}^{\infty} \left[\left(P_m^{(3)} + D_m^{(3)} \ln(x) \right) x^m \right] \right\}, \\ v_0^{(4)}(x) = \frac{\sqrt{x(1-x)}}{[(\alpha-1)S-2](\alpha+\alpha^2)} \left\{ \sum_{m=0}^{\infty} \left[\left(P_m^{(4)} + D_m^{(4)} \cdot \ln(x) \right) x^{(m-1)} \right] \right\}, \tag{11.25}$$

where the coefficients $P_m^{(1)}$, $P_m^{(2)}$, $P_m^{(3)}$, $P_m^{(4)}$, $D_m^{(3)}$, $D_m^{(4)}$ are determined by the expressions posted on https://www.icmm.ru/compcoeff/.

The general solution for u_0 and v_0 are as follows:

$$\begin{aligned} u_0(x) &= C_1 \cdot u_0^{(1)}(x) + C_2 \cdot u_0^{(2)}(x) + C_3 \cdot u_0^{(3)}(x) + C_4 \cdot u_0^{(4)}(x), \\ v_0(x) &= C_1 \cdot v_0^{(1)}(x) + C_2 \cdot v_0^{(2)}(x) + C_3 \cdot v_0^{(3)}(x) + C_4 \cdot v_0^{(4)}(x), \end{aligned} \tag{11.26}$$

where C_1, C_2, C_3, C_4 are the constants determined from a preset combination of boundary conditions (11.3)–(11.5).

To construct partial solutions to the system of equations (11.10), we perform a series of transformations [18], and obtain, as a result, a system of two differential equations with respect to w_k, v_l:

$$\begin{aligned} f_4(x)\frac{d^4 w_k(x)}{dx^4} + f_3(x)\frac{d^3 w_k(x)}{dx^3} + f_2(x)\frac{d^2 w_k(x)}{dx^2} + \\ + f_1(x)\frac{d w_k(x)}{dx} + f_0(x)\, w_k(x) = 0, \end{aligned} \tag{11.27a}$$

$$\begin{aligned} \psi_2(x)\frac{d^2 v_k(x)}{dx^2} + \psi_0(x)\, v_k(x) = \phi_3(x)\frac{d^3 w_k(x)}{dx^3} + \\ + \phi_2(x)\frac{d^2 w_k(x)}{dx^2} + \phi_1(x)\frac{d w_k(x)}{dx} + \phi_0(x)\, w_k(x), \end{aligned} \tag{11.27b}$$

where $f_0, f_1, f_2, f_3, f_4, \psi_0, \psi_2, \phi_0, \phi_1, \phi_2, \phi_3$ are written as:

$$\begin{aligned} f_0(x) &= \frac{1}{2} x\alpha(\alpha+1)(x-1)\left[2x(\alpha+3)(\alpha-2)(x-1) + k^2 - 1\right] + \\ &+ \frac{1}{16}(k-1)^2(k+1)^2, \\ f_1(x) &= x(1-x)(2x-1)\left[4x\left(\alpha^2+\alpha-3\right)(x-1) + \frac{1}{2}k^2 - \frac{1}{2}\right], \\ f_2(x) &= -\frac{1}{2}x^2(x-1)^2\left[4x\left(\alpha^2+\alpha-18\right)(x-1) + \frac{1}{2}k^2 - 13\right], \\ f_3(x) &= 6x^3(x-1)^3(2x-1), \\ f_4(x) &= x^4(x-1)^4, \\ \psi_0(x) &= x\alpha(\alpha+1)(x-1) + \frac{1}{4}(1-k^2), \\ \psi_2(x) &= x^2(x-1)^2, \\ \phi_0(x) &= -\frac{1}{2}x\left[4x\alpha(\alpha+1)(x-1) - k^2 + 1\right]\frac{(2x-1)}{k}, \\ \phi_1(x) &= \frac{1}{2}x\left[4x\left(\alpha^2+\alpha-4\right)(x-1) - k^2 + 1\right]\frac{(x-1)}{k}, \\ \phi_2(x) &= -5x^2(2x-1)\frac{(x-1)^2}{k}, \\ \phi_3(x) &= 2x^3\frac{(x-1)^3}{k}. \end{aligned} \tag{11.28}$$

Furthermore, the performed transformations results in the relation that establishes the dependence of the function u_k on the functions w_k, v_k and their derivatives:

$$u_k(x) = \frac{\sqrt{x(1-x)}}{2x(x-1)k(S\alpha+2S+2)}\left\{4x^2(x-1)\frac{d^2w_k(x)}{dx^2}+4x(2x-1)\frac{dw_k(x)}{dx}-\right.$$
$$-\left[4\alpha x(x-1)(1+\alpha)+k^2(S+1)+1\right]\cdot w_k(x)-$$
$$\left.-\left[2kSx(x-1)\frac{dv_k(x)}{dx}+k(2x-1)(S+2)v_k(x)\right]\right\}. \tag{11.29}$$

Equation (11.27a) is independent of equation (11.27b) and is a fourth-order linear differential equation with respect to the function w_k. Equation (11.27b) can be considered as a second-order differential equation with respect to v_k with the right-hand side depending on w_k. This specific feature of differential equations (11.27) and the resulting relation (11.29) allow us to define a sequence of partial solutions for the functions ω_k, v_k, u_k. The concept of this sequence is as follows. At the first stage, from the solution of equation (11.27a) we get four partial solutions $w_k^{(1)}, w_k^{(2)}, w_k^{(3)}, w_k^{(4)}$ written in the following form

$$\omega_k^{(1)}(x) = \sum_{m=0}^{\infty}\left[A_m^{(1)}x^{\left(m+\frac{k+1}{2}\right)}\right], \qquad \omega_k^{(2)}(x) = \sum_{m=0}^{\infty}\left[A_m^{(2)}x^{\left(m+\frac{k-1}{2}\right)}\right],$$
$$\omega_k^{(3)}(x) = \sum_{m=0}^{\infty}\left[\left(A_m^{(3)}+B_m^{(3)}\ln(x)\right)x^{\left(m-\frac{k-1}{2}\right)}\right], \tag{11.30}$$
$$\omega_k^{(4)}(x) = \sum_{m=0}^{\infty}\left[\left(A_m^{(4)}+B_m^{(4)}\ln(x)\right)x^{\left(m-\frac{k+1}{2}\right)}\right],$$

where the coefficients $A_m^{(1)}, A_m^{(2)}, A_m^{(3)}, A_m^{(4)}, B_m^{(3)}, B_m^{(4)}$ are determined by the relations posted on https://www.icmm.ru/compcoeff/.

Sequentially substituting the obtained partial solutions into the right-hand side of equation (11.27b) and solving it as the inhomogeneous equation, we find four partial solutions $v_k^{(1)}, v_k^{(2)}, v_k^{(3)}, v_k^{(4)}$ written as

$$v_k^{(1)}(x) = \sum_{m=0}^{\infty}\left[P_m^{(1)}x^{\left(m+\frac{k+1}{2}\right)}\right],$$
$$v_k^{(2)}(x) = \sum_{m=0}^{\infty}\left[P_m^{(2)}x^{\left(m+\frac{k-1}{2}\right)}\right],$$
$$v_k^{(3)}(x) = \sum_{m=0}^{\infty}\left\{\left[P_m^{(3)}+D_m^{(3)}\ln(x)\right]x^{\left(m-\frac{k-1}{2}\right)}\right\}, \tag{11.31}$$
$$v_k^{(4)}(x) = \sum_{m=0}^{\infty}\left\{\left[P_m^{(4)}+D_m^{(4)}\ln(x)\right]x^{\left(m-\frac{k+1}{2}\right)}\right\},$$

where the coefficients $P_m^{(1)}, P_m^{(2)}, P_m^{(3)}, P_m^{(4)}, D_m^{(3)}, D_m^{(4)}$ are determined by the relations available on https://www.icmm.ru/compcoeff/.

Then, solving equation (11.27b) as a homogeneous one, we find two more partial solutions $v_k^{(5)}$, $v_k^{(6)}$. The form of this differential equation indicates that the point $x = 0$ is a regular singular point. The construction of partial solutions in the form of generalized power series is accomplished in the framework of the above approach. These partial solutions are written as

$$
\begin{aligned}
v_k^{(5)}(x) &= \sum_{m=0}^{\infty}\left[P_m^{(5)} x^{\left(m+\frac{k+1}{2}\right)}\right],\\
v_k^{(6)}(x) &= \sum_{m=0}^{\infty}\left[\left(P_m^{(6)} + D_m^{(6)} \ln(x)\right) x^{\left(m+\frac{k-1}{2}\right)}\right],
\end{aligned}
\tag{11.32}
$$

where $P_m^{(5)}, P_m^{(6)}, D_m^{(6)}$ are defined on https://www.icmm.ru/compcoeff/.

Then, using partial solutions $w_k^{(1)}, w_k^{(2)}, w_k^{(3)}, w_k^{(4)}, v_k^{(1)}, v_k^{(2)}, v_k^{(3)}, v_k^{(4)}, v_k^{(5)}, v_k^{(6)}$, and the obtained relation (30), we determine six partial solutions $u_k^{(1)}$, $u_k^{(2)}$, $u_k^{(3)}$, $u_k^{(4)}, u_k^{(5)}, u_k^{(6)}$, represented as

$$
\begin{aligned}
u_k^{(1)} &= \frac{2\sqrt{x(1-x)}}{kx(x-1)(S(\alpha+2)+2)} \sum_{m=0}^{\infty}\left[E_m^{(1)} x^{\left(m+\frac{k+1}{2}\right)}\right],\\
u_k^{(2)} &= \frac{2\sqrt{x(1-x)}}{kx(x-1)(S(\alpha+2)+2)} \sum_{m=0}^{\infty}\left[E_m^{(2)} x^{\left(m+\frac{k-1}{2}\right)}\right],\\
u_k^{(3)} &= \frac{2\sqrt{x(1-x)}}{kx(x-1)(S(\alpha+2)+2)} \sum_{m=0}^{\infty}\left[\left(E_m^{(3)} + G_m^{(3)} \ln(x)\right) x^{\left(m-\frac{k-1}{2}\right)}\right],\\
u_k^{(4)} &= \frac{2\sqrt{x(1-x)}}{kx(x-1)(S(\alpha+2)+2)} \sum_{m=0}^{\infty}\left[\left(E_m^{(4)} + G_m^{(4)} \ln(x)\right) x^{\left(m-\frac{k+1}{2}\right)}\right],\\
u_k^{(5)} &= \frac{\sqrt{x(1-x)}}{kx(x-1)(S(\alpha+2)+2)} \sum_{m=0}^{\infty}\left[E_m^{(5)} x^{\left(m+\frac{k+1}{2}\right)}\right],\\
u_k^{(6)} &= \frac{\sqrt{x(1-x)}}{kx(x-1)(S(\alpha+2)+2)} \sum_{m=0}^{\infty}\left[\left(E_m^{(6)} + G_m^{(6)} \ln(x)\right) x^{\left(m-\frac{k-1}{2}\right)}\right],
\end{aligned}
\tag{11.33}
$$

where the coefficients $E_m^{(1)}$, $E_m^{(2)}$, $E_m^{(3)}$, $E_m^{(4)}$, $E_m^{(5)}$, $E_m^{(6)}$, $G_m^{(3)}$, $G_m^{(4)}$, $G_m^{(6)}$ for any value of $m \geq 0$ are determined on https://www.icmm.ru/compcoeff/index2.html.

General solutions for u_k, v_k, w_k take the following form

$$
\begin{aligned}
u_k(x) &= C_1 \cdot u_k^{(1)}(x) + C_2 \cdot u_k^{(2)}(x) + C_3 \cdot u_k^{(3)}(x) +\\
&\quad + C_4 \cdot u_k^{(4)}(x) + C_5 \cdot u_k^{(5)}(x) + C_6 \cdot u_k^{(6)}(x),\\
v_k(x) &= C_1 \cdot v_k^{(1)}(x) + C_2 \cdot v_k^{(2)}(x) + C_3 \cdot v_k^{(3)}(x) +\\
&\quad + C_4 \cdot v_k^{(4)}(x) + C_5 \cdot v_k^{(5)}(x) + C_6 \cdot v_k^{(6)}(x),\\
w_k(x) &= C_1 \cdot w_k^{(1)}(x) + C_2 \cdot w_k^{(2)}(x) + C_3 \cdot w_k^{(3)}(x) + C_4 \cdot w_k^{(4)}(x),
\end{aligned}
\tag{11.34}
$$

where $C_1, C_2, C_3, C_4, C_5, C_6$ are the constants determined from a preset combination of boundary conditions (11.3)–(11.5).

For the examined variant of a conical body, the constructed general solutions for $k = 0$, $k \geq 1$ and the preset combination of boundary conditions are used to derive a homogeneous system of linear algebraic equations for the constants C_i. The coefficients of this system of equations depend on the vertex angles of conical bodies, elastic characteristics of materials, and the characteristic exponent α. From the condition of existence of a nonzero solution to the system of linear algebraic equations we find the exponents α, determining the nature of stress singularity at the vertices of conical bodies.

Let us consider numerical results for a solid cone ($0 \leq r \leq \infty$, $0 \leq \varphi \leq 2\pi$, $0 \leq \theta \leq \theta_0$). Here we use partial solutions, for which the regularity condition is identically fulfilled at $x = 0$ (or $\theta = 0$): $w_0^{(1)}$ is used for axisymmetric rotation; $u_0^{(1)}$, $v_0^{(1)}$, $u_0^{(2)}$, $v_0^{(2)}$ — for axisymmetric deformation without rotation; $u_k^{(1)}$, $v_k^{(1)}$, $w_k^{(1)}$, $u_k^{(2)}$, $v_k^{(2)}$, $w_k^{(2)}$, $u_k^{(5)}$, $v_k^{(5)}$ — for nonaxisymmetric deformation. All results in this work were obtained for Poisson's ratio $\nu = 0.3$. Figure 11.2 presents the values $\mathrm{Re}\,\alpha_n < 1$, determining singular solutions for a solid cone with stress and displacement boundary conditions. These values are identical to the results of [19, 15]. It should be noted that for a solid cone with stress boundary conditions, the singular solutions appear at the zero, first and second harmonics of the Fourier series, whereas for a cone with displacement boundary conditions — at the zero and first harmonics of the Fourier series. Figure 11.3 shows new results disclosing the nature of stress singularity at the vertex of a solid cone with boundary conditions of ideal sliding prescribed on its lateral surface. Here, singular solutions are possible at the zero, first and second harmonics of the Fourier series and at the angle θ_0 smaller than π.

The proposed method has proved to be effective in determining the region of singular solutions for a hollow cone with two conical boundary surfaces $\theta = \theta_0$ and

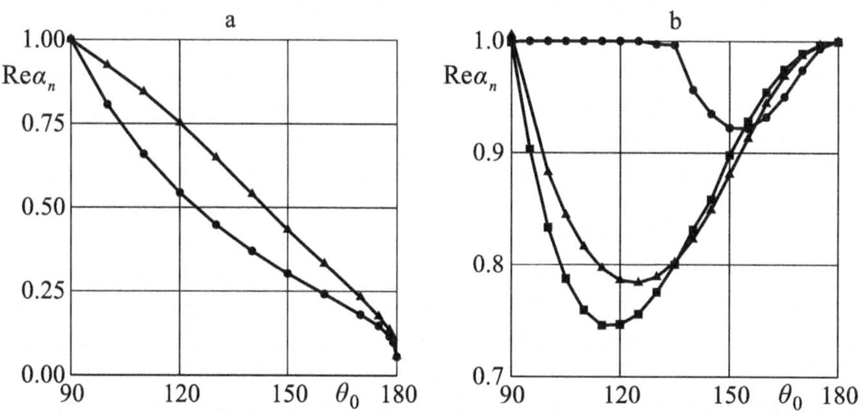

Fig. 11.2 Dependence of $\mathrm{Re}\,\alpha_n$ on the vertex angle of the solid cone with boundary conditions on the lateral surface for displacements (a) and stresses (b) (▲ — $k = 0$, • — $k = 1$, ■ — $k = 2$)

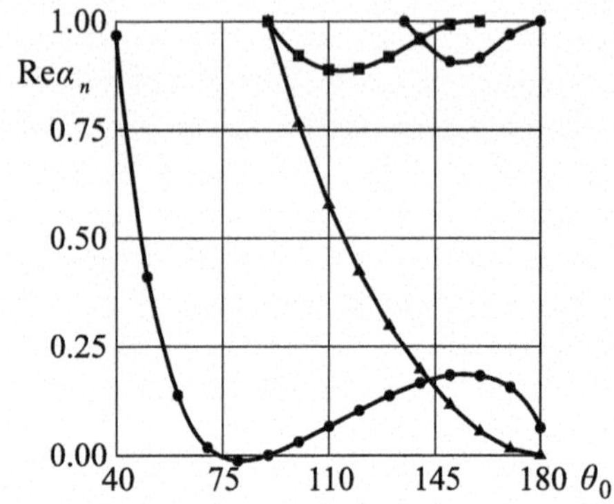

Fig. 11.3 Dependence of Reα_n on the vertex angle of the cone with boundary conditions on the lateral surface corresponding to an ideal sliding (▲ — $k = 0$, • — $k = 1$, ■ — $k = 2$)

$\theta = \theta_1$ (hollow cone) under different variants of boundary conditions. In this case, it is necessary to use all partial solutions (11.18), (11.19), (11.24), (11.25), (11.30), (11.31), (11.32) to ensure the fulfillment of boundary conditions on the two conical surfaces. As an example, Fig. 11.4 shows the dependence of eigenvalues Re$\alpha_n < 1$ on the angle of the outer conical surface θ_0 for different internal cone angles θ_1. Zero stress boundary conditions are prescribed on the conical surfaces. Here, the solid line corresponds to the actual eigenvalues and the dashed line — to the complex ones.

In the case of a hollow cone, different combinations of boundary conditions on the inner and outer conical surfaces can be used. Here we consider two variants. In the first variant, zero stresses are specified on the inner surface and zero displacements — on the outer surface. In the second variant, zero displacements are preset on the inner surface and zero stresses — on the outer surface. The variation of the stress singularity exponent Re$\alpha_n < 1$ as a function of the outer conical surface angle θ_0 at different values of the inner surface angle is shown in Fig. 11.5 for the first variant of boundary conditions. The eigenvalues, leading to the occurrence of stress singularity, appear at the values of θ_0 higher than 80°. For the second variant of the boundary conditions the dependence of eigenvalues Re$\alpha_n < 1$ is shown in Fig. 11.6.

11.3 Numerical-analytical Method of Stress Singularity Analysis at the Vertices of Circular and Non-circular Conical Bodies

We consider a semi-infinite circular or non-circular cone, whose vertex coincides with the center of spherical coordinates r, θ, φ, and the base is perpendicular to the axis $\theta = 0$. To analyze the character of the stress singularity, we need to construct eigensolutions, which will be similar in form to the asymptotic representation of solution [12],

$$u_k(r,\theta,\varphi) = r^{\lambda} \xi_k(\theta,\varphi), \qquad k = 1,2,3 \tag{11.35}$$

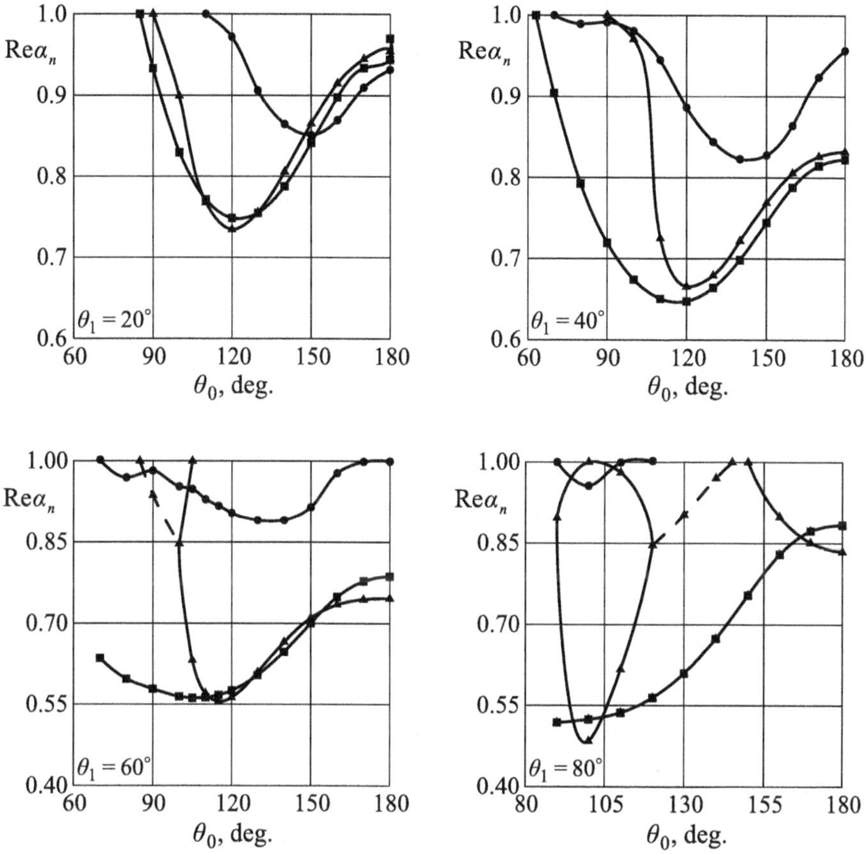

Fig. 11.4 Dependence of Reα_n on the angle θ_0 at fixed angles θ_1 of the hollow cone and zero stresses on the lateral surfaces for different values of k (▲ — $k = 0$, • — $k = 1$, ■ — $k = 2$)

Fig. 11.5 Dependence of Reα_n on the angle θ_0 for different values of θ_1 at zero stresses on the inner surface and zero displacements on the outer lateral surface (▲ — $k = 0$, • — $k = 1$, ■ — $k = 2$)

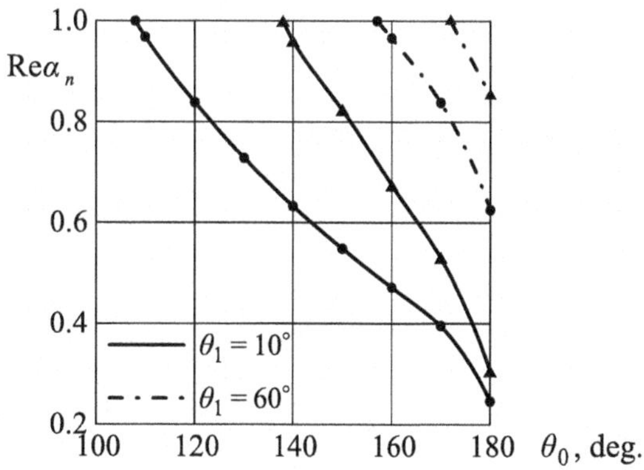

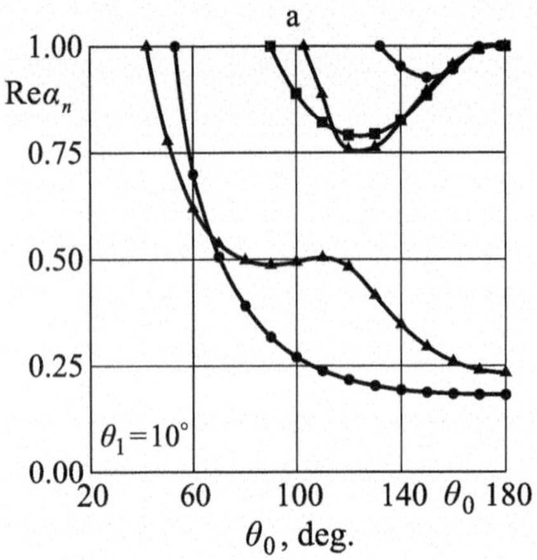

Fig. 11.6 Dependence of Reα_n on the angle θ_0 for different values of θ_1 of the hollow cone with zero displacements on the inner surface and zero stresses on the outer lateral surface (▲ — $k = 0$, • — $k = 1$, ■ — $k = 2$)

and satisfy in the examined domain the equilibrium equations

$$\frac{1}{1-2\nu}\operatorname{grad}\operatorname{div}\mathbf{u} + \nabla^2\mathbf{u} = 0 \tag{11.36}$$

and uniform boundary conditions prescribed on the lateral surface of the cone, namely, zero displacements

$$\mathbf{u} = 0 \tag{11.37}$$

or zero stresses

$$\frac{\nu}{1-2\nu}\mathbf{n}\operatorname{div}\mathbf{u} + \mathbf{n}\cdot\nabla\mathbf{u} + \frac{1}{2}\mathbf{n}\times\operatorname{rot}\mathbf{u} = 0. \tag{11.38}$$

Here $\mathbf{u}$ is the displacement vector, $\mathbf{n}$ is the unit vector of the external normal, ν is Poisson's ratio.

A variant of boundary conditions corresponding to the ideal sliding conditions on the lateral surface may be also of interest. These conditions are as follows:

$$u_\theta = 0, \quad \tau_{r\theta} = 0, \quad \tau_{\varphi\theta} = 0. \tag{11.39}$$

On the lateral surface of the cone, mixed boundary conditions can be prescribed, that is, conditions (11.38) are set at $0 \le \varphi \le \varphi_1$ and conditions (11.39) are specified at $\varphi_1 \le \varphi \le 2\pi$.

In addition to a solid cone, the study can be conducted for a hollow cone with two lateral surfaces. For a circular cone, the domain occupied by this body is defined as follows: $0 \le r \le \infty$, $0 \le \varphi \le 2\pi$, $\theta_1 \le \theta \le \theta_2$ ($\theta_1 = 0$ corresponds to a solid cone). In this case, one of the variants of boundary conditions (11.37)–(11.39) can be imposed on the lateral surfaces.

To construct eigenvalues, we substitute expressions (11.35) into Eqs. (11.36), to obtain a system of partial differential equations with respect to functions $\xi_k(\theta,\varphi)$ and parameter λ

$$\begin{aligned}
L_1(\lambda,\xi_k) &= 2(1-\nu)(k_1-2)\xi_1 + k_2(\xi_2 \operatorname{ctg}\theta + \xi_{2\theta} + \frac{1}{\sin\theta}\xi_{3\varphi}) + \\
&+ (1-2\nu)(\operatorname{ctg}\theta\, \xi_{1\theta} + \xi_{1\theta\theta} + \frac{1}{\sin^2\theta}) = 0, \\
L_2(\lambda,\xi_k) &= \left[(1-2\nu)k_1 - \frac{2(1-\nu)}{\sin^2\theta}\right]\xi_2 + k_3\xi_{1\theta} - (3-4\nu)\frac{\operatorname{ctg}\theta}{\sin\theta}\xi_{3\varphi} + \\
&+ \frac{(1-2\nu)}{\sin^2\theta}\xi_{2\varphi\varphi} + \frac{1}{\sin\theta}\xi_{3\theta\varphi} + 2(1-\nu)(\operatorname{ctg}\theta\, \xi_{2\theta} + \xi_{2\theta\theta}) = 0, \\
L_3(\lambda,\xi_k) &= (1-2\nu)(k_1 - \frac{1}{\sin^2\theta})\xi_3 + k_3\frac{1}{\sin\theta}\xi_{1\varphi} - (3-4\nu)\frac{\operatorname{ctg}\theta}{\sin\theta}\xi_{2\varphi} + \\
&+ \frac{2(1-\nu)}{\sin^2\theta}\xi_{3\varphi\varphi} + \frac{1}{\sin\theta}\xi_{2\theta\varphi} + (1-2\nu)(\operatorname{ctg}\theta\, \xi_{3\theta} + \xi_{3\theta\theta}) = 0.
\end{aligned} \tag{11.40}$$

Here, $k_1 = \lambda^2 + \lambda$, $k_2 = \lambda - 3 + 4\nu$, $k_3 = \lambda + 4 - 4\nu$, $\xi_{k\theta} = \partial\xi_k/\partial\theta$, $\xi_{k\varphi} = \partial\xi_k/\partial\varphi$, $\xi_{k\theta\theta} = \partial^2\xi_k/\partial\theta^2$, etc.

Based on the asymptotic expression (11.35), boundary conditions (11.37), (11.38) are transformed to the following form:

$$M_1(\lambda,\xi_k) \equiv \xi_1 = 0, \qquad M_2(\lambda,\xi_k) \equiv \xi_2 = 0, \qquad M_3(\lambda,\xi_k) \equiv \xi_3 = 0. \tag{11.41}$$

$$\begin{aligned}
&M_1(\lambda,\xi_k) \equiv \xi_{1\theta} + \xi_2(\lambda - 1) = 0, \\
&M_2(\lambda,\xi_k) \equiv (1-\nu)\xi_{2\theta} + (1+\nu\lambda)\xi_1 + \nu \operatorname{ctg}\theta\, \xi_2 + \frac{\nu}{\sin\theta}\xi_{3\varphi} = 0, \\
&M_3(\lambda,\xi_k) \equiv \xi_{3\theta} + \frac{1}{\sin\theta}\xi_{2\varphi} - \operatorname{ctg}\theta\, \xi_3 = 0.
\end{aligned} \tag{11.42}$$

Here L_k and M_k are the differential operators.

In addition to a homogeneous cone, as an object of study we can also consider a composite cone, e.g., a circular cone occupying the domain $V = V_1 + V_2$, where the subdomain V_1 (subdomain V_2) is a segment of the cone made of the material with shear modulus G_1 (G_2) and Poisson's ratio ν_1 (ν_2). The subdomain geometry is defined by the relations $0 \le r \le \infty$, $0 \le \varphi \le 2\pi$, $\theta_1 \le \theta \le \theta_2$ ($\theta_2 \le \theta \le \theta_3$). In particular cases, θ_1 and θ_3 can be respectively equal to 0 and π.

For a composite cone, eigensolutions (11.35) in each of the subdomains V_1 and V_2, must satisfy the equations of equilibrium (11.36), which will differ only in the values of the elastic material constants. In this case, one of the three variants of boundary conditions (11.37), (11.38) and (11.39) can be used for the surfaces $\theta = \theta_1$ ($\theta \ne 0$) and $\theta = \theta_3$ ($\theta \ne \pi$), while the condition on a contact surface is that of ideal bonding of layers

$$
\begin{aligned}
&u_r^{(1)} = u_r^{(2)}, \quad u_\varphi^{(1)} = u_\varphi^{(2)}, \quad u_\theta^{(1)} = u_\theta^{(2)}, \\
&\sigma_\theta^{(1)} = \sigma_\theta^{(2)}, \quad \tau_{r\theta}^{(1)} = \tau_{r\theta}^{(2)}, \quad \tau_{\varphi\theta}^{(1)} = \tau_{\varphi\theta}^{(2)},
\end{aligned}
\tag{11.43}
$$

or ideal sliding

$$
u_\theta^{(1)} = u_\theta^{(2)}, \quad \sigma_\theta^{(1)} = \sigma_\theta^{(2)}, \quad \tau_{r\theta}^{(1)} = \tau_{r\theta}^{(2)} = \tau_{\varphi\theta}^{(1)} = \tau_{\varphi\theta}^{(2)} = 0. \tag{11.44}
$$

We propose the following scheme of problem solution. Let us represent Eqs. (11.40) in a weak form [34], for which purpose we multiply them by the appropriate variations $\delta\xi_k(\theta,\varphi)$ and integrate over the region S cut by the cone from the sphere. As a result we get

$$
\int_S \left[\sum_{k=1}^{3} L_k(\lambda,\xi_1,\xi_2,\xi_3)\,\delta\xi_k(\theta,\varphi) \right] dS = 0. \tag{11.45}
$$

Equations (11.45) are solved using the finite element method (FEM). The finite-element implementation of these equations is a rather complicated procedure, since it requires the use of two-dimensional elements to ensure the continuity of the functions ξ_k, as well as the continuity of their first derivatives. Without going into details, we simply note that in FEM, there are no effective procedures for constructing such elements. In this regard, after performing identity transformations with the aim to reduce the order of derivatives of functions in the solutions of Eq. (11.45) and considering boundary conditions (11.42), we obtain the following equation

$$
\begin{aligned}
&\iint_S \{ [2(1-\nu)(k_1-2)\sin\theta\xi_1 + k_1(\cos\theta\xi_2 + \sin\theta\xi_{2\theta} + \xi_{3\varphi})]\,\delta\xi_1 - \\
&\quad -(1-2\nu)(\sin\theta\,\xi_{1\theta}\,\delta\xi_{1\theta} + \frac{1}{\sin\theta}\,\xi_{1\varphi}\,\delta\xi_{1\varphi}) - \frac{2(1-\nu)}{\sin\theta}\,\xi_{3\varphi}\,\delta\xi_{3\varphi} + \\
&+\left[(1-2\nu)k_1\sin\theta\,\xi_2 - \frac{2(1-\nu)}{\sin\theta}\xi_2 + k_3\sin\theta\,\xi_{1\theta} - (3-4\nu)\mathrm{ctg}\theta\,\xi_{3\varphi}\right]\delta\xi_2 - \\
&\qquad -2(1-\nu)\sin\theta\,\xi_{2\theta}\,\delta\xi_{2\theta} - 2\nu\,\xi_{3\varphi}\,\delta\xi_{2\theta} - \\
&\qquad -(1-2\nu)\left(\frac{1}{\sin\theta}\,\xi_{2\varphi}\,\delta\xi_{2\varphi} + \xi_{3\theta}\,\delta\xi_{2\varphi}\right) + \\
&\quad +\left[(1-2\nu)k_1\sin\theta\,\xi_3 - \frac{1-2\nu}{\sin\theta}\xi_3 + k_3\,\xi_{1\varphi} + (3-4\nu)\,\mathrm{ctg}\,\theta\,\xi_{2\varphi}\right]\delta\xi_3 - \\
&\qquad -2\nu\,\xi_{2\theta}\,\delta\xi_{3\varphi} - (1-2\nu)\left(\sin\theta\,\xi_{3\theta}\,\delta\xi_{3\theta} + \xi_{2\varphi}\,\delta\xi_{3\theta}\right)\}\,d\theta d\varphi + \\
&+\int_l \{(1-2\nu)(1-\lambda)\sin\theta\,\xi_2\,\delta\xi_1 - 2[(1+\nu\lambda)\sin\theta\,\xi_1 + \nu\cos\theta\,\xi_2]\delta\xi_2 + \\
&\qquad +(1-2\nu)\cos\theta\,\xi_3\,\delta\xi_3 - (1-2\nu)\}\,dl = 0,
\end{aligned}
\tag{11.46}
$$

where l is the boundary of the surface S with prescribed stresses.

Reduction of the order of derivatives allows us to use such finite elements that ensure only the continuity of the functions ξ_k. In our simulation we used finite elements in the form of triangles and the Lagrangian linear polynomial approximation of the functions ξ_k.

In the numerical analysis of circular conical bodies with unmixed boundary conditions imposed on the lateral conical surfaces, the functions $\xi_k(\theta,\varphi)$ can be represented as a Fourier series in the circumferential coordinate φ

$$
\begin{aligned}
\xi_1 &= \sum_{n=0}^{\infty} \beta_1^{(n)}(\theta) \cos n\varphi, \\
\xi_2 &= \sum_{n=0}^{\infty} \beta_2^{(n)}(\theta) \cos n\varphi, \\
\xi_3 &= \sum_{n=0}^{\infty} \beta_3^{(n)}(\theta) \sin n\varphi.
\end{aligned}
\tag{11.47}
$$

In view of expansion (11.47), Eqs. (11.45) and boundary conditions (11.41), (11.42) for each of the harmonics of the Fourier series can be written in the following form (dashed line indicates the derivative with respect to θ, the upper index (n) for $\beta_1, \beta_2, \beta_3$ is omitted):

$$
\begin{aligned}
&\int_{\theta_1}^{\theta_2} \{[2(1-\nu)(k_1-2)\sin^2\theta\,\beta_1 + k_2(\cos\theta\sin\theta\,\beta_2 + \sin^2\theta\,\beta'_2 + n\sin\theta\,\beta_3) + \\
&\quad + (1-2\nu)(\cos\theta\sin\theta\,\beta_1' + \sin^2\theta\,\beta_1'' - n^2\beta_1)]\,\delta\xi_1 + [(1-2\nu)k_1\sin^2\theta\,\beta_2 + \\
&\qquad + k_3\sin^2\theta\,\beta_1' - 2(1-\nu)\beta_2 - (1-2\nu)n^2\beta_2 + n\sin\theta\,\beta_3' - \\
&\qquad - (3-4\nu)n\cos\theta\,\beta_3 + 2(1-\nu)(\cos\theta\sin\theta\,\beta_2' + \sin^2\theta\,\beta_2'')]\,\delta\xi_2 + \\
&\quad + [(1-2\nu)(k_1\sin^2\theta\,\beta_3 - \beta_3 + \cos\theta\sin\theta\,\beta_3' + \sin^2\theta\,\beta_3'') - k_3 n\sin\theta\,\beta_1 - \\
&\qquad - (3-4\nu)n\cos\theta\,\beta_2 - n\sin\theta\,\beta_2' - 2(1-\nu)n^2\beta_3]\,\delta\xi_3\}\,d\theta = 0.
\end{aligned}
\tag{11.48}
$$

$$
\begin{aligned}
M_1(\lambda,\beta_k) &\equiv \beta_1' + \beta_2(\lambda-1) = 0, \\
M_2(\lambda,\beta_k) &\equiv (1-\nu)\beta_2' + (1+\nu\lambda)\beta_1 + \nu\,\mathrm{ctg}\,\theta\,\beta_2 + \frac{\nu n}{\sin\theta}\beta_3 = 0, \\
M_3(\lambda,\beta_k) &\equiv \beta_3' + \frac{n}{\sin\theta}\beta_2 - \mathrm{ctg}\,\theta\,\beta_3 = 0.
\end{aligned}
\tag{11.49}
$$

In the numerical implementation, the use of expansion (11.47) allows us to change from a two-dimensional problem to a set of separate one-dimensional problems for each of the harmonics of the Fourier series. In the finite element implementation of one-dimensional problems, in contrast to that of two-dimensional problems, the presence of well-tried finite elements ensures continuity of approximating functions and their first derivatives between two adjacent elements. It means that in this case we can directly carry out the finite element implementation of Eqs. (11.48). As finite elements, we used one-dimensional two-node elements, in which the functions $\beta_i^{(n)}(\theta)$ are approximated with a cubic polynomial defined by the values of the function and its derivatives $d\beta_i^{(n)}/d\theta$ at the ends of the segment (one-dimensional element).

As in a two-dimensional version, we can employ the procedure of reducing the order of derivatives in Eq. (11.48). Then, in the case of applying the finite element method of solution to these equations, it becomes possible to use one-dimensional elements ensuring continuity of only approximated functions, in particular, one-dimensional two-node elements with linear approximation of functions $\beta_i^{(n)}(\theta)$.

The application of the Bubnov procedure together with the finite element method reduces the formulated problem to a search for eigenvalues (EV) and eigenvectors of an algebraic asymmetric band matrix. To find complex eigenvalues, the obtained algebraic problem is solved using the algorithm based on the application of the Muller method and the argument principle [17], which allowed us to obtain acceptable numerical results.

The reliability and efficiency of the proposed method and the algorithm for its numerical implementation can be substantiated by the results of two numerical experiments. The first experiment is designed to realize the possibility of comparing the numerical and analytical results for a homogeneous continuous circular cone $(0 \le r \le \infty,\ 0 \le \varphi \le 2\pi,\ 0 \le \theta \le \theta_2)$ [15]. In a two-dimensional variant with the number of nodal variables equal to $\sim 10^3$, the difference between the numerical and analytical results is less than one percent. The second computational experiment is based on the analysis of the convergence of the numerical method depending on the degree of discretization of the computational domain. As an example, Fig. 11.7a shows a numerical solution (solid curve) depending on the number of nodal variables N and analytical results (dashed curve) at $\theta_2 = 2\pi/3$, $\nu = 0.3$. The results of such experiments demonstrate not only the convergence of the numerical procedure, but also make it possible to choose a variant of discretization of the computational domain, which can provide acceptable accuracy.

Let us consider the results of solving a number of new problems. Figure 11.7b presents the eigenvalues calculated for a solid circular cone ($\nu = 0.3$) at boundary conditions (11.39) corresponding to the ideal sliding conditions. Figure 11.7c displays the eigenvalues for one of the variants of a continuous circular cone ($\theta_2 = 2\pi/3$, $\nu = 0.3$) at mixed boundary conditions prescribed on the lateral surface: zero displacements at and zero stresses at $\varphi_1 \le \varphi \le 2\pi$. It should be noted that in this problem, the representation of the desired solution as a Fourier series in the angular coordinate φ is not allowed. Hereinafter, the solid curve corresponds to real eigenvalues, while the dashed curve corresponds to the complex eigenvalues.

Calculations were performed for a composite cone which allowed us to evaluate the effect of ratios of mechanical characteristic on the stress singularity exponents. In Fig. 11.8, for the composite cone under boundary conditions (11.43) and $\theta_1 = 0$, $\theta_2 = \pi/3$, $\theta_3 = 2\pi/3$, $\nu_1 = \nu_2 = 0.3$ the values of $\mathrm{Re}\lambda_k < 1$ are plotted against the ratio G_1/G_2.

The method under consideration allows us to obtain numerical results for different cone shapes, including a cone whose base is an ellipse. The geometry of the boundary of the surface (11.46), which is cut by a cone from a sphere, is defined by

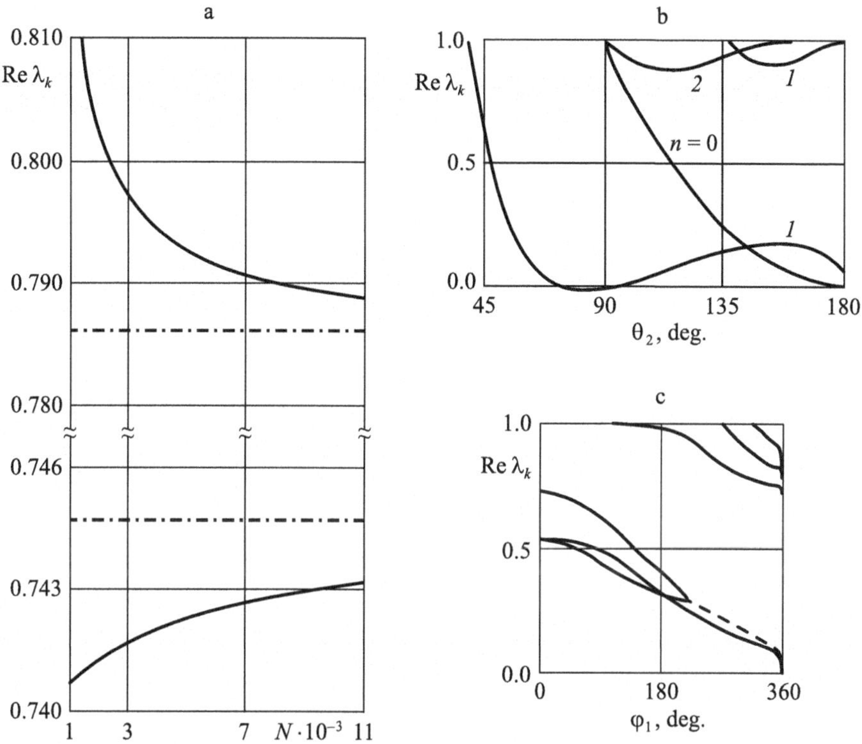

Fig. 11.7 Dependence of Reλ_k on the value of N (a). Dependences of Reλ_k on the angle θ_2 (b) and on the angle φ_1 (c)

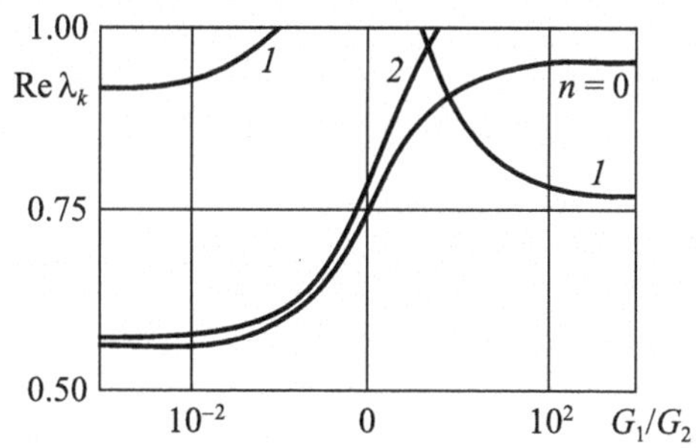

Fig. 11.8 Dependence of Reλ_k on the ratio G_1/G_2

the relation

$$\mathrm{tg}\theta = \mathrm{tg}\theta_2 \left(\frac{1}{\left(\cos^2\varphi + æ^{-2}\sin^2\varphi\right)^{-/2}} \right), \qquad æ = \frac{\mathrm{a}}{\mathrm{b}}. \tag{11.50}$$

Here a and b are the semi-axes of the ellipse, $2\theta_2$ is the vertex angle of the cone in the plane passing through the cone vertex and the semi-axis a. Figure 11.9 shows

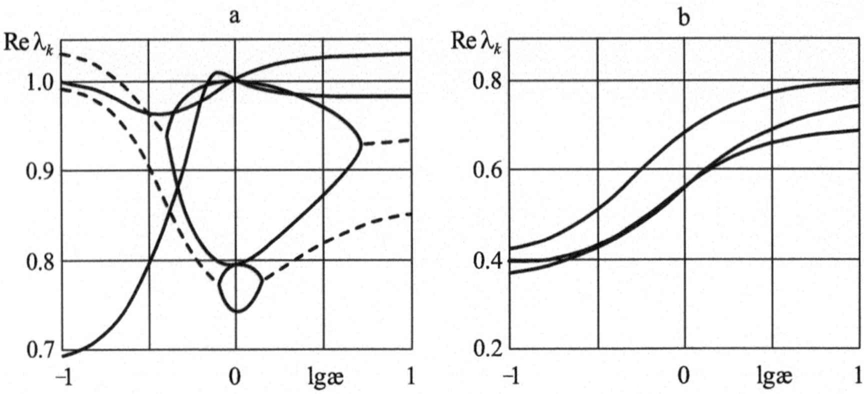

Fig. 11.9 Dependence of Reλ_k on the value of æ at zero stress (a) and at zero displacement (b)

the results of calculations of eigenvalues at zero stress (a) and at zero displacement (b) specified on the lateral surface of the cone.

The proposed method has proved to be effective in calculating all eigenvalues of interest. Furthermore, within the error of the numerical method, it allows one to calculate multiple eigenvalues. For example, in [15] analytical results on the multiplicity of the eigenvalue $\lambda = 1$ were presented. In particular, at $\theta_2 = \pi/2$ the multiplicity is found to be 6 and at $\theta_2 = \pi$ the multiplicity is 9. The method under consideration can be used to find all multiple eigenvalues within the accuracy of the third place with the number of finite elements being equal to about three thousand.

11.4 Finite Element Analysis of Stress Singularity in Three-dimensional Problems of Elasticity Theory

To determine the power law relationship of stresses in the vicinity of singular points, a numerical technique [29] is proposed. It is based on the statement that the stress distribution along the radial line, originating from a singular point, can be expressed as [4, 35]

$$\sigma = A_1 r^{\lambda-1} + O(r^{\lambda}), \tag{11.51}$$

where r is the distance from the singular point, A_1 is some constant, λ is the parameter, characterizing the degree of stress singularity, and $O(r^{\lambda})$ represents all terms of the order r^{λ} and higher. For small distances r, the singular term dominates and equation (11.51) can be approximated by

$$\sigma \simeq A_1 r^{\lambda-1},$$

or

$$\log\sigma = \log A_1 + (\lambda - 1)\log r, \tag{11.52}$$

where λ is the smallest eigenvalue [4]. The parameter λ is determined using the FEM procedure with finite element meshes refined towards the singular points (Fig. 11.10). To establish the relationship (11.52) via numerical experiments, it is necessary to find the discretization, such that in the vicinity of a singular point at a number of nodal points on the radial line originating from the singular point the following relations will be fulfilled with sufficient accuracy:

$$\lambda - 1 \approx \frac{\log\left(\frac{\sigma_1}{\sigma_2}\right)}{\log\left(\frac{r_1}{r_2}\right)} \approx \frac{\log\left(\frac{\sigma_2}{\sigma_3}\right)}{\log\left(\frac{r_2}{r_3}\right)} \approx \ldots \approx \frac{\log\left(\frac{\sigma_{n-1}}{\sigma_n}\right)}{\log\left(\frac{r_{n-1}}{r_n}\right)}, \tag{11.53}$$

where $r_1, r_2, \ldots, r_n$ are the distances from the singular point, $\sigma_1, \sigma_2, \ldots, \sigma_n$ are stresses at the corresponding nodal points $r_1, r_2, \ldots, r_n$, respectively. λ is the required stress singularity exponent. The derivation of this relationship makes it possible to calculate the value of λ, which determines the stress behavior (including that of stress singularity) in the vicinity of a singular point.

The algorithm is tested by solving two- and three-dimensional problems of elasticity theory and comparing the stress singularity exponents found by the proposed numerical algorithm with those obtained from the known analytical and numerical solutions. As two-dimensional problems, we considered a plate with notches (Fig. 11.11a), a plate with a fixed edge (Fig. 11.11b), and a composite plate (Fig. 11.11c), which contained singular points associated, respectively, with breaking of surface smoothness, a change in the type of boundary conditions, and a contact of dissimilar materials. For all problems, the obtained numerical results agree with the analytical results up to the third decimal place.

The proposed numerical algorithm for computing the stress singularity exponents in the vicinity of singular points is of considerable independent significance for problems, which cannot be solved analytically in the vicinity of singular points. To problem of crack propagation, whose front is perpendicular to the surface xOy of an

Fig. 11.10 The example of finite-element mesh with gradual refinement near singular point

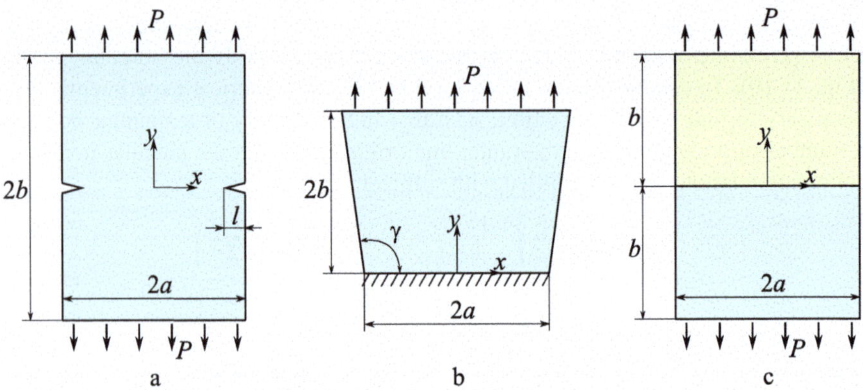

Fig. 11.11 Plate with V-notches on lateral edges (a); plate with a fixed edge (b); composite plate (c)

elastic half-space (Fig. 11.12a). The stress singularity exponent is evaluated at the tip of the crack with coordinates $x = y = z = 0$. For this problem, work [23] presents the results of numerical calculation of stress singularity exponents for an isotropic material ($\nu = 0.3$) and an orthotropic material, the elastic characteristics of which are summarized in Table 11.1.

As a computational scheme for this problem, we use a cube (Fig. 11.12b). The conditions of opening mode (the mode I) are simulated by the normal displacements applied parallel to the xOz-plane, and the conditions of sliding mode (the mode II) are simulated by the tangential displacements applied parallel to the x-axis and in the opposite directions.

Table 11.2 presents the values of stress singularity exponents for a crack tip under loads of mode I and II obtained in [23] and calculated with the proposed numerical

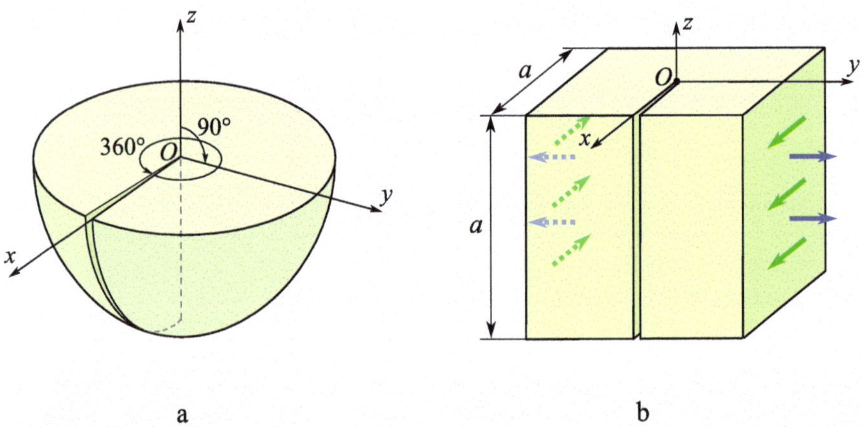

Fig. 11.12 Crack, the front of which is perpendicular to the surface of an elastic half-space (a); its computational scheme (b)

Table 11.1 Elastic characteristics of carbon fiber reinforced plastic [23]

Material	E_i, GPa	G_{ij}, GPa	ν_{ij}
	$E_x = 130.3$	$G_{xy} = 4.502$	$\nu_{xy} = 0.33$
Carbon fiber reinforced plastic	$E_y = 9.377$	$G_{xz} = 4.502$	$\nu_{xz} = 0.33$
	$E_z = 9.377$	$G_{yz} = 2.865$	$\nu_{yz} = 0.33$

Table 11.2 Comparison between stress singularity exponents calculated by formula (11.53) and obtained in [23] for a crack whose front is perpendicular to the surface of an elastic half-space (three-dimensional problem)

	Isotropic ($\nu = 0.3$)		Anisotropic (Table 11.1)	
	λ_1 (mode II)	λ_2 (mode I)	λ_1 (mode II)	λ_2 (mode I)
Numerical algorithm	0.40	0.55	0.46	0.52
Numerical result from [23]	0.3929	0.5483	0.4543	0.5227

algorithm, which uses the finite element method to determine the stress asymptotics based on relations (11.53). In this case, the difference between the stress singularity exponents calculated by formula (11.53) and those presented in [23] is less than 1.8%.

Hence, the effectiveness and high accuracy of the proposed numerical algorithm for calculating the stress singularity exponents in the vicinity of singular points for homogeneous and piecewise homogeneous bodies, including those with anisotropic properties have been substantiated by the results of solution of two- and three-dimensional problems of elasticity theory.

11.5 Conclusion

The analytical method for constructing eigenvalues for circular cones has been considered. The relations developed in this study can be used to construct solutions, and estimate the character of stress singularity for different variants of conical bodies (solid, hollow, composite cones) under different types of boundary conditions set on the lateral surfaces and contact surfaces of different materials. Numerical results have been presented on the nature of stress singularity at the vertex of a solid cone under boundary conditions specified in terms of displacements, stresses, mixed type boundary conditions and at the vertex of a hollow cone under different variants of boundary conditions specified on the lateral surfaces. A numerical algorithm for evaluating the nature of stress singularity in the vicinity of singular points of elastic bodies has been considered. It is based on the derivation of a power law relationship for stresses from the numerically determined stress-strain state in the vicinity of a singular point. The efficiency and high accuracy of the proposed numerical algorithm for calculating stress singularity exponents in the vicinity of singular points for homogeneous and piecewise homogeneous bodies, including those with anisotropic properties, have been demonstrated.

Acknowledgements Sections 11.2 and 11.3 were performed within the state assignment of ICMM UB RAS (topic No. AAAA-A19-19012290100-8). Section 11.4 was funded by RFBR and Perm Territory (project No. 20-41-596007).

References

1. T. Apel, V. Mehrmann, and D. Watkins. Structured eigenvalue methods for the computation of corner singularities in 3d anisotropic elastic structures. *Computer Methods in Applied Mechanics and Engineering*, 191(39):4459–4473, 2002.
2. V.A. Babeshko, E.V. Glushkov, N.V. Glushkova, O.N. Lapina. Stress singularity in the vicinity of the vertex of an elastic trihedron. *Dokl. Akad. Nauk SSSR*, 318(5):1113–1116, 1991.
3. Z.P. Bazant, L.M. Keer. Singularities of elastic stresses and harmonic functions at conical notches and inclusions. *International Journal of Solids and Structures*, 10(9):957–965, 1974.
4. E.B. Becker, R.S. Dunham, and M. Stern. Some stress intensity calculations using finite elements. in V.A. Pulmans, A.P. Kabaila, editors, *Finite element methods in engineering : proceedings of the 1974 International Conference on Finite Element Methods in Engineering*, pages 117–138, Kensington, May 1974. the School of Civil Engineering, the University of New South Wales, Unisearch Ltd.
5. A. Carpinteri, M. Paggi. Asymptotic analysis in linear elasticity: From the pioneering studies by Wieghardt and Irwin until today. *Engineering Fracture Mechanics*, 76(12):1771–1784, 2009.
6. A. Dimitrov, H. Andrä, E. Schnack. Singularities near three-dimensional corners in composite laminates. *International Journal of Fracture*, 115(4):361–375, 2002.
7. J.W. Eischen. Fracture of nonhomogeneous materials. *International Journal of Fracture*, 34(1):3–22, 1987.
8. A.Yu. Fedorov, V.P. Matveenko. Investigation of stress behavior in the vicinity of singular points of elastic bodies made of functionally graded materials. *Journal of Applied Mechanics*, 85(6):061008, apr 2018.
9. C.S. Huang, A.W. Leissa. Stress singularities in bimaterial bodies of revolution. *Composite Structures*, 82(4):488–498, 2008.
10. E. Kamke. *Differentialgleichungen: Lösungsmethoden und Lösungen*. Vieweg+Teubner Verlag Wiesbaden, Leipzig, 1977.
11. H. Koguchi. Stress singularity analysis in three-dimensional bonded structure. *International Journal of Solids and Structures*, 34:461–480, 1997.
12. V.A. Kondrat'ev. Boundary value problems for elliptic equations in domains with conical or angular points. *Transactions of the Moscow Mathematical Society*, 16:227–313, 1967.
13. T.O. Korepanova, V.P. Matveenko, N.V. Sevodina. Numerical analysis of stress singularity at singular points of three-dimensional elastic bodies. *Acta Mechanica*, 224(9):2045–2063, 2010.

14. T.O. Korepanova, V.P. Matveenko, I.N. Shardakov. Analytical constructions of eigensolutions for isotropic conical bodies and their applications for estimating stress singularity. *Doklady Physics*, 59(7):335–340, 2014.
15. V.A. Kozlov, V. Maz'ya, and J.S. Rossmann. *Spectral Problems Associated with Corner Singularities of Solutions to Elliptic Equations*. American Mathematical Society, Rhode Island, 2001.
16. V.P. Matveenko, T.O. Nakaryakova, N.V. Sevodina, I.N. Shardakov. Investigation of singularity of stresses at the apex of a cone with an elliptic base. *Doklady Physics*, 51(11):630–633, 2006.
17. V.P. Matveenko, M.A. Sevodin, N.V. Sevodina. Applications of Muller's method and the argument principle to eigenvalue problems in solid mechanics. *Computational Continuum Mechanics*, 7(3):331–336, 2014.
18. N.M. Matveev. *Integration Techniques of Ordinary Differential Equations*. Higher School, Moscow, 1967.
19. V.P. Matveyenko, T.O. Nakaryakova, N.V. Sevodina, I.N. Shardakov. Stress singularity at the vertex of homogeneous and composite cones for different boundary conditions. *Journal of Applied Mathematics and Mechanics*, 72(3):331–337, 2008.
20. S.E. Mikhailov. Stress singularity in the vicinity of an angle edge in an anisotropic composite and some applications to fibrous composites. *Izv. Akad. Nauk SSSR, Mekh. Tverd. Tela*, (5):103–110, 1979.
21. N.F. Morozov. Brittle fracture problems and their solution using elasticity methods. In *Mechanics and Progress in Science and Engineering, Vol. 3: Solid Mechanics [in Russ.]*, pages 54–63. Nauka, Moscow, 1988.
22. N.F. Morozov, B.N. Semenov. Using the Novozhilov criterion of brittle fracture to determine fracture loads for V-shaped notches in complex stressed states. *Izv. Akad. Nauk SSSR, Mekh. Tverd. Tela*, (1):122–126, 1986.
23. S.S. Pageau, S.B. Biggers. Finite element evaluation of free-edge singular stress fields in anisotropic materials. *International Journal for Numerical Methods in Engineering*, 38(13):2225–2239, 1995.
24. S.S. Pageau, S.B. Biggers. A finite element approach to three-dimensional singular stress states in anisotropic multi-material wedges and junctions. *International Journal of Solids and Structures*, 33(1):33–47, jan 1996.
25. M. Paggi, A. Carpinteri. On the stress singularities at multimaterial interfaces and related analogies with fluid dynamics and diffusion. *Applied Mechanics Reviews*, 61(2):020801, 2008.
26. K.S. Parihar, L.M. Keer. Elastic stress singularities at conical inclusions. *International Journal of Solids and Structures*, 14(4):261–263, 1978.
27. C.R. Picu, V. Gupta. Three-dimensional stress singularities at the tip of a grain triple junction line intersecting the free surface. *Journal of the Mechanics and Physics of Solids*, 45(9):1495–1520, 1997.
28. L.P. Pook. A 50-year retrospective review of three-dimensional effects at cracks and sharp notches. *Fatigue & Fracture of Engineering Materials & Structures*, 36(8):699–723, 2013.
29. I.S. Raju, J.H. Crews. Interlaminar stress singularities at a straight free edge in composite laminates. *Computers & Structures*, 14(1):21–28, 1981.
30. M.P. Savruk, S. Shkarayev. Stress singularities for three-dimensional corners using the boundary integral equation method. *Theoretical and Applied Fracture Mechanics*, 36:263–275, 2001.
31. G.B. Sinclair. Stress singularities in classical elasticity – I: Removal, interpretation, and analysis. *Applied Mechanics Reviews*, 57(4):251–297, 2004.
32. G.B. Sinclair. Stress singularities in classical elasticity – II: Asymptotic identification. *Applied Mechanics Reviews*, 57(5):385–439, 2004.
33. N. Somaratna and T.C.T. Ting. Three-dimensional stress singularities at conical notches and inclusions in transversely isotropic materials. *Journal of Applied Mechanics*, 53(1):89–96, 1986.
34. G.Strang and G.J. Fix. *An Analysis of the Finite Element Method*. Prentice-Hall, Inc., Englewood, N.J., 1973.

35. M.L. Williams. Stress singularities resulting from various boundary conditions in angular corners of plates in extension. *Journal Applied Mechanics*, 19(4):526–528, 1952.

Chapter 12
On Estimating Prestress State in an Elastic Cylinder

Rostislav D. Nedin, Alexander O. Vatulyan, and Victor O. Yurov

Abstract There is a lack of studies devoted to modeling and identification of a complex prestress-strain state in cylindrical bodies. We consider a problem for a finite cylinder in the presence of prestress fields of an inhomogeneous structure. The prestress tensor is given by four nonzero components that are functions of the coordinates; the cases of initial inflation, pre-torsion and pre-tension are considered. The influence of the three considered prestress types on the field of small superimposed displacements was analyzed. The inverse problem on recovering the intensities of the prestress types is studied, given the additional data on the measured displacements in the entire region.

Key words: Cylinder, Prestress, Torsion, Inflation, Tension, Inhomogeneous, Weak Statement, Inverse Problem, Sensitivity Analysis

12.1 Intro

One of the urgent and promising tasks of modern mechanics of deformable solid body is the development of theoretical and numerical models of objects made of modern functionally graded materials (FGM) with complex inhomogeneous physical and mechanical properties, which are characterized by the presence of prestress state. Present-day FGM manufacturing technologies, such as 3D printing, make it possible to design objects of complex geometry without employing standard technologies, including casting, which requires additional production of molds. For the

Rostislav D. Nedin · Alexander O. Vatulyan · Victor O. Yurov
Institute of Mathematics, Mechanics and Computer Sciences named after I.I. Vorovich, Southern Federal University, 8a Mil'chakova Str., 344090 Rostov-on-Don & Southern Mathematical Institute, Vladikavkaz Scientific Center of Russian Academy of Sciences, 22 Markusa Str., 362027 Vladikavkaz, Russian Federation,
e-mail: rdn90@bk.ru, aovatulyan@sfedu.ru, vyurov@sfedu.ru

H. Altenbach et al. (eds.), *Advances in Solid and Fracture Mechanics*,
Advanced Structured Materials 180,
https://doi.org/10.1007/978-3-031-18393-5_12

manufacture of FGM, high-temperature technologies are usually used (e.g., surfacing, sintering, work hardening, etc.), due to which, after cooling, prestress fields are formed in the samples, reaching levels of significant effect on their dynamic characteristics [1].

As in the case of material inhomogeneity, there is a lack of studies devoted to the identification of significantly inhomogeneous factors of prestress state in the literature. This is mainly due to the complexity of the mathematical apparatus and computational difficulties that arise at the stage of formulating and solving the accompanying inverse problems. As for the studies of a homogeneous prestress state, in the literature they often use an approach to solving problems of restoring the prestress field in various structures based on a priori information that the prestress state is formed as a result of subjecting to some mechanical load. For instance, in [2], the authors investigate waves propagation in double cylindrical rods in the presence of prestress. The interaction between two rods is modeled within the framework of the Hertzian contact problem under the action of a static load; subsequently, this state is interpreted as the initial stress state and is used further when considering the analysis of wave propagation in a double cylindrical structure.

[3] presents a multiscale model for studying hollow cylinders with an arbitrary functional gradient, with fibers, particles or disc-shaped reinforcing elements subjected to harmonic loads. Stress analysis is performed by dividing the cylinders into several layers, each with uniform properties, which functionally differ in the thickness. To describe the effective properties of each layer reinforced with fibers, particles or disk-shaped inclusions, averaging is used in the framework of the Mori-Tanaka approach.

In [4], a prestress state in the welded joint of an annular structure modeled by a cylindrical hoop was studied. The blind hole method was used to check the distribution of residual stress in the ring structure, and the test results, in order to check the validity, were compared with the results of the finite element method (FEM) calculation. This made it possible to formulate recommendations on the optimal mode of welding ring structures.

Currently, dental implants are often cylindrical or conical with a screw structure, and most screw-retained implants have a uniform solid structure [5]. In this case, the Young modulus of the implants is significantly higher than that of the surrounding tissues, and the implants fail due to the stress shielding effect. Thus, dental implants with a porous FGM structure distribute stress to surrounding tissues and provide long-term stability. Such materials have attracted considerable interest from researchers due to their light weight, excellent energy absorption capacity, and fine thermal stability properties.

It should be noted that the approaches used in practice to determine a homogeneous prestress state and homogeneous material parameters in most situations are inapplicable to modern composite and FGM materials in which a prestress state with a clearly inhomogeneous structure is usually formed. To study cylinders and plates made of such materials, it is necessary to use full-fledged two-dimensional and three-dimensional models with variable mechanical characteristics and prestress state factors. In turn, to study the corresponding inverse reconstruction problems, new methods are required that allow one to qualitatively estimate not only the level, but also the nature of the distribution of field characteristics that depend on three variables. Currently, one can find many works devoted to the development of algebraization methods, including projection methods, aimed at solving multidimensional direct and inverse problems for elastic bodies in the presence of coupled fields. We also note the importance of using variational interpretations and weak formulations of direct and inverse problems for deriving operator relations and obtaining efficient numerical solutions. In [6], a new numerical technique for solving problems of the linear elasticity theory based on a combination of the Galerkin method and the FEM is presented, based on the use of weak formulations for the corresponding differential operators.

There is still a lack of research devoted to the identification of planar or volumetric distributions of prestress fields in solids based on a number of surface measurements within the nondestructive concepts. As such measurements, one can use the displacements amplitudes measured in a set of points of some part of the body's surface for several vibration frequencies; in such a statement, the inverse problem is nonlinear and essentially ill-posed. In contrast to that, sometimes it is possible to consider a simpler, linear, inverse problem on prestress reconstruction based on the measured displacement data inside the entire region. In [7] a few techniques for solving such inverse problem were proposed, and some solution uniqueness aspects were discussed.

Nowadays several approaches to modeling prestress in solids are used in literature; some of them are based on the linearization of nonlinear relations in continuum mechanics, for instance, in [8, 9, 10, 11]. A number of linearized approaches to prestress modelling are reviewed in [12]. On the basis of such theoretical modeling approaches, in [13, 14, 15, 16, 17, 18], the methodologies for determining inhomogeneous prestress fields in elastic bodies (including those made of functionally graded composites) were proposed. The methods are mainly based on iterative-regularization schemes for solving the corresponding ill-posed coefficient inverse problems. Such approaches can be useful when dealing with incomplete measurement data.

In this paper, we study the problem for a finite cylinder in the presence of prestress fields of an inhomogeneous structure. A study of the influence of several prestress types on the displacement field under the action of an external load was made. The inverse problem of finding the intensities of three types of prestress states is studied in the presence of additional information on the displacement field measured in the region.

12.2 Linearized Deformation Model for a Prestressed Body

To formulate the problems on vibrations of prestressed cylindrical bodies, we will start with the approach of linearizing the original nonlinear continuum mechanics motion equations and boundary conditions by perturbing the original body's configuration (taken as the reference one) and superimposing small deformations on it [8, 11, 12, 19].

We shall proceed from the standard nonlinear dynamic problem statement written for the 1st Piola-Kirchhoff stress tensor $\mathbf{T}'$ [8]:

$$\nabla \cdot \mathbf{T}' = \rho \ddot{\mathbf{u}}', \tag{12.1}$$

$$\mathbf{n} \cdot \mathbf{T}'|_{S_\sigma} = \mathbf{P}', \quad \mathbf{u}'|_{S_u} = 0, \tag{12.2}$$

Here, the body is rigidly clamped on the surface part S_u and subjected to the load $\mathbf{P}'$ on S_σ, ρ is the body density, $\mathbf{n}$ — unit vector of the outer normal to the body's surface, a dash corresponds to the total nonlinearized components (e.g., $\mathbf{u}'$ is the total displacement vector); the nabla-operator ∇ is defined in the metrics of the natural undeformed configuration of the body. Representing $\mathbf{T}'$ in terms of the symmetric total 2nd Piola–Kirchhoff stress tensor σ', we get $\mathbf{T}' = \sigma' \cdot \mathbf{F}$, where $\mathbf{F} = \nabla\mathbf{R} = \nabla(\mathbf{r}+\mathbf{u}') = \mathbf{E} + \nabla\mathbf{u}'$ – material deformation gradient; $\mathbf{r}$, $\mathbf{R}$ – radius-vectors referred to the undeformed and perturbed (actual) configurations, respectively; $\mathbf{E}$ – identity tensor.

Now let us use the classical linearization approach (corresponding to superimposing small deformation on the finite one) and decompose all the total components (displacements, stress, strain, loads, etc.) by those from the initially-stressed and perturbed states: $\mathbf{u}' = \mathbf{u}^0 + \mathbf{u}$, $\sigma' = \sigma^0 + \sigma$, $\boldsymbol{\varepsilon}' = \boldsymbol{\varepsilon}^0 + \boldsymbol{\varepsilon}$, $\mathbf{P}' = \mathbf{P}^0 + \mathbf{P}$, etc.

For instance, the linearization of the total 1st Piola-Kirchhoff stress tensor $\mathbf{T}'$ (omitting detailed tensor transformations) will result in

$$\mathbf{T}' = \sigma' \cdot (\mathbf{E} + \nabla\mathbf{u}') = (\sigma^0 + \sigma) \cdot (\mathbf{E} + \nabla\mathbf{u}^0 + \nabla\mathbf{u}) = \mathbf{T}^0 + \mathbf{T},$$

$$\mathbf{T}^0 = \sigma^0 + \sigma^0 \cdot \nabla\mathbf{u}^0, \quad \mathbf{T} = \sigma + \sigma^0 \cdot \nabla\mathbf{u} + \sigma \cdot \nabla\mathbf{u}^0 \tag{12.3}$$

Here, in view of smallness of the superimposed deformation, we have also neglected the term $\sigma \cdot \nabla\mathbf{u}$; $\mathbf{T}^0$ and $\mathbf{T}$ represent initial and perturbed components of the tensor $\mathbf{T}'$, respectively. In the same way, the linearized initial and incremental strain tensors take forms

$$\boldsymbol{\varepsilon}^0 = \frac{1}{2}(\nabla\mathbf{u}^0 + \nabla\mathbf{u}^{0\mathrm{T}} + \nabla\mathbf{u}^0 \cdot \nabla\mathbf{u}^{0\mathrm{T}}), \ \boldsymbol{\varepsilon} = \frac{1}{2}(\nabla\mathbf{u} + \nabla\mathbf{u}^{\mathrm{T}} + \nabla\mathbf{u}^0 \cdot \nabla\mathbf{u}^{\mathrm{T}} + \nabla\mathbf{u} \cdot \nabla\mathbf{u}^{0\mathrm{T}}),$$

respectively (note that both components contain the initial displacement gradients explicitly).

Let us consider perturbing the initially-stressed state in the form of small steady-state vibrations and represent $\mathbf{P}(t) = \mathbf{P}e^{-i\omega t}$, $\mathbf{u}(t) = \mathbf{u}e^{-i\omega t}$, and then eliminate the time factor from the governing equations in the common way. After linearizing the motion equations (12.1) and boundary conditions (12.2), we get two boundary-problem statements: for the initial, self-balanced, state (also neglecting the difference between the normal vectors in the initial and actual configurations) in the form

$$\nabla \cdot \mathbf{T}^0 = 0, \quad \mathbf{u}^0|_{S_u} = 0, \quad n \cdot \mathbf{T}^0|_{S_\sigma} = \mathbf{P}^0. \tag{12.4}$$

and for the perturbed configuration, written for small incremental values, as follows:

$$\nabla \cdot \mathbf{T} + \rho\omega^2\mathbf{u} = 0, \quad \mathbf{u}|_{S_u} = 0, \quad n \cdot \mathbf{T}|_{S_\sigma} = \mathbf{P}. \tag{12.5}$$

where $\mathbf{T}^0$ and $\mathbf{T}$ were introduced in (12.3). In such a formulation, the statements (12.4) and (12.5) can be found, for instance, in the works of A.N. Guz [11] (the review with an extensive list of names of scientists who contributed to the development of this theory may be found in [12])). As it was done in most of the works in literature, we shall consider a particular case of the described statement (12.5), when the changes of volumes and areas can be neglected, and the initial deformed state can be determined by geometrically linear theory, ignoring the initial displacement gradient: $\nabla\mathbf{u}^0 \approx 0$. In this case, one may set the incremental objective stress tensor $\boldsymbol{\sigma}$ in the most standard way, satisfying, for example, Hooke's law for the linear elastic material. For an isotropic material, we take $\boldsymbol{\sigma} = \lambda \mathbf{E}\,\mathrm{tr}\,\boldsymbol{\varepsilon} + 2\mu\boldsymbol{\varepsilon}$ for the infinitesimal linear strain tensor $\boldsymbol{\varepsilon} = \frac{1}{2}\left(\nabla\mathbf{u} + \nabla\mathbf{u}^{\mathrm{T}}\right)$, where λ and μ are the Lamé coefficients. In this way, the incremental 1st Piola-Kirchhoff stress tensor $\mathbf{T}$ gets the simplified form $\mathbf{T} = \boldsymbol{\sigma} + \boldsymbol{\sigma}^0 \cdot \nabla\mathbf{u}$.

More details on the technique of deriving such linearized models of prestress elastic and electroelastic bodies can be found in [12, 20]. The model described above makes it possible to simulate inhomogeneous prestress fields of various nature without taking into account initial deformation explicitly and is of certain convenience when treating inverse problems on prestress identification.

The general weak statement of the described problem has the form [12, 19]:

$$\int_V \mathbf{T} \odot \nabla\mathbf{v} \mathrm{d}V - \omega^2 \int_V \rho\mathbf{u} \cdot \mathbf{v} \mathrm{d}V - \int_{S_\sigma} \mathbf{P} \cdot \mathbf{v} \mathrm{d}S = 0, \tag{12.6}$$

where $\mathbf{v}$ — test vector function that satisfies the main boundary conditions from Eq. (12.6), V — volume occupied by the body.

12.3 Weak Problem Statement for Prestressed Cylinder

Consider the problem of vibrations of an inhomogeneous cylinder in the presence of four nonzero components of the prestress tensor: $\sigma^0_{rr}, \sigma^0_{\phi\phi}, \sigma^0_{zz}, \sigma^0_{r\phi}$ in the cylindrical coordinate system r, ϕ, z. The cylinder volume is defined as

$$V = \{r \in [R_1, R_2], \phi \in [0, 2\pi), z \in [0, L]\}.$$

We will consider an axisymmetric problem, assuming that the test vector function and the displacement vector have the form $\mathbf{v} = v_r(r,z)\mathbf{e}_r + v_\phi(r,z)\mathbf{e}_\phi + v_z(r,z)\mathbf{e}_z$ and $\mathbf{u} = u_r(r,z)\mathbf{e}_r + u_\phi(r,z)\mathbf{e}_\phi + u_z(r,z)\mathbf{e}_z$, respectively, and

$$\frac{\partial u_r}{\partial \phi} = \frac{\partial u_\phi}{\partial \phi} = \frac{\partial u_z}{\partial \phi} = 0.$$

To formulate a weak statement in the considered particular case, we shall write down the components of the tensor **T** based on the constitutive relations from (12.5), assuming the cylinder's material be isotropic with variable properties:

$$
\begin{aligned}
T_r &= \lambda\left(\frac{\partial u_r}{\partial r}+\frac{u_r}{r}+\frac{\partial u_z}{\partial z}\right)+2\mu\frac{\partial u_r}{\partial r}+\sigma^0_{rr}\frac{\partial u_r}{\partial r}-\sigma^0_{r\phi}\frac{u_\phi}{r},\\
T_\phi &= \lambda\left(\frac{\partial u_r}{\partial r}+\frac{u_r}{r}+\frac{\partial u_z}{\partial z}\right)+2\mu\frac{u_r}{r}+\sigma^0_{\phi\phi}\frac{u_r}{r}+\sigma^0_{r\phi}\frac{\partial u_\phi}{\partial r},\\
T_z &= \lambda\left(\frac{\partial u_r}{\partial r}+\frac{u_r}{r}+\frac{\partial u_z}{\partial z}\right)+2\mu\frac{\partial u_z}{\partial z}+\sigma^0_{zz}\frac{\partial u_z}{\partial z},\\
T_{rz} &= \mu\left(\frac{\partial u_z}{\partial r}+\frac{\partial u_r}{\partial z}\right)+\sigma^0_{rr}\frac{\partial u_z}{\partial r},\\
T_{zr} &= \mu\left(\frac{\partial u_z}{\partial r}+\frac{\partial u_r}{\partial z}\right)+\sigma^0_{zz}\frac{\partial u_r}{\partial z},\\
T_{r\phi} &= \mu\left(\frac{\partial u_\phi}{\partial r}-\frac{u_\phi}{r}\right)+\sigma^0_{rr}\frac{\partial u_\phi}{\partial r}+\sigma^0_{r\phi}\frac{u_r}{r},\\
T_{z\phi} &= \mu\frac{\partial u_\phi}{\partial z}+\sigma^0_{zz}\frac{\partial u_\phi}{\partial z},\\
T_{\phi z} &= \mu\frac{\partial u_\phi}{\partial z}+\sigma^0_{r\phi}\frac{\partial u_z}{\partial r},\\
T_{\phi r} &= \mu\left(\frac{\partial u_\phi}{\partial r}-\frac{u_\phi}{r}\right)-\sigma^0_{\phi\phi}\frac{u_\phi}{r}+\sigma^0_{r\phi}\frac{\partial u_r}{\partial r}.
\end{aligned}
\tag{12.7}
$$

Let, for the sake of generality, be $S_\sigma=\partial V=S_-\cup S_+\cup S_1\cup S_2$, were

$$
\begin{aligned}
S_- &= \{r\in[R_1,R_2],\phi\in[0,2\pi),z=0\},\\
S_+ &= \{r\in[R_1,R_2],\phi\in[0,2\pi),z=L\},\\
S_1 &= \{r=R_1,\phi\in[0,2\pi),z\in[0,L]\},\\
S_2 &= \{r=R_2,\phi\in[0,2\pi),z\in[0,L]\}.
\end{aligned}
$$

By integrating over volume and surface in (12.6) and performing standard tensor transformations, taking into account the boundary conditions from Eq. (12.5) and the independence of all the considered functions from the circumferential coordinate ϕ, we get

$$
\begin{aligned}
&\int_{\Omega}\left(\lambda K_{\lambda}^{uv}+\mu K_{\mu}^{uv}+\sigma_{r\phi}^{0}K_{r\phi}^{uv}+\sigma_{rr}^{0}K_{rr}^{uv}+\sigma_{\phi\phi}^{0}K_{\phi\phi}^{uv}+\sigma_{zz}^{0}K_{zz}^{uv}\right)\mathrm{d}\Omega\\
&\qquad-\omega^{2}\int_{\Omega}\rho\left(u_{r}v_{r}+u_{\phi}v_{\phi}+u_{z}v_{z}\right)\mathrm{d}\Omega\\
&\qquad-\int_{R_{1}}^{R_{2}}\left(T_{zr}v_{r}+T_{z\phi}v_{\phi}+T_{zz}v_{z}\right)\Big|_{z=\{0;L\}}r\mathrm{d}r\\
&\qquad-\int_{0}^{L}\left(T_{rr}v_{r}+T_{r\phi}v_{\phi}+T_{rz}v_{z}\right)r\Big|_{r=\{R_{1};R_{2}\}}\mathrm{d}z=0
\end{aligned}
\tag{12.8}
$$

where the following functions are introduced

$$
\begin{aligned}
K_{\lambda}^{uv}&=\left(u_{r,r}+\frac{u_{r}}{r}+u_{z,z}\right)\left(v_{r,r}+\frac{v_{r}}{r}+v_{z,z}\right),\\
K_{\mu}^{uv}&=2\left(u_{r,r}v_{r,r}+\frac{u_{r}}{r}\frac{v_{r}}{r}+u_{z,z}v_{z,z}\right)+\left(u_{r,z}+u_{z,r}\right)\left(v_{r,z}+v_{z,r}\right)\\
&+\left(u_{\phi,r}-\frac{u_{\phi}}{r}\right)\left(v_{\phi,r}-\frac{v_{\phi}}{r}\right)+u_{\phi,z}v_{\phi,z},\\
K_{r\phi}^{uv}&=\left(u_{\phi,r}v_{r}+u_{r}v_{\phi,r}-u_{\phi}v_{r,r}-u_{r,r}v_{\phi}\right)r^{-1},\\
K_{rr}^{uv}&=u_{r,r}v_{r,r}+u_{\phi,r}v_{\phi,r}+u_{z,r}v_{z,r},\\
K_{\phi\phi}^{uv}&=\left(u_{r}v_{r}\right)r^{-2}+\left(u_{\phi}v_{\phi}\right)r^{-2},\\
K_{zz}^{uv}&=u_{r,z}v_{r,z}+u_{\phi,z}v_{\phi,z}+u_{z,z}v_{z,z}.
\end{aligned}
$$

In this case, the region of a flat longitudinal section of the cylinder is considered:

$$
\Omega=\{r\in[R_{1},R_{2}],z\in[0,L]\},\mathrm{d}\Omega=r\mathrm{d}r\mathrm{d}z.
$$

Note that the test vector function $\mathbf{v}$ satisfies the same essential boundary conditions as the displacement vector $\mathbf{u}$.

Remark 12.1. To verify the computational FE scheme based on the proposed weak statement (12.8), a series of computational experiments was carried out: FE solutions were compared with Saint-Venant solutions for the cases of axial tension and torsion; a comparison was also made with the solution for an annular region with a uniformly distributed internal pressure. Verification within steady oscillations is carried out for a special case for which a semi-analytical solution exists; at both cylinder's ends, the sliding boundary conditions are set

$$u_z = 0, \mu\left(\frac{\partial u_z}{\partial r} + \frac{\partial u_r}{\partial z}\right) = \mu\frac{\partial u_\phi}{\partial z} = 0,$$

on the outer side S_2 the normal load is distributed according to the law $\cos\left(\pi z L^{-1}\right)$. In this case, the problem is reduced to the one-dimensional one (by separating the axial coordinate) and investigated using the shooting method. The results obtained quite accurately repeat the results of calculations published in the work [18] (the relative error when comparing the solutions was no more than 1%).

12.4 Sensitivity Analysis

Consider the deformation of the hollow cylinder with the parameters

$$R_1 = 0.5R_2, L = 10R_2, R_2 = 1\text{ m}, \mu = 70\text{ GPa}, \lambda = 105\text{ GPa}, \rho = 7950\ \frac{\text{kg}}{\text{m}^3}$$

(steel #10X17H13M2T), in case of cantilever conditions ($u_r = u_\phi = u_z = 0$, $v_r = v_\phi = v_z = 0$ on S_-) under the action of an external load of three following types (a unit load is indicated in each type):

1. Torsion of the cylinder by a tangential load applied to the free end $z = L$:
   $T_{zr} = 0, T_{z\phi} = r, T_{zz} = 0.$
2. Inflation by internal pressure for $r = R_1$:
   $T_{rr} = 1, T_{r\phi} = 0, T_{rz} = 0.$
3. Stretching by axial load applied to the end $z = L$:
   $T_{zr} = 0, T_{z\phi} = 0, T_{zz} = 1.$

For each type of loading, the deformation characteristics of the cylinder are studied in the presence and absence of inhomogeneous prestress fields. The sensitivity of cylinder displacements under the action of the considered static loads to three prestress types is analyzed (similar prestress laws for an infinite cylinder may be found in [18]):

- Prestress state I (torsion):

$$\sigma^{\text{I}}_{r\phi} = \varepsilon_{\text{I}}\frac{R_2^2}{r^2}, \quad \sigma^{\text{I}}_{rr} = 0, \quad \sigma^{\text{I}}_{\phi\phi} = 0, \quad \sigma^{\text{I}}_{zz} = 0 \tag{12.9}$$

- Prestress state II (inflation):

$$\sigma^{\text{II}}_{r\phi} = 0, \quad \sigma^{\text{II}}_{rr} = \varepsilon_{\text{II}}\frac{R_1^2\left(r^2 - R_2^2\right)}{r^2\left(R_2^2 - R_1^2\right)}, \quad \sigma^{\text{II}}_{\phi\phi} = \varepsilon_{\text{II}}\frac{R_1^2\left(r^2 + R_2^2\right)}{r^2\left(R_2^2 - R_1^2\right)}, \quad \sigma^{\text{II}}_{zz} = 0 \tag{12.10}$$

- Prestress state III (tension):

$$\sigma^{\text{III}}_{r\phi} = 0, \quad \sigma^{\text{III}}_{rr} = 0, \quad \sigma^{\text{III}}_{\phi\phi} = 0, \quad \sigma^{\text{III}}_{zz} = \varepsilon_{\text{III}} \tag{12.11}$$

Here the constant values ε_{I}, $\varepsilon_{\mathrm{II}}$, $\varepsilon_{\mathrm{III}}$ represent prestress amplitudes. Below in the paper, we present the results of assessing the possibility of using fields of analytical type (12.9)-(12.11) in the finite cylinder problem. Let us represent the prestress tensor in the general case as $\boldsymbol{\sigma}_0 = \boldsymbol{\sigma}_{\mathrm{I}} + \boldsymbol{\sigma}_{\mathrm{II}} + \boldsymbol{\sigma}_{\mathrm{III}}$.

We studied the problems in the presence of prestress of the three described types and without it, with the corresponding solutions

$$\mathbf{u} = \left(u_r, u_\phi, u_z\right)$$

and

$$\mathbf{u}_0 = \left(u_{r0}, u_{\phi 0}, u_{z0}\right),$$

respectively. We built 2D graphs (Fig. 12.1) showing the function of the prestress "influence" on the deformation of the cylinder – the relative change in the length of the displacement vector

$$\delta = \frac{|\mathbf{u} - \mathbf{u}_0|}{d} 100\%,$$

where

$$d = \max_{r,z} |\mathbf{u}_0| = \max_{r,z} \sqrt{(u_{r0})^2 + \left(u_{\phi 0}\right)^2 + (u_{z0})^2}$$

is the largest characteristic length of the displacement vector for the problem without prestress. The figure also shows the respective maximum influence values

$$\delta_{\max} = \max_{r,z} \delta$$

in each case considered.

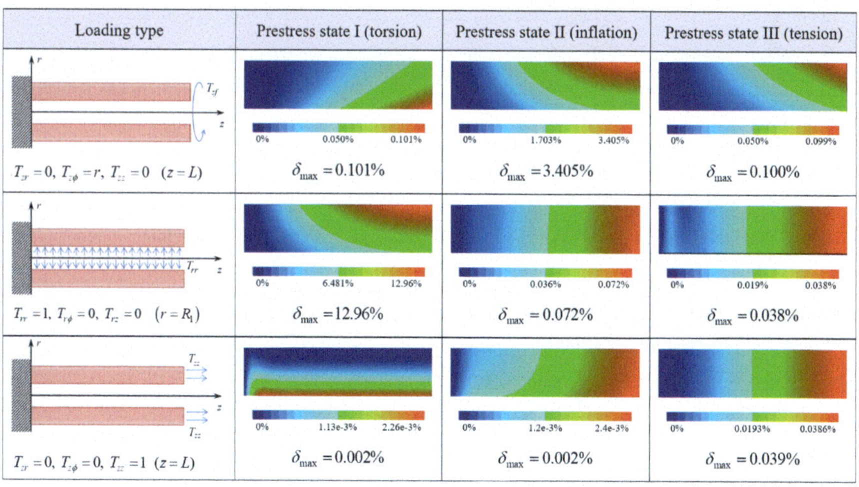

Fig. 12.1 Influences of 3 prestress types on the cylinder displacement vector $\mathbf{u}$ under 3 modes of static loading

Remark 12.2. Further, everywhere in the work on the images of 2D fields, the longitudinal coordinate $z \in [0, L]$ is plotted along the abscissa axis, and the radial coordinate $r \in [R_1, R_2]$ – along the ordinate axis. FE calculations were carried out on the grid $[200 \times 20]$ (with FE providing quadratic approximation); in some cases local FE mesh refinement was used in the vicinity of the clamped end $z = 0$.

In this case, we considered the prestress of the maximum level 0.001μ (where μ is the shear modulus), for example, the prestress proportionality coefficients of the first type have the form $\varepsilon_{\rm I} = 0.001\mu$, $\varepsilon_{\rm II} = \varepsilon_{\rm III} = 0$.

Based on the results of studying the influence of prestress, some recommendations are formulated below for their restoration in the course of solving the inverse problem with additional information about the measured deformation characteristics of the cylinder. For efficient level reconstruction $\varepsilon_{\rm I}$ (for prestress state I) it is rational to implement the cylinder inflation mode (loading of the second type $T_{rr} = 1, T_{r\phi} = 0, T_{rz} = 0$ for $r = R_1$); to recover $\varepsilon_{\rm II}$ (prestress II) it is advisable to implement the loading of the first type ($T_{zr} = 0, T_{z\phi} = r, T_{zz} = 0$ for $z = L$). The level $\varepsilon_{\rm III}$ has the same maximum effect under loadings of both second and third types. Note that under loading of the first type (torsion), for prestress states II-III, it is characteristic $u_r = u_z \equiv 0$. When considering loading of the second and third types (inflation, tension) and prestress states II-III, there is $u_\phi \equiv 0$. Prestress I is characterized by the presence of all three components of the displacement vector.

Remark 12.3. Note that the laws of change in prestress (12.9)-(12.11) do not satisfy the boundary conditions at the ends of the cylinder, since they are solutions to some model problems for an infinite cylinder and a ring. However, as shown below, laws (12.9)-(12.11) can nevertheless be used as prestress when considering small deformations of an elongated prestressed cylinder; the greatest difference takes place in the area of the cantilever-pinched edge.

Below we present the results of comparing the given prestress laws with the prestress fields calculated by the FE in the course of solving an auxiliary statics problem for a cylinder under the influence of the corresponding initial mechanical loads.

12.5 Comparison of Analytical and FE Prestress Fields

This section presents the results of analyzing the proximity of the analytical prestress laws (12.9)-(12.11) to the corresponding numerical FE solutions for a finite cantilever-clamped cylinder under similar loading modes. Everywhere below, the prestress obtained as a result of FE calculations will be denoted as $\hat{\sigma}^{\alpha}_{ij}$, $\alpha = {\rm I}, {\rm II}, {\rm III}$.

12.5.1 Initial Inflation

A numerical FE solution is constructed for the auxiliary statics problem on initial inflation by internal pressure of a final cantilever-clamped cylinder that satisfies the following boundary conditions:

$$u_r = u_\phi = u_z = 0 \quad \text{on} \quad S_-, \qquad T_{zr} = 0, T_{z\phi} = 0, T_{zz} = 0 \quad \text{on} \quad S_+,$$
$$T_{rr} = q_r^0, T_{r\phi} = 0, T_{rz} = 0 \quad \text{on} \quad S_1, \qquad T_{rr} = 0, T_{r\phi} = 0, T_{rz} = 0 \quad \text{on} \quad S_2.$$

Figure 12.2 presents some characteristic distributions of the calculated stress components for the internal pressure value $q_r^0 = 0.001\mu$. At that, the two components $\hat{\sigma}_{r\phi}^{\rm II} = \hat{\sigma}_{\phi z}^{\rm II} = 0$, and the stresses $\hat{\sigma}_{rz}^{\rm II}$, $\hat{\sigma}_{zz}^{\rm II}$ are localized in the clamp vicinity; the stresses $\hat{\sigma}_{rr}^{\rm II}$, $\hat{\sigma}_{\phi\phi}^{\rm II}$ for $z = [0.2L, L]$ are practically indistinguishable from $\sigma_{rr}^{\rm II}$, $\sigma_{\phi\phi}^{\rm II}$. The maximum and minimum values in the color gradient are hereinafter indicated below the field images.

In Fig. 12.3 section graphs $\hat{\sigma}_{zz}^{\rm II}$ (for the component $\hat{\sigma}_{rz}^{\rm II}$ we got similar tendencies) are given in the vicinity of the cylinder's clamped end. The figure reflects the presence of stress concentration in the vicinity of the end S_-. In the point $(r = R_1, z = 0)$ on the edge, it is most pronounced (which follows from the analysis of the solution by means of distinguishing regular parts). Let us further consider the relative difference between the analytical and FE fields of the prestresses $\sigma_{rr}^{\rm II}$, $\sigma_{\phi\phi}^{\rm II}$ and $\hat{\sigma}_{rr}^{\rm II}$, $\hat{\sigma}_{\phi\phi}^{\rm II}$, respectively (as the most characteristic components). To do this, we introduce into consideration the functions

$$\delta_r = \frac{\left|\sigma_{rr}^{\rm II} - \hat{\sigma}_{rr}^{\rm II}\right|}{\max\left|\sigma_{rr}^{\rm II}\right|} 100\%, \delta_\phi = \frac{\left|\sigma_{\phi\phi}^{\rm II} - \hat{\sigma}_{\phi\phi}^{\rm II}\right|}{\max\left|\sigma_{\phi\phi}^{\rm II}\right|} 100\%, \delta_z = \frac{\left|\sigma_{zz}^{\rm III} - \hat{\sigma}_{zz}^{\rm III}\right|}{\max\left|\sigma_{zz}^{\rm III}\right|} 100\%,$$

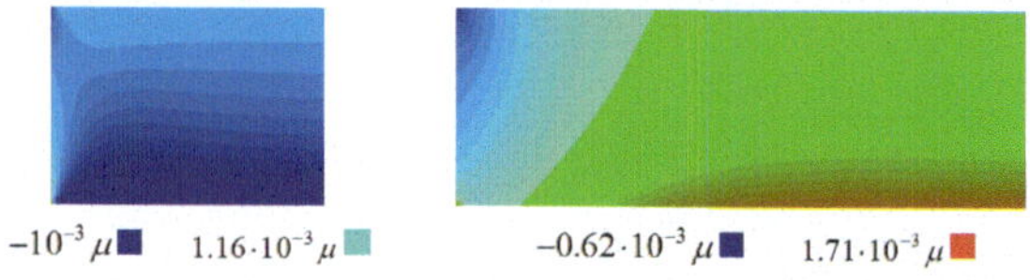

Fig. 12.2 On the left: $\sigma_{rr}^{\rm II}$ for $z = [0, 0.1L]$. On the right: $\sigma_{\phi\phi}^{\rm II}$ for $z = [0, 0.2L]$; $r \in [R_1, R_2]$.

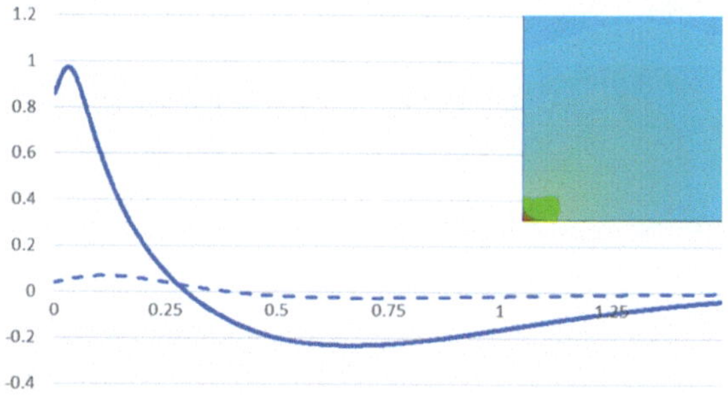

Fig. 12.3 The stress $\hat{\sigma}_{zz}^{\rm II}/q_r^0$ in two sections: $r = 0.75R_2$ – dotted line, $r = 0.55R_2$ – solid line; $z \in [0, 0.15L]$.

see Figs. 12.4-12.6.

12.5.2 Initial Axial Tension

Similarly, we consider the problem of axial pretensioning of the finite cantilever-clamped cylinder by means of the load of magnitude $q_z^0 = 0.001\mu$, applied to the free end S_+. In this case, we obtain that $\hat{\sigma}_{r\phi}^{\mathrm{III}} = \hat{\sigma}_{\phi z}^{\mathrm{III}} = 0$, the stresses $\hat{\sigma}_{rr}^{\mathrm{III}}$, $\hat{\sigma}_{\phi\phi}^{\mathrm{III}}$, $\hat{\sigma}_{rz}^{\mathrm{III}}$ are localized in the clamp vicinity, the stress $\hat{\sigma}_{zz}^{\mathrm{III}}$ for $z = [0.2L, L]$ is close to the corresponding analytical law $\sigma_{zz}^{\mathrm{III}}$ (Figs. 12.5-12.6). Relative difference in the FE approximation of the function $\hat{\sigma}_{zz}^{\mathrm{III}}$ and its analytical representation (12.11) is shown in Fig. 12.6.

12.5.3 Initial Torsion

Let us consider the problem of initial torsion of the finite cylinder subjected to the tangential load $\tau_\phi^0 = 0.001\mu$, applied to the inner and outer boundaries. The calculations show that $\hat{\sigma}_{rr}^{\mathrm{I}} = \hat{\sigma}_{\phi\phi}^{\mathrm{I}} = \hat{\sigma}_{zz}^{\mathrm{I}} = \hat{\sigma}_{rz}^{\mathrm{I}} = 0$, the stress component $\hat{\sigma}_{\phi z}^{\mathrm{I}}$ is localized in the clamp vicinity, and the stress $\hat{\sigma}_{r\phi}^{\mathrm{I}}$ for $z = [0.2L, L]$ coincides with $\sigma_{zz}^{\mathrm{I}} = \varepsilon_{\mathrm{I}} \frac{R_2^2}{r^2}$ with 0.05% precision (see Fig. 12.7).

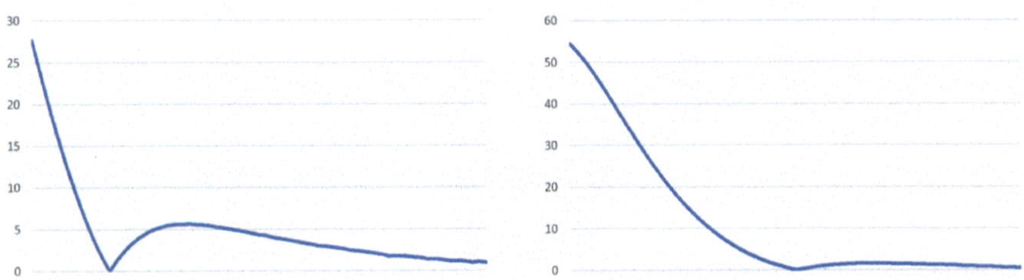

Fig. 12.4 On the left: δ_r for $z \in [0, 0.1L]$. To the right: δ_ϕ for $z \in [0, 0.2L]$ in section $r = 0.75R_2$.

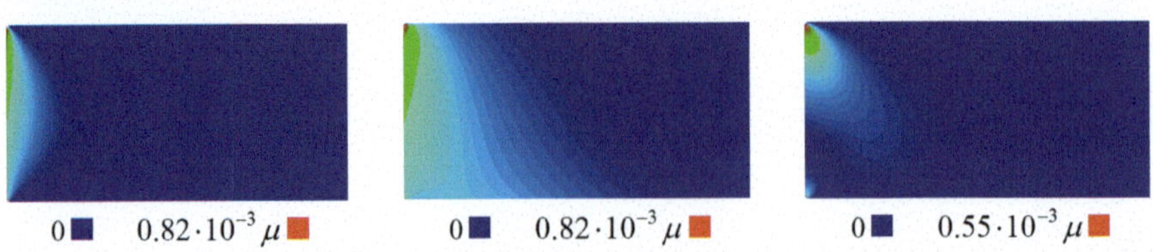

Fig. 12.5 From left to right: $\hat{\sigma}_{rr}^{\mathrm{III}}$, $\hat{\sigma}_{\phi\phi}^{\mathrm{III}}$, $|\hat{\sigma}_{rz}^{\mathrm{III}}|$ for $z \in [0, 0.1L]$.

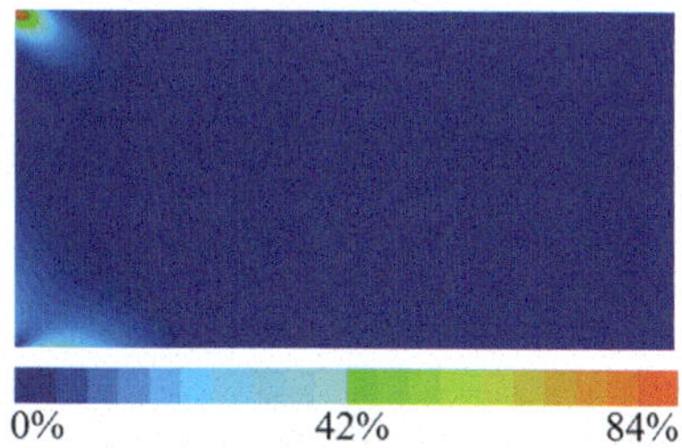

Fig. 12.6 Function δ_z for $z \in [0, 0.1L]$

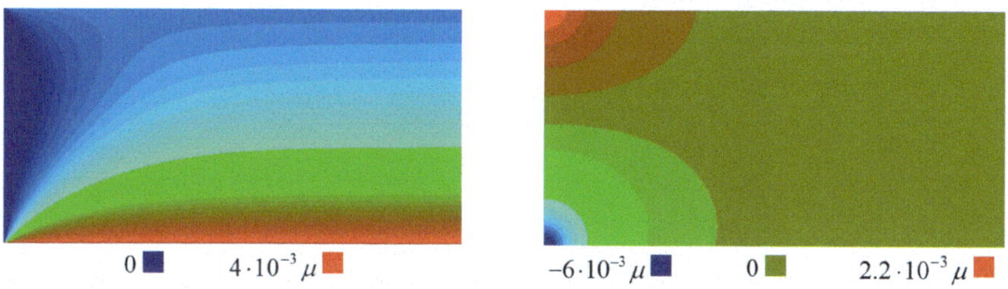

Fig. 12.7 On the left: $\hat{\sigma}^{\mathrm{I}}_{r\phi}$. On the right: $\hat{\sigma}^{\mathrm{I}}_{\phi z}$; $z \in [0, 0.1L]$.

The comparison results provided reveal that significant differences in the considered analytical and FE stresses are observed in the clamp vicinity; in the working area of the cylinder, the difference does not exceed 1%. This makes it possible to use the analytical prestress laws of change as a model simplification instead of providing additional calculation of FE solutions.

12.6 Inverse Problem of Prestress Identification

12.6.1 Problem Statement

Let us consider the inverse problem of restoring the prestress amplitudes in a cylinder in the presence of additional data on the displacements measured in the cylinder region Ω. Within such a formulation, the inverse problem is linear and does not require the use of special iterative-regularization approaches [19]. To investigate it, we shall use the weak statement (12.8), replacing the test functions with the components of the displacement field: $v_j = u_j, j = r, \phi, z$; we also put $\omega = 0$ and rearrange the equation terms:

$$\int\limits_{\Omega} \left(\sigma^0_{r\phi} K^{uu}_{r\phi} + \sigma^0_{rr} K^{uu}_{rr} + \sigma^0_{\phi\phi} K^{uu}_{\phi\phi} + \sigma^0_{zz} K^{uu}_{zz} \right) \mathrm{d}\Omega = B^u \qquad (12.12)$$

where we denote

$$B^u = F^u - \int\limits_{\Omega} K^{uu}_{\lambda\mu} \mathrm{d}\Omega, \qquad K^{uu}_{\lambda\mu} = \lambda K^{uu}_{\lambda} + \mu K^{uu}_{\mu},$$

$$F^u = \int\limits_{R_1}^{R_2} \left(T_{zr}u_r + T_{z\phi}u_\phi + T_{zz}u_z\right)\big|_{z=L} r\mathrm{d}r + R_1 \int\limits_{0}^{L} \left(T_{rr}u_r + T_{r\phi}u_\phi + T_{rz}u_z\right)\big|_{r=R_1} \mathrm{d}z$$

12.6.2 Problem Solution

After substituting the prestress in the form

$$\sigma^{\mathrm{I}}_{r\phi} = \varepsilon_{\mathrm{I}} \frac{R_2^2}{r^2}, \sigma^{\mathrm{II}}_{rr} = \varepsilon_{\mathrm{II}} \frac{R_1^2\left(r^2 - R_2^2\right)}{r^2\left(R_2^2 - R_1^2\right)}, \sigma^{\mathrm{II}}_{\phi\phi} = \varepsilon_{\mathrm{II}} \frac{R_1^2\left(r^2 + R_2^2\right)}{r^2\left(R_2^2 - R_1^2\right)}, \sigma^{\mathrm{III}}_{zz} = \varepsilon_{\mathrm{III}}$$

(on the basis of the laws (12.9)-(12.11)) into (12.12), we get the system of linear algebraic equations $a_{ji}\varepsilon_i = b_j, \quad i,j = \overline{1,3}$, whose coefficients are calculated according to

$$
\begin{aligned}
a_{j1} &= \int\limits_{\Omega} \frac{R_2^2}{r^2} K^{uu}_{r\phi}\left(u_r^{(j)}, u_\phi^{(j)}, u_z^{(j)}\right) \mathrm{d}\Omega, \\
a_{j2} &= \int\limits_{\Omega} \left[\frac{R_1^2\left(r^2 - R_2^2\right)}{r^2\left(R_2^2 - R_1^2\right)} K^{uu}_{rr}\left(u_r^{(j)}, u_\phi^{(j)}, u_z^{(j)}\right) + \frac{R_1^2\left(r^2 + R_2^2\right)}{r^2\left(R_2^2 - R_1^2\right)} K^{uu}_{\phi\phi}\left(u_r^{(j)}, u_\phi^{(j)}, u_z^{(j)}\right) \right] \mathrm{d}\Omega, \\
a_{j3} &= \int\limits_{\Omega} K^{uu}_{zz}\left(u_r^{(j)}, u_\phi^{(j)}, u_z^{(j)}\right) \mathrm{d}\Omega, \\
b_j &= F^u\left(u_r^{(j)}, u_\phi^{(j)}, u_z^{(j)}\right) - \int\limits_{\Omega} K^{uu}_{\lambda\mu}\left(\lambda, \mu, u_r^{(j)}, u_\phi^{(j)}, u_z^{(j)}\right) \mathrm{d}\Omega,
\end{aligned}
$$

where $u_r^{(j)}, u_\phi^{(j)}, u_z^{(j)}$ – the corresponding solutions of the direct problem (simulating the measurements data) for the jth loading type.

12.6.3 Computational Experiments

In Table 12.1 we present the results of identification of various combinations of values $\varepsilon_{\mathrm{I}}, \varepsilon_{\mathrm{II}}, \varepsilon_{\mathrm{III}}$, characterizing the amplitudes of the considered types of inhomogeneous prestress in the cylinder. From the table it can be seen that the described procedure for identifying prestress with amplitudes not lower than 10^{-5} with respect to the shear modulus gives a sufficiently high recovery accuracy. In this case, the restored amplitudes can also be used to judge the type of prestressed state in the cylinder.

Table 12.1 Reconstruction of the inhomogeneous prestress amplitudes (*tilde* means an approximate value calculated with an accuracy of at least 10^{-6})

Test	Original prestress amplitudes $\times 10^{-3}$			Reconstruction $\times 10^{-3}$		
	ε_{I}	$\varepsilon_{\mathrm{II}}$	$\varepsilon_{\mathrm{III}}$	$\tilde{\varepsilon}_{\mathrm{I}}$	$\tilde{\varepsilon}_{\mathrm{II}}$	$\tilde{\varepsilon}_{\mathrm{III}}$
1	1	1	1	~ 1	~ 1	~ 1
2	0.1	1	1	~ 0.1	~ 1	~ 1
3	10^{-2}	1	1	$0.96 \cdot 10^{-2}$	~ 1	~ 1
4	10^{-3}	1	1	$0.61 \cdot 10^{-3}$	~ 1	~ 1
5	1	10^{-6}	1	~ 1	$0.96 \cdot 10^{-6}$	~ 1
6	1	1	10^{-6}	~ 1	~ 1	$1.07 \cdot 10^{-6}$

12.7 Conclusion

- A weak problem statement is formulated for a finite prestressed cylinder. The prestress tensor is given by four non-zero components depending on the coordinates.
- There are two ways to set prestress fields for a cylinder clamped by one end. The first one is based on constructing the prestress field as a finite element solution of the corresponding initial statics problem on the deformation of a cylinder under some preload; the cases of initial inflation, torsion and tension are considered. The second way is based on using the analytical solutions for a ring section and infinite cylinder. A significant difference between the analytical and finite element solutions is observed in a fairly small neighborhood of the clamped end of the cylinder. The possibility of using the analytical solutions instead of numerical ones is demonstrated by analyzing the prestress effect on the cylinder's deformation characteristics. The calculated displacement fields in the problems with analytical and numerical prestress differ insignificantly due to the cylinder fixation conditions.
- The influence of the considered prestress types on the field of small superimposed displacements is studied; sensitivity analysis was carried out for tensile, inflating and twisting load regimes. The computational experiments performed have shown that the highest prestress sensitivity is observed mainly in the vicinity of the free edge.
- The inverse problem on the reconstruction of the intensities of the three considered prestress types is studied, when the additional data on the displacements in the entire area is provided. The system of linear equations for finding the prestress amplitudes is obtained based on the weak formulation equations. In the absence of input noise, a sufficiently high accuracy of the inverse solution is observed.

Acknowledgements The study was supported by the Russian Science Foundation, grant #18-71-10045, https://rscf.ru/project/18-71-10045/.

References

1. G.S. Schajer. *Practical Residual Stress Measurement Methods*. John Wiley & Sons, Ltd., 2013.
2. L. Chunlei, H. Qiang, L. Yijie, L. Xiucheng, W. Bin. Investigation of wave propagation in double cylindrical rods considering the effect of prestress. *Journal of Sound and Vibration*. 353:164-180, 2015.
3. W. Guannan, D. Leiting, N.A. Satya. Direct and inverse multi-scale analyses of arbitrarily functionally graded layered hollow cylinders (discs), with different shaped reinforcements, under harmonic loads. COMPOSITE STRUCTURES. 188:425-437, 2018
4. W. Ma, H. Zhang, W. Zhu, F. Xu, C. Yang. Study on Residual Stress of Welded Hoop Structure. *Applied Sciences*. 10(8):2838, 2020.
5. M. Jażdżewska, M. Bartmański. Nanotubular oxide layer formed on helix surfaces of dental screw implants. *Coatings*. 11(2):115, 2021.
6. W. Chunmei, W. Junping, W. Ruishu, Z. Ran. A locking-free weak Galerkin finite element method for elasticity problems in the primal formulation. *Journal of Computational and Applied Mathematics*. 307:346-366, 2016.
7. R.D. Nedin, A.O. Vatulyan. Advances in Modeling and Identification of Prestresses in Modern Materials, in: *Advanced Materials Modelling for Mechanical, Medical and Biological Applications* (ed. by H. Altenbach, V.A. Eremeyev, A. Galybin, A. Vasiliev), pp. 357-374, 2020 (Advanced Structured Materials, Vol. 155).
8. C.A. Truesdell. *A First Course in Rational Continuum Mechanics, Vol. 1: General Concepts*. Academic Press, 1977.
9. A. Hoger. On the determination of residual stress in an elastic body. *Journal of Elasticity*. 16:303-324, 1986.
10. R.L. Robertson. Determining residual stress from boundary. Measurements: A linearized approach. *Journal of Elasticity*. 52:63-73, 1998.
11. A.N. Guz, F.G. Makhort, O.I. Gushcha. *Introduction to Acoustoelasticity (in Russ.)*. Kiev: Naukova Dumka, 1977.
12. R.D. Nedin, V.V. Dudarev, A.O. Vatulyan. Some aspects of modeling and identification of inhomogeneous residual stress. *Engineering Structures*. 151:391-405, 2017.
13. R.D. Nedin, A.O. Vatulyan, V.V. Dudarev, I.V. Bogachev. Detection of nonuniform residual strain in a pipe. *Int J Solids Struct*. 139-140:121-128, 2018.
14. R.D. Nedin, A.O. Vatulyan. Inverse problem of non-homogeneous residual stress identification in thin plates. *Int J Solids Struct*. 50:2107-2114, 2013.
15. R.D. Nedin, A.O. Vatulyan. Concerning one approach to the reconstruction of heterogeneous residual stress in plate. *ZAMM J. Appl. Math. Mech.* 94(1-2):142-149, 2014.
16. R.D. Nedin, A.O. Vatulyan, I.V. Bogachev. Direct and inverse problems for prestressed functionally graded plates in the framework of the Timoshenko model. *Mathematical Methods in the Applied Sciences*. 41(4):1600-1618, 2018.
17. A.O. Vatulyan, R.D. Nedin. On the Reconstruction of the Characteristics of the Plane Initial Stress State. *Mechanics of Solids*. 55(5):624-632, 2020.
18. V.O. Yurov, R.D. Nedin, A.O. Vatulyan. Oscillations of a non-uniform finite hollow cylinder under conditions of complex prestressed state. *Engineering Structures* 221:111019, 2020.
19. A.O. Vatulyan, V.V. Dudarev, R.D. Nedin. *Initial stresses: modeling and identification (in Russ.)*. Monograph. Rostov-on-Don: SFedU Publishing House, 2015.
20. R.D. Nedin, V.V. Dudarev, A.O. Vatulyan. Vibrations of inhomogeneous piezoelectric bodies in conditions of residual stress-strain state. *Appl. Math. Modelling*. 63:219–242, 2018

Chapter 13
A Method of the J_R-curve Determination Using Linear Normalization

Oxana G. Rybakina and Olga A. Strogonova

Abstract An analytical method for the J_R-curve determination is proposed. The method is based on the principle of load normalization by J. Landes and the linear relationship between the normalized load gradient and the crack length increment established in the paper by E. Reese and K. Schwalbe. As long as the final values of the load and the crack length increment are known, the method makes it possible to calculate the crack length increment for each load value throughout the entire loading process. Further, using the experimental relationship between the load and the displacement of the point of application of the load, the J_R–curve is determined. An example of calculation and comparison with the results obtained by other methods is given.

Key words: Crack, J_R-curve, Load normalization, Load gradient

13.1 Introduction

Crack extension conditions based on the concepts of linear fracture mechanics and related structural requirements should be applied when the behavior of the material is predominantly elastic and the fracture is brittle. In cases where the material is capable of large plastic deformations, the use of linear fracture mechanics provides a conservative calculation, as a result of which the performance of the material is not fully exploited. Quantification of various stages of the ductile fracture process can be based on the use of the J–integral, which is a characteristic parameter of the crack front. At present, methods have been developed for calculating this parameter, which is a function of the applied load, crack geometry, and mechanical properties of the material. To assess the critical state of a structural element containing a crack,

---

Oxana G. Rybakina · Olga A. Strogonova
Krylov State Research Centre, 44 Moskovskoe shosse, St.Petersburg, Russian Federation,
e-mail: rybakina37@mail.ru,o.a.strogonova@gmail.com

H. Altenbach et al. (eds.), *Advances in Solid and Fracture Mechanics*,
Advanced Structured Materials 180,
https://doi.org/10.1007/978-3-031-18393-5_13

as in conventional strength calculations, the value of this parameter is compared with the critical value characterizing a specific material. In the presence of large plastic deformations, critical situations become more diverse, namely, we can talk about crack initiation, its stable growth, and unstable propagation.

The J_R–curve is the dependence of the values of the J–integral on the values of the increment of the crack length, obtained as a result of testing the samples and is a characteristic of the crack resistance of the material. Standards have been developed with the help of which J_R–curves are constructed [1]. The simplest, but time-consuming method is to test a series of samples. Each sample of the series with the initial crack length a_0 is loaded once and then unloaded in order to obtain, after testing the entire series, two data arrays: an array of load values P corresponding to the beginning of unloading, and an array of values of the plastic displacement component υ_{pl} . After unloading, the length of the crack a is fixed on each sample (using heat treatment or some other method), and then the sample is loaded again until the final failure. Thus, the third data array is obtained — an array of crack length increment values $\Delta a = a - a_0$. These arrays contain as many points as there are samples tested; according to these data, a J_R–curve is constructed. A less laborious method uses a single sample that is repeatedly partially unloaded and reloaded to continue the test. The "load–crack opening" diagram in the elastic section of unloading and subsequent loading makes it possible to determine the length of the crack using the relations of linear fracture mechanics (this method is called the elastic compliance method). However, as practice shows, this method gives reliable results for moderately ductile metals, in which the region of plastic deformations in the vicinity of the crack front is small. For highly plastic materials, this method cannot be used to determine the length of the crack with the required accuracy.

To assess the crack resistance of highly plastic materials, the efforts of researchers were directed to the creation of other methods for determining the crack length increment when testing samples of various geometries. The papers [2, 3] introduced the load separation principle. The load P applied to the sample is represented as a product of two functions, one of which depends only on the geometry of the sample and the size of the crack, and the other only on the plastic component of the displacement of the point of application of the load. The form of these functions, based on experimental studies, is specified for the most common sample geometries. In [4], the concept of a normalized load gradient was introduced and, based on experimental data, it was shown that there is a linear relationship between the normalized load gradient and the crack length increment. Based on these results, a graphical procedure was proposed that allows plotting the dependence of the normalized load on the crack length increment. To determine it, it is necessary to have a diagram "$P - \upsilon_{pl}$", obtained on one sample, and unloading is carried out after the load reaches the maximum value $P = P_{max}$ and some decrease to the value $P = P_f < P_{max}$ occurred. The corresponding values $P = P_f$ and $a = a_f$ should be recorded.

In this paper, analytical dependencies have been obtained that make it possible to determine the J_R-curve in the presence of the above diagram "$P - \upsilon_{pl}$ ", obtained on one standard sample, and the values of the load and crack length corresponding to the moment of unloading.

13.2 Statement of the Problem

We assume that the load applied to the sample can be represented as:

$$P(a, v_{pl}) = G(a/W)H(v_{pl}/W) \quad (W \text{ is the sample width}).$$

Based on numerous experimental studies carried out on standard samples used to construct the J$_R$–curve, in [2] the form of the function was proposed:

$$G(a/W) = BW\left(1 - \frac{a}{W}\right)^{\eta_{pl}} \quad (B \text{ is the sample thickness}),$$

where the parameter η_{pl} depends on the sample geometry and does not depend on the crack size. In [3], specific numerical values η_{pl} were obtained for various sample geometries: $\eta_{pl} = 2,13$ for tensile tests, $\eta_{pl} = 1,94$ for three-point bending tests, $\eta_{pl} = 0,96$ for tests of a compact sample. Following [2], we introduce the concept of normalized load P_N:

$$P_N = \frac{P}{BW\left(1 - \frac{a}{W}\right)^{\eta_{pl}}}.$$

Obviously, if the normalized load $P_N = H(v_{pl})$ is known, then the initial load $P(a, v_{pl})$ is also known. Let us introduce the gradient of the normalized load ΔP_N, corresponding to a fixed level of load P and due to the increment of the crack length Δa , [4]:

$$\Delta P_N(\Delta a) = P_N(a_0 + \Delta a) - P_N(a_0) = \frac{P}{BW}\left(\frac{1}{\left(1 - \frac{a_0 + \Delta a}{W}\right)^{\eta_{pl}}} - \frac{1}{\left(1 - \frac{a_0}{W}\right)^{\eta_{pl}}}\right). \quad (13.1)$$

For fixed values of a_0, based on this formula a series of curves $\Delta P_N = \Delta P_N(\Delta a)$ can be obtained. In [4], based on a large amount of experimental data obtained on various materials, it was shown that the dependence $\Delta P_N = \Delta P_N(\Delta a)$ is linear over the entire range of P and Δa . The corresponding straight line on the plane $(\Delta a, \Delta P_N)$ passes through the point $(\Delta a_f, \Delta P_N(\Delta a_f))$ and touches the curve $\Delta P_N = \Delta P_N(\Delta a)$ corresponding to the load $P = P_{max}$. This graphical construction completely determines the parameters of the linear dependence. With a known dependence $\Delta P_N = \Delta P_N(\Delta a)$, using formula (13.1), for any value of the crack length increment Δa, the load value P is determined. In this paper, the J$_R$- curve is determined analytically on the basis of the experimental diagram "$P - v_{pl}$" and the values P_f and a_f obtained on one standard sample.

13.3 Obtaining the Solution

The dependence of the normalized load gradient on the crack increment has the form:

$$\Delta P_{\mathrm{N}}(\Delta a) = \Delta P_{\mathrm{N}}(\Delta a_{\mathrm{f}}) + (\Delta P_{\mathrm{N}})'|_{P=P_{\max}}(\Delta a - \Delta a_{\mathrm{f}}). \tag{13.2}$$

For $(\Delta P_{\mathrm{N}})'|_{P=P_{\max};\,\Delta a(P=P_{\max})}$ from (13.1) we get:

$$(\Delta P_{\mathrm{N}})'|_{P=P_{\max};\,\Delta a(P=P_{\max})} = \frac{\eta_{\mathrm{pl}} P_{\max}}{BW^2\left(1 - \frac{a_0 + \Delta a|_{P=P_{\max}}}{W}\right)^{\eta_{\mathrm{pl}}+1}}. \tag{13.3}$$

Relations (13.1) and (13.2) are satisfied at the point corresponding to the maximum load

$$\Delta a|_{P=P_{\max}}, \Delta P_{\mathrm{N}}(\Delta a|_{P=P_{\max}}),$$

which leads to the equations:

$$\Delta P_{\mathrm{N}}(\Delta a|_{P=P_{\max}}) = \frac{P_{\max}}{BW}\left(\frac{1}{\left(1 - \frac{\Delta a|_{P=P_{\max}}}{W}\right)^{\eta_{\mathrm{pl}}}} - \frac{1}{\left(1 - \frac{a_0}{W}\right)^{\eta_{\mathrm{pl}}}}\right); \tag{13.4}$$

$$\Delta P_{\mathrm{N}}(\Delta a|_{P=P_{\max}}) = \Delta P_{\mathrm{N}}(\Delta a_{\mathrm{f}}) + (\Delta P_{\mathrm{N}})'|_{P=P_{\max};\,\Delta a(P=P_{\max})}(\Delta a|_{P=P_{\max}} - \Delta a_{\mathrm{f}}) \tag{13.5}$$

Substituting (13.3) into (13.5) leads to a system of equations for determining $\Delta a|_{P=P_{\max}}$ and $\Delta P_{\mathrm{N}}(\Delta a|_{P=P_{\max}})$. After transformations, we obtain a transcendental equation for $\Delta a|_{P=P_{\max}}$; then from relation (13.3) we find $(\Delta P_{\mathrm{N}})'|_{P=P_{\max};\,\Delta a(P=P_{\max})}$. For an arbitrary value of the crack length increment $\Delta a \leq \Delta a_{\mathrm{f}}$ from (13.2) we determine $\Delta P_{\mathrm{N}}(\Delta a)$ and from (13.1) the value of the load P on the experimental diagram "$P - v_{\mathrm{pl}}$".

Further in this section, to simplify the notation, we will use the following:

$$\frac{a_0}{W} \sim a_0; \quad \frac{a_0 + \Delta a_{\mathrm{f}}}{W} \sim a_{\mathrm{f}}; \quad \frac{a_0 + \Delta a|_{P=P_{\max}}}{W} \sim a_{\mathrm{m}};$$

$$\frac{\Delta a}{W} \sim \Delta a; \quad \frac{\Delta a|_{P=P_{\max}}}{W} \sim \Delta a_{\mathrm{m}}; \quad \frac{\Delta a_{\mathrm{f}}}{W} \sim \Delta a_{\mathrm{f}}.$$

The transcendental equation for the crack length increment at maximum load Δa_{m} is written as:

$$\frac{P_{\max}}{BW(1-(a_0+\Delta a_{\mathrm{m}}))^{\eta_{\mathrm{pl}}}} - \frac{P_{\max}}{BW(1-a_0)^{\eta_{\mathrm{pl}}}} =$$

$$= \frac{P_{\mathrm{f}}}{BW(1-a_{\mathrm{f}})^{\eta_{\mathrm{pl}}}} - \frac{P_{\mathrm{f}}}{BW(1-a_0)^{\eta_{\mathrm{pl}}}} + \frac{\eta_{\mathrm{pl}} P_{\max}}{BW^2(1-(a_0+\Delta a_{\mathrm{m}}))^{\eta_{\mathrm{pl}}+1}}(\Delta a_{\mathrm{m}} - \Delta a_{\mathrm{f}})$$

or after transformations and simplifications

$$\frac{1}{(1-(a_0+\Delta a_m))^{\eta_{pl}+1}} - \frac{\eta_{pl}+1}{\eta_{pl}(1-a_f)(1-(a_0+\Delta a_m))^{\eta_{pl}}} + \\ + \left(1-\frac{P_f}{P_{max}}\right)\frac{1}{\eta_{pl}(1-a_f)(1-a_0)^{\eta_{pl}}} + \frac{P_f/P_{max}}{\eta_{pl}(1-a_f)^{\eta_{pl}+1}} = 0 \tag{13.6}$$

and solved numerically.

Next, we consider the determination of a J$_R$–curve using tests of a standard sample for a three-point bend; this type of test is one of the most common. As mentioned earlier, in this case $\eta_{pl} = 1{,}94$. However, taking into account the scatter of experimental data, we round off the value of the coefficient η_{pl} and then take $\eta_{pl} = 2$. Then Eq. (13.6) reduces to an algebraic equation of the 3rd degree with respect to $z = \frac{1}{1-(a_0+\Delta a_m)}$:

$$z^3 + az^2 + bz + c = 0, \tag{13.7}$$

where

$$a = -\frac{3}{2(1-a_f)}; \quad b = 0; \quad c = \left(1-\frac{P_f}{P_{max}}\right)\frac{1}{2(1-a_f)(1-a_0)^2} + \frac{P_f/P_{max}}{2(1-a_f)^3}.$$

The solution of Eq. (13.7) in accordance with [5], taking into account $c > 0, a < 0$, gives:

$$\cos\alpha = 2\left(1-\frac{P_f}{P_{max}}\right)\left[1-\left(\frac{1-a_f}{1-a_0}\right)^2\right] - 1.$$

Given the expression for z, we get three roots:

$$\frac{1}{1-(a_0+\Delta a_m)} = \frac{1}{1-a_f}\left(\frac{1}{2}+\cos\frac{\alpha}{3}\right),$$

$$\frac{1}{1-(a_0+\Delta a_m)} = \frac{1}{1-a_f}\left(\frac{1}{2}-\cos\left(\frac{\alpha}{3}\pm\frac{\pi}{3}\right)\right).$$

From the condition $\Delta a_m < \Delta a_f$ we have

$$\frac{1}{1-(a_0+\Delta a_m)} < \frac{1}{1-a_f}.$$

The solution that satisfies the indicated inequality is written as:

$$\frac{1}{1-(a_0+\Delta a_m)} = \frac{1}{1-a_f}\left(\frac{1}{2}+\cos\beta\right),$$

where

$$\cos\beta = \cos\left\{\frac{2\pi}{3} - \frac{1}{3}\arccos\left[2\left(1-\frac{P_f}{P_{max}}\right)\left[1-\left(\frac{1-a_f}{1-a_0}\right)^2\right]-1\right]\right\}.$$

Finally,

$$\Delta a_{\mathrm{m}} = 1 - a_0 - \frac{1 - a_{\mathrm{f}}}{\frac{1}{2} + \cos\beta}.$$

Note that from the condition

$$-\frac{1}{2} + \frac{1 - a_{\mathrm{f}}}{1 - a_0} < \cos\beta < \frac{1}{2}$$

follows the inequality

$$\frac{P_{\mathrm{f}}}{P_{\max}} > \frac{2\left(\frac{1-a_{\mathrm{f}}}{1-a_0}\right)^2}{1 + \frac{1-a_{\mathrm{f}}}{1-a_0}}$$

that is equivalent to

$$a_{\mathrm{f}} > 1 - (1 - a_0)\frac{\frac{P_{\mathrm{f}}}{P_{\max}} + \sqrt{\left(\frac{P_{\mathrm{f}}}{P_{\max}}\right)^2 + \frac{8P_{\mathrm{f}}}{P_{\max}}}}{4},$$

imposing some restrictions on the values $P_{\mathrm{f}}/P_{\max}$, a_0 and a_{f}. Before starting the processing of experimental data, it is necessary to make sure that these conditions are met. Otherwise, the proposed method cannot be used.

Let us return to consideration of relation (13.2). It is easy to see that the plot of dependence $\Delta P_{\mathrm{N}}(\Delta a)$, generally speaking, does not pass through the origin of the coordinates, which does not allow using the results obtained for small values of Δa. Let us represent the dependence $\Delta P_{\mathrm{N}}(\Delta a)$ on the section $0 \le \Delta a \le \Delta a_{\mathrm{m}}$ in the form

$$\Delta P_{\mathrm{N}}(\Delta a) = A_1 \Delta a + A_2 (\Delta a)^n, \tag{13.8}$$

moreover, the coefficients A_1 and A_2 are determined from the condition that at the transition point from dependence (13.8) to linear ($\Delta a = \Delta a_{\mathrm{m}}$) the function $\Delta P_{\mathrm{N}}(\Delta a)$ and its derivative with respect Δa are continuous, n satisfies the condition $1 < n \le 2$.

From these conditions we get:

$$A_1 = (\Delta P_{\mathrm{N}})'|_{P=P_{\max}} + \frac{n\left(\Delta P_{\mathrm{N}}(\Delta a_{\mathrm{f}}) - (\Delta P_{\mathrm{N}})'|_{P=P_{\max}} \Delta a_{\mathrm{f}}\right)}{(n-1)\Delta a_{\mathrm{m}}},$$

$$A_2 = -\frac{n\left(\Delta P_{\mathrm{N}}(\Delta a_{\mathrm{f}}) - (\Delta P_{\mathrm{N}})'|_{P=P_{\max}} \Delta a_{\mathrm{f}}\right)}{(n-1)\Delta a_{\mathrm{m}}^2}.$$

At $\Delta a \to 0$,

$$A_1 = \frac{dP_{\mathrm{N}}}{d(\Delta a)} = \frac{2P_0}{BW^2(1-a_0)^3},$$

where P_0 is the load at which the crack starts to move. From here we get the expression for P_0:

$$\frac{2P_0}{BW^2(1-a_0)^3} = (\Delta P_N)'|_{P=P_{max}} + \frac{n\left(\Delta P_N(\Delta a_f) - (\Delta P_N)'|_{P=P_{max}}\Delta a_f\right)}{(n-1)\Delta a_m} \quad (13.9)$$

The value P_0 is obtained experimentally, for example, using the acoustic emission method, after which the value of n is determined from relation (13.9).

Now the dependence $\Delta P_N(\Delta a)$ is built on the entire range $0 \le \Delta a \le \Delta a_f$. Next, from (13.3) $(\Delta P_N)'|_{P=P_{max};\,\Delta a(P=P_{max})}$ is determined. For an arbitrary Δa from (13.2) and (13.8) we find $\Delta P_N(\Delta a)$, then using (13.1) the load P.

13.4 Examples of Application of the Method and Analysis of Results

An experimental study was carried out on a steel sample of the SENB type (single edge notch bending) [1], yield strength $\sigma_{0,2} = 560$ MPa. The characteristics of the sample required for the calculation are given in Table 13.1, the results of calculating the parameters that determine thc normalized load and its gradient are presented in Table 13.2.

Figures 13.1 and 13.2 show the dependencies $\Delta P_N(\Delta a)$ and $P_N(\Delta a)$, and the dotted lines in the range $0 \le \Delta a \le \Delta a_m$ correspond to dependence (13.8) at $n = 1.5$, and the solid lines in the range $\Delta a_m \le \Delta a \le \Delta a_f$ correspond to the linear dependence of the normalized load gradient on the crack length increment. In Fig. 13.2, the dashdotted line shows the dependence of the reduced load

$$\bar{P} = \frac{P}{BW\left(1 - \frac{a_0}{W}\right)^2}$$

on the increment in the length of the crack.

Table 13.1 The characteristics of the sample required for the calculation

a_0, mm	W, mm	B, mm	a_f, mm	P_f, N	Δa_f, mm	$P_N(\Delta a_f)$, N/mm^2	$\Delta P_N(\Delta a_f)$, N/mm^2	P_{max}, N	P_f/P_{max}
24.73	50.1	22.85	29.71	65377	4.98	344.8	122.1	78409	0.83

Table 13.2 The results of calculating the parameters that determine the normalized load and its gradient

Δa_m, mm	$\Delta P'_N$ N/mm^3	A_1 N/mm^3	A_2 N/mm$^{3.5}$	P_0/P_{max}
1.45	25.13	18.78	3.51	0.89

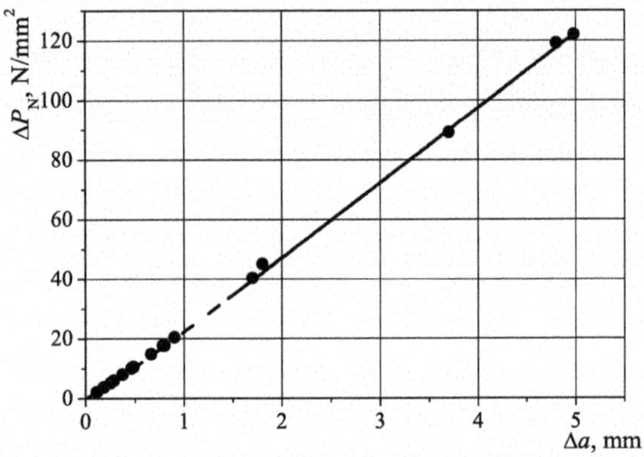

Fig. 13.1 Dependence of the normalized load gradient on the crack length increment

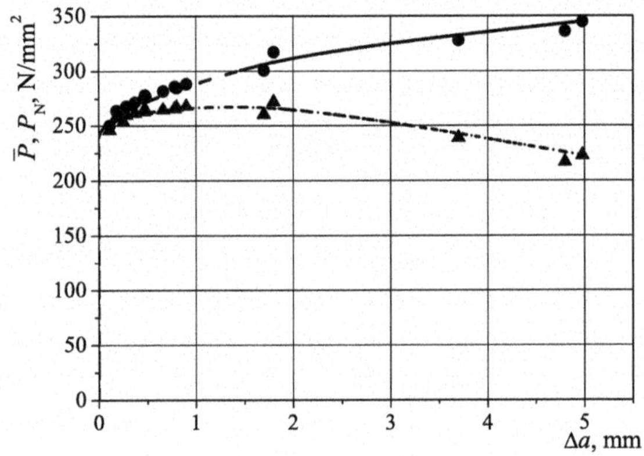

Fig. 13.2 Dependence of the normalized load and the reduced load on the crack length increment.

Another six specimens made from the same steel were tested to different load levels and after unloading, the crack length a_f was measured on them. The relevant data are presented in Table 13.3.

Table 13.3 Test data for different load levels

No.	a_0, mm	W, mm	B, mm	a_f, mm	Δa, mm	P, N	$P_N(\Delta a)$, N/mm^2	$\Delta P_N(\Delta a)$, N/mm^2	$\bar{P}(\Delta a)$, N/mm^2
1	24.90	50.10	22.85	25.70	0.80	77324	284.8	17.79	267.0
2	25.20	50.00	22.85	26.10	0.90	75268	288.3	20.5	267.8
3	25.70	50.10	22.90	27.40	1.70	70821	300.9	40.4	260.2
4	25.80	50.10	22.90	27.60	1.80	73404	317.2	45.3	272.0
5	24.90	50.10	22.85	28.60	3.70	69142	328.0	89.2	238.7
6	25.70	50.10	22.90	30.50	4.80	59008	336.0	119.2	216.8

In Figs. 13.1 and 13.2, dots show the results obtained on samples 1–6. The experimental data obtained on samples 1–6 correlate well with the calculated curves. Good agreement between the results obtained by methods that differ significantly from each other indicates the applicability of the load separation principle, the possibility of using the parameter value $\eta_{pl} = 2$ and confirms the presence of a linear relationship between the normalized load gradient $\Delta P_N(\Delta a)$ and the crack length increment Δa.

After the dependence between the load and the increment of the crack length is established, it is not difficult to calculate the values of the J-integral along the entire path of loading the sample. At the point with the number i, which corresponds to the values of the load $P_{(i)}$, the plastic component of the displacement $\upsilon_{pl(i)}$ and the length of the crack $a_{(i)}$, the value of the J-integral $J_{(i)}$ is determined as:

$$J_{(i)} = \frac{K_{I(i)}^2}{E'} + J_{pl(i)},$$

$$J_{pl(i)} = \left[J_{pl(i-1)} + \frac{2}{(W - a_{(i-1)})B} \int_{\upsilon_{pl(i-1)}}^{\upsilon_{pl(i)}} P(\upsilon_{pl}) d\upsilon_{pl} \right] \frac{W - a_{(i)}}{W - a_{(i-1)}},$$

where $K_{I(i)}$ is the stress intensity factor and E' is the modulus of normal elasticity. The results are presented in Fig. 13.3. The resulting J_R–curve is a convenient tool in the calculation of crack extension and evaluation of the stability of crack propagation.

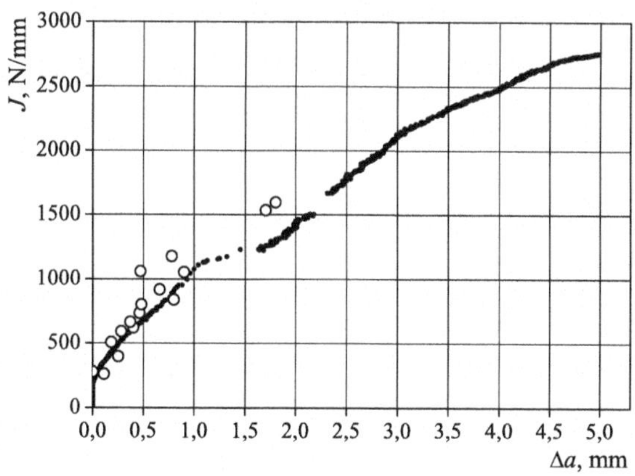

Fig. 13.3 J_R–curve.

13.5 Conclusion

The presented results make it possible to determine the J_R–curve with the minimum amount of necessary experimental data using analytical methods for processing these data. To obtain a fracture curve, it is sufficient to test one sample while recording the load and the displacement of the load application point and evaluate the final load and crack length at the stage after the load reaches its maximum value. The method is based on experimental facts (the load separation principle, the numerical value of the parameter η_{pl}, linear dependence $\Delta P_{\text{N}}(\Delta a)$), the limits of applicability of which are subject to further research.

References

1. BS 7448-4: Fracture mechanics toughness tests — Part 4: Method for determination of fracture resistance curves and initiation values for stable crack extension in metallic materials, 1997.
2. J.D. Landes, Z. Zhou, K. Lee and, R. Herrera. Normalization method for developing J-R curves with LMN function. *Journal of Testing and Evaluation*. 19(4):305–311, 1991.
3. M.H. Sharobeam, J.D. Landes. The load separation criterion and methodology in ductile fracture mechanics. *International Journal of Fracture*. 47(2):81–104, 1991.
4. E.D. Reese, K.H. Schwalbe. The linear normalization technique — an alternative procedure for determining J-R curves from a single specimen test record based on Landes' normalization method. *Fatigue & Fracture of Engineering Materials & Structures*. 16(3):271–280, 1993.
5. G.A. Korn, T.M. Korn. *Mathematical Handbook for Scienists and Engineers*. Dover Publications, Inc., New York, 2000.

Chapter 14
Application of Galerkin's Method to Buckling of Functionally Graded or Stepped Columns

Joel Storch and Isaac Elishakoff

Abstract We address several buckling problems for nonuniform columns obtaining exact solutions as well as approximate solutions based on Galerkin's method. It is shown that faster convergence is obtained using the higher order self-adjoint form of the governing differential equation. A detailed derivation is given for the buckling of heavy nonuniform columns. For the case of a heavy stepped column, in which the classical Galerkin method fails, we present a modified method employing generalized functions which displays excellent convergence to the exact solution.

Key words: Buckling, Galerkin method, Stepped columns, Functionally graded columns, Generalized functions

14.1 Introduction

The exact solutions for buckling loads of a homogenous uniform beam can be found in most textbooks about the mechanics of solids. The number of closed form solutions for non-uniform columns are extremely limited. In the 18th century, Leonhard Euler (1759) [1] gave the first closed-form solution in which the flexural rigidity is given as a polynomial: $(a+bx/L)^m$, where a,b are constants, L is the column's length, m is a positive integer, and x is an axial coordinate. The next closed form solution was provided by Engesser (1899) [2]. In the 20th century, Duncan (1937) [3] and Elishakoff (2000) [4] contributed additional closed-form solutions. Novel

Joel Storch
Department of Mechanical Engineering, California StateUniversity, Northridge CA 91330, USA,
e-mail: jas405@caa.columbia.edu

Isaac Elishakoff
Department of Ocean and Mechanical Engineering, Florida Atlantic University, Boca Raton, FL 33431-0991, USA,
e-mail: elishako@fau.edu

H. Altenbach et al. (eds.), *Advances in Solid and Fracture Mechanics*,
Advanced Structured Materials 180,
https://doi.org/10.1007/978-3-031-18393-5_14

solutions were reported in the monographs by Elishakoff (2005) [5] and Elishakoff, Pentaras and Gentilini (2010) [6] devoted to mechanics of functionally graded material structures. Associated works inspired by [5, 6] are those by Elishakoff, Eisenberger and Delmas (2016) [7], Ayadoglu (2008) [8], Li (2009) [9], Maalavi (2009) [10], Singh and Li (2009) [11], Coskun (2009) [12], Darbandi, Firouz-Abadi and Haddapour (2010) [13], Huang and Li (2011) [14], Huang and Luo (2011) [15], Babilio (2013) [16], Shan and Chen (2013) [17], Krutyi (2016) [18] and Ioakimidis (2017) [19], *inter alia.*

This study is committed to attain new closed form solutions for a column that is simply supported-clamped for inhomogeneous, functionally graded, or stepped columns. An assumption was made that the column has a uniform cross sectional area throughout its length, but the modulus of elasticity varies along the column. Papers by Byskov [20], Fox and Kapoor [21], Vinogradov [22], Bert et al [23], Livne [24], Canfield [25], Elishakoff et al. [26], Storch and Elishakoff [27], Elishakoff and Boutur [28] all deal with different aspects of associated eigenvalue problems.

Carroll [29] writes in his paper on foundations of solid mechanics, "...there has been a considerable shift in emphasis from analytical solutions toward numerical solutions of initial and boundary value problems. Nevertheless, closed form analytical solutions are still of considerable utility and should be pursued." In view of this remark, this study is committed to attain new closed form solutions for the buckling of nonuniform columns in addition to approximate solutions based on the Galerkin method.

14.2 The Clamped-Free Column

In the paper by Elishakoff and Boutur [28], the problem of determining the first critical buckling load of a clamped-free column is addressed using the Galerkin method. Three sets of trial functions (Eqs.(3)-(5)) are considered and all fail to converge to the true solution even though they satisfy all the boundary conditions. For reference, we resolve this problem after reducing the original 4th order equation to 2nd order.

The governing equation for the buckling mode $v(x,t)$ of a column with bending stiffness EI under an axial load P is given by

$$EI\frac{\mathrm{d}^4 v}{\mathrm{d}x^4}+P\frac{\mathrm{d}^2 v}{\mathrm{d}x^2}=0 \tag{14.1}$$

If the beam is clamped at $x=0$ and free at $x=L$, then the corresponding boundary conditions are: $v(0)=v'(0)=0$, $v''(L)=0$ and $Pv'(L)+EIv'''(L)=0$. Due to the last boundary condition, Eq. (14.1) can be replaced with the third order equation

$$\frac{\mathrm{d}^3 v}{\mathrm{d}\xi^3}+\lambda\frac{\mathrm{d}v}{\mathrm{d}\xi}=0 \tag{14.2}$$

where we have introduced the non-dimensional coordinate $\xi = x/L$ and buckling load $\lambda = PL^2/EI$. A further reduction in order can be made by setting $w = \mathrm{d}v/\mathrm{d}\xi$ resulting in the eigenvalue problem

$$\frac{\mathrm{d}^2 w}{\mathrm{d}\xi^2} + \lambda w = 0, \quad w(0) = 0, \ \frac{\mathrm{d}w}{\mathrm{d}\xi}(1) = 0. \tag{14.3}$$

To apply Galerkin's method, we set

$$w \cong \sum_{j=1}^{N} a_j \varphi_j(\xi)$$

and choose

$$\varphi_j(\xi) = \sin\left[(2j-1)\pi\xi/2\right].$$

Rendering the error residual orthogonal to each of the comparison functions, we arrive at the generalized algebraic eigenvalue problem

$$(\mathbf{K} + \lambda\mathbf{M})\,\mathbf{a} = \mathbf{0} \tag{14.4}$$

where $\mathbf{K}$ and $\mathbf{M}$ are diagonal matrices given by

$$k_{ii} = -\frac{\pi^2}{8}(2i-1)^2, \quad m_{ii} = \frac{1}{2} \quad (i = 1, 2, \ldots) \tag{14.5}$$

Setting the determinant of $\mathbf{K} + \lambda\mathbf{M}$ to zero, we find an explicit expression for the critical buckling loads

$$P_n = \frac{\pi^2 EI}{4L^2}(2n-1)^2 \tag{14.6}$$

Note that these values agree with those given on pp. 47-48 of Timoshenko and Gere [30]. The reason we recover the exact eigenvalues is due to the fact that our trial solution is a linear combination of the exact mode shapes.

14.3 Buckling of a Heavy Simply Supported-Sliding Column

Consider now a uniform column of length L, bending stiffness EI and weight per unit length p which is simply supported at its base and free to slide at its tip (see Fig. 2 in Elishakoff and Boutur [28]). This problem was previously studied by Bürgermeister and Steup [31], pp.195-197. The governing differential equation is

$$EI\frac{\mathrm{d}^3 v}{\mathrm{d}x^3} + px\frac{\mathrm{d}v}{\mathrm{d}x} = 0 \tag{14.7}$$

with boundary conditions

$$v'(0) = v(L) = v''(L) = 0 \tag{14.8}$$

We first provide an exact solution. Following Timoshenko and Gere [30], we let

$$z = \frac{2}{3}\sqrt{\frac{p}{EI}x^3}$$

so that

$$\frac{\mathrm{d}}{\mathrm{d}x}() = \left(\frac{3pz}{2EI}\right)^{1/3}\frac{\mathrm{d}}{\mathrm{d}z}().$$

The differential equation (14.7) is transformed to

$$\frac{\mathrm{d}^2u}{\mathrm{d}z^2} + \frac{1}{z}\frac{\mathrm{d}u}{\mathrm{d}z} + \left(1 - \frac{1}{9z^2}\right)u = 0 \tag{14.9}$$

where $u = \mathrm{d}v/\mathrm{d}z$. The general solution to the differential equation (14.9) can be expressed in the form

$$u = c_1 J_{1/3}(z) + c_2 J_{-1/3}(z) \tag{14.10}$$

where $J_\nu(z)$ denotes the Bessel function of the first kind of order ν while c_1 and c_2 are arbitrary constants. The first boundary condition in (14.8) requires that $z^{1/3}u = 0$ at $z = 0$ from which we find $c_2 = 0$. The third boundary condition in (14.8) requires that

$$3z\frac{\mathrm{d}u}{\mathrm{d}z} + u = 0$$

at $z = \frac{2}{3}\sqrt{\frac{pL^3}{EI}}$ from which we obtain the characteristic equation for the critical loads.

$$J_{-2/3}\left(\frac{2}{3}\sqrt{\frac{pL^3}{EI}}\right) = 0 \tag{14.11}$$

The first critical value of p is approximately $3.4766EI/L^3$.

The buckling mode is given by $u = \int J_{1/3}(z)\mathrm{d}z$ where the constant of integration is obtained from the second boundary condition in (14.8) i.e. $v = 0$ at $z = \frac{2}{3}\sqrt{pL^3/EI}$. Introducing the parameter $\lambda = pL^3/EI$ and expressing z in terms of x, we obtain

$$v = {}_1F_2\left(\frac{2}{3};\frac{4}{3};\frac{5}{3};-\frac{\lambda}{9}\right) - \xi^2\,{}_1F_2\left(\frac{2}{3};\frac{4}{3};\frac{5}{3};-\frac{\lambda\xi^3}{9}\right) \tag{14.12}$$

where $\xi = x/L$ and ${}_1F_2$ denotes the generalized hypergeometric function (see Rainville [32]). Figure 14.1 shows a plot of the first buckling mode.

We now proceed to solve for the critical values of p using the Galerkin method. First note that Eq. (14.7) can be written in the form

$$\frac{\mathrm{d}^3v}{\mathrm{d}\xi^3} + \lambda\xi\frac{\mathrm{d}v}{\mathrm{d}\xi} = 0 \tag{14.13}$$

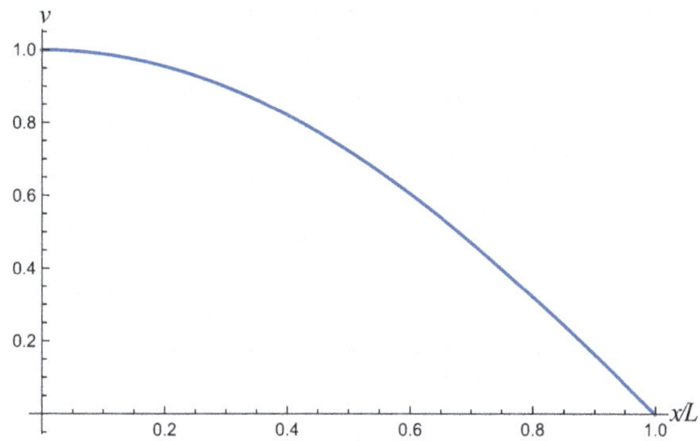

Fig. 14.1 First buckling mode of simply supported-sliding heavy column

We choose as comparison functions $\varphi_i(\xi) = \cos[(2i-1)\pi\xi/2]$ which satisfy all the boundary condition (14.8). Inserting the trial solution $v = \sum_{j=1}^{N} a_j \varphi_j(\xi)$ into Eq. (14.13) and demanding that the error residual be orthogonal to $\varphi_i(\xi), i = 1, 2, \ldots, N$, we obtain the generalized eigenvalue problem

$$(\mathbf{M} + \lambda \mathbf{K})\mathbf{a} = \mathbf{0} \tag{14.14}$$

where

$$\begin{aligned}
m_{ij} &= \int_0^1 \varphi_i \varphi_j''' \, d\xi = \frac{\pi^2 (2j-1)^3 \left[(2i-1)(-1)^{i+j} - 2j + 1\right]}{16(i-j)(i+j-1)} \quad (i \neq j) \\
m_{ii} &= \int_0^1 \varphi_i \varphi_i''' \, d\xi = \frac{\pi^2}{8}(2i-1)^2 \\
k_{ij} &= \int_0^1 \xi \varphi_i \varphi_j' d\xi = \frac{(-1)^{i+j+1}(2i-1)(2j-1)}{4(i-j)(i+j-1)} \quad (i \neq j) \\
k_{ii} &= \int_0^1 \xi \varphi_i \varphi_i' \, d\xi = -1/4
\end{aligned} \tag{14.15}$$

The second column in Table 14.1 shows the estimates for the first critical value of λ for several values of N. We see that retaining only the first term in the trial solution results if a 42% relative error and 25 terms are needed to match the exact result to 4 significant digits.

It is interesting to compare the above solution to that based on the fourth order equation obtained by differentiating Eq. (14.13) (constituting Eq.(7) in Elishakoff and Boutur [28])

$$\frac{d^4 v}{d\xi^4} + \lambda \frac{d}{d\xi}\left(\xi \frac{dv}{d\xi}\right) = 0 \tag{14.16}$$

Table 14.1 Convergence of Galerkin method to first critical buckling load for a simply supported-sliding heavy column

N	3rd Order	4th Order
1	4.9348	3.5116
2	3.5786	3.4770
3	3.5554	3.4767
4	3.5041	3.4766
10	3.4815	
15	3.4791	
20	3.4779	
25	3.4775	
60	3.4768	
100	3.4767	
exact	3.4766	3.4766

The eigenvalue problem defined by the above differential equation and homogeneous boundary conditions (14.8) along with the additional boundary condition $v'''(0) = 0$, is self-adjoint, so we expect better convergence to the eigenvalue λ than in the previous solution. The same choice of comparison functions will again satisfy all boundary conditions. Again, we obtain the eigenvalue problem (14.14), where now

$$\begin{aligned} m_{ij} &= \int_0^1 \varphi_i \frac{\mathrm{d}^4\varphi_j}{\mathrm{d}\xi^4} = 0 \quad (i \neq j) \\ m_{ii} &= \frac{\pi^4}{32}(2i-1)^4 \\ k_{ij} &= \int_0^1 \varphi_i \left(\xi\varphi_j'\right)' \mathrm{d}\xi \\ &= -\frac{(2i-1)(2j-1)i(2-4j)+(-1)^{i+j}(2(i-1)i+2(j-1)j+1)+2j-1}{8(i-j)^2(i+j-1)^2} \quad (i \neq j) \\ k_{ii} &= -\frac{1}{16}\left[4+(2i-1)^2\pi^2\right] \end{aligned} \tag{14.17}$$

Based on the 3rd column in Table 14.1, we see that retaining only the first term in the trial solution results if a 1% relative error and only 4 terms are needed to match the exact result to 5 significant digits.

14.4 Buckling of a Stepped Column Under Axial Load

We investigate the bucking of a stepped column fixed at its base and free at its tip (see Fig. 4 in Elishakoff and Boutur [29]). The analytical solution can be found in Timoshenko and Gere [30], pp. 113-114, in which the critical buckling load P is shown to satisfy the transcendental equation

$$\tan(k_1 L_1)\tan(k_2 L_2) = k_1/k_2 \tag{14.18}$$

where $k_1 = \sqrt{P/EI_1}$ and $k_2 = \sqrt{P/EI_2}$. Numerical solutions can be computed once the two ratios L_1/L_2 and I_1/I_2 are specified. To make this transparent, we define the dimensionless buckling load λ^2 by

$$\lambda^2 = \frac{PL_2^2}{EI_2} \tag{14.19}$$

after which Eq. (14.18) assumes the form

$$\tan\lambda\tan\left(\frac{L^*}{\sqrt{I^*}}\lambda\right) = \frac{1}{\sqrt{I^*}} \tag{14.20}$$

where

$$L^* = L_1/L_2,\ I^* = I_1/I_2 \tag{14.21}$$

In order to solve Eq. (14.20), we need an approximation for the desired root. An approximate value for the first critical buckling load can be obtained by the method of virtual work in which we assume the following form for the first buckling mode

$$v = \delta(1 - \cos\frac{\pi x}{2L}) \tag{14.22}$$

where δ denotes the tip deflection. Note that this form satisfies both boundary conditions at the fixed end ($v(0) = 0$, $v'(0) = 0$) and the natural boundary condition at the free end ($v''(L) = 0$). The bending moment M at any point along the column is given by $M(x) = P(\delta - v)$ while the increment in strain energy due to bending is given by the expression

$$\Delta U = \frac{1}{2}\int_0^L \frac{M^2(x)}{EI(x)}\mathrm{d}x \tag{14.23}$$

Splitting the integral over the two sections of the column and employing Eq. (14.22), we obtain

$$\Delta U = \frac{P^2\delta^2}{2E}\left[\frac{1}{I_2}\int_0^{L_2}\cos^2\frac{\pi x}{2L}\mathrm{d}x + \frac{1}{I_1}\int_{L_2}^{L}\cos^2\frac{\pi x}{2L}\mathrm{d}x\right] \tag{14.24}$$

Performing the integrations and simplifying, we find

$$\Delta U = \frac{P^2\delta^2\left[\pi\left(I_1L_2+I_2L_1\right)+L\left(I_1-I_2\right)\sin\frac{\pi L_2}{L}\right]}{4\pi EI_1I_2} \tag{14.25}$$

The virtual work done by the applied axial load P is given by the expression

$$\Delta W = \frac{P}{2}\int_0^L (\mathrm{d}v/\mathrm{d}x)^2\,\mathrm{d}x = \frac{\pi^2\delta^2P}{16L} \tag{14.26}$$

Equating ΔW to ΔU and employing Eq. (14.19), we obtain the following expression for the dimensionless buckling load.

$$\lambda^2 = \frac{\pi^3 I^*}{4\left(L^*+1\right)\left[\pi\left(I^*+L^*\right)+\left(I^*-1\right)\left(L^*+1\right)\sin\frac{\pi}{L^*+1}\right]} \tag{14.27}$$

Figure 14.2 shows the variation of the first critical buckling load as a function of the inertia ratio I^* for several values if the length ratio L^*. The solid curves are based on the approximate formula (14.27) while the dashed curves were obtained from the exact formulation (14.20).

14.4.1 Galerkin Solution: First Version

The governing differential equation for the buckling of the fixed-free stepped column is given by

$$\frac{\mathrm{d}}{\mathrm{d}x}\left[EI(x)\frac{\mathrm{d}^2v}{\mathrm{d}x^2}\right]+P\frac{\mathrm{d}v}{\mathrm{d}x}=0 \tag{14.28}$$

where $EI(x) = EI_2$ for $0 \le x < L_2$ and $EI(x) = EI_1$ for $L_2 < x \le L$ The associated boundary conditions read: $v(0) = v'(0) = 0, \quad v''(L) = 0$. We assume a trial function of the form $v = \sum_{j=1}^{N} a_j\varphi_j(x)$ with $\varphi_j(x) = 1-\cos\frac{(2j-1)\pi x}{2L}$. Multiplying Eq. (14.28)

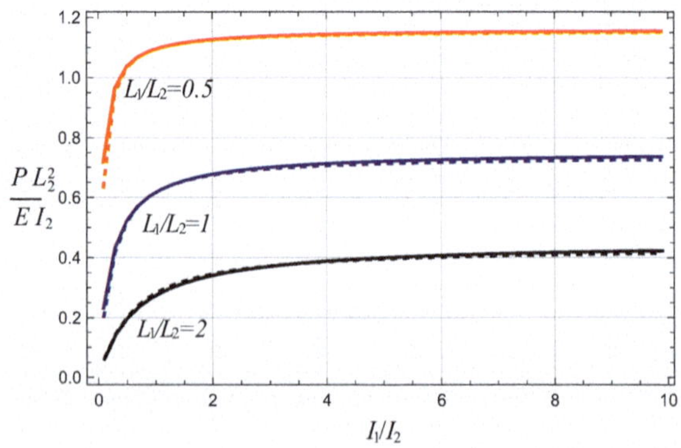

Fig. 14.2 First critical buckling load of a stepped column

by $\varphi_i(x)$ and integrating over the interval$(0, L)$, we obtain

$$P\sum_{j=1}^{N} a_j \int_0^L \varphi_i(x)\varphi_j'(x)\,dx - \sum_{j=1}^{N} a_j \int_0^L EI(x)\varphi_i'(x)\varphi_j''(x)\,dx = 0 \tag{14.29}$$

where we have used integration by parts (observing the boundary conditions) to simplify the second term. Multiplying Eq. (14.29) by L_2^2/EI_2 and recalling Eq. (14.19), we arrive at the generalized eigenvalue problem

$$\mathbf{Ma} = \lambda^2\mathbf{Ka} \tag{14.30}$$

where

$$\begin{aligned}
k_{ij} &= \int_0^L \varphi_i(x)\varphi_j'(x)\mathrm{d}x = \frac{(2i-1)\left[(2j-1)(-1)^{i+j+1}+2i-1\right]}{4(i-j)(i+j-1)} \qquad (i \neq j) \\
k_{ii} &= \frac{1}{2} \\
m_{ij} &= \frac{L_2^2}{EI_2}\int_0^L EI(x)\varphi_i'(x)\varphi_j''(x)\mathrm{d}x \\
&= \alpha_{ij}\frac{I^*(2j-1)(-1)^{i+j+1}+2i-1+(I^*-1)\beta_{ij}}{16(L^*+1)^2(i-j)(i+j-1)} \quad (i \neq j) \\
m_{ii} &= \frac{(2i-1)^2\pi^2\left[(I^*-1)\cos\left[\frac{\pi(2i-1)}{L^*+1}\right]+I^*+1\right]}{16(L^*+1)^2}
\end{aligned} \tag{14.31}$$

where

$$\begin{aligned}
\alpha_{ij} &= (2i-1)(2j-1)^2\pi^2 \\
\beta_{ij} &= (i+j-1)\cos\left[\frac{\pi(i-j)}{L^*+1}\right]+(i-j)\cos\left[\frac{\pi(i+j-1)}{L^*+1}\right]
\end{aligned} \tag{14.32}$$

Table 14.2 (second column) shows the estimates to the first nondimensional buckling load for L^*=1 and I^*=0.25. Retaining two terms in the trial function, yields an error of 31%, while retaining the first 20 terms brings the error down to below 1%.

14.4.2 Galerkin Solution: Second Version

Differentiating Eq. (14.28) with respect to x we obtain the self-adjoint form

$$\frac{\mathrm{d}^2}{\mathrm{d}x^2}\left[EI(x)\frac{\mathrm{d}^2 v}{\mathrm{d}x^2}\right]+P\frac{\mathrm{d}^2 v}{\mathrm{d}x^2}=0 \tag{14.33}$$

In addition to the three boundary conditions following Eq. (14.28) we have

Table 14.2 Convergence of Galerkin method to first critical buckling load for a clamped—free stepped column (L^*=1, I^*=0.25)

N	PL^2/EI_2 3rd Order	PL^2/EI_2 4th Order
2	1.9805	1.8275
3	1.7153	1.6440
5	1.5963	1.5946
10	1.5800	1.5606
20	1.5300	1.5371
exact	1.5153	1.5153

$$EI_1 v'''(L) + Pv'(L) = 0 \tag{14.34}$$

reflecting the zero shear force condition at the column tip. Employing the same set of basis functions $\varphi_i(x)$ as in the previous section we find (integrating by parts)

$$\begin{gathered}\int_0^L \varphi_i v'' \mathrm{d}x = v'(L) - \int_0^L v' \varphi_i{}' \mathrm{d}x \\ \int_0^L \varphi_i \frac{\mathrm{d}^2}{\mathrm{d}x^2}\left[EI(x)\frac{\mathrm{d}^2 v}{\mathrm{d}x^2}\right]\mathrm{d}x = EI_1 v'''(L) - \int_0^L \frac{\mathrm{d}}{\mathrm{d}x}\left[EI(x)v''\right]\varphi_i{}' \mathrm{d}x\end{gathered} \tag{14.35}$$

Multiplying Eq. (14.33) by $\varphi_i(x)$, integrating over $(0, L)$ and employing Eqs. (14.34)-(14.35), we obtain

$$P\int_0^L \varphi_i{}' v' \mathrm{d}x + \int_0^L \varphi_i{}' \frac{\mathrm{d}}{\mathrm{d}x}\left[EI(x)v''\right]\mathrm{d}x = 0 \tag{14.36}$$

Integrating the second term by parts and observing the boundary conditions, we finally obtain

$$P\int_0^L \varphi_i{}' v' \mathrm{d}x - \int_0^L EI(x) v'' \varphi_i{}'' \mathrm{d}x = 0 \tag{14.37}$$

Inserting the expansion

$$v(x) = \sum_{j=1}^{N} a_j \varphi_j(x)$$

into Eq. (14.37) and multiplying thru by LL_2^2/EI_2,we obtain the generalized eigenvalue problem $\mathbf{Ma} = \lambda^2 \mathbf{Ka}$ where

$$k_{ij} = L \int_0^L \varphi_i{}'(x)\varphi_j{}'(x)\mathrm{d}x$$
$$m_{ij} = \frac{LL_2^2}{I_2} \int_0^L I(x)\varphi_i{}''(x)\varphi_j{}''(x)\mathrm{d}x \tag{14.38}$$

Performing the integrations, we find

$$\begin{aligned} k_{ij} &= 0 \; (i \neq j) \\ k_{ii} &= \frac{\pi^2}{8}(2i-1)^2 \\ m_{ij} &= \frac{\pi^3(1-2i)^2(1-I^*)(1-2j)^2\left[(i+j-1)\alpha_{ij}+(i-j)\beta_{ij}\right]}{32(L^*+1)^2(i-j)(i+j-1)} \quad (i \neq j) \\ m_{ii} &= \frac{\pi^3(2i-1)^3\left[\pi(2i-1)(I^*L^*+1)+(I^*-1)(L^*+1)\sin\frac{\pi-2\pi i}{L^*+1}\right]}{32(L^*+1)^3} \end{aligned} \tag{14.39}$$

where $\alpha_{ij} = \sin\frac{\pi(i-j)}{L^*+1}$ and $\beta_{ij} = \sin\frac{\pi(i+j-1)}{L^*+1}$

Table 14.2 (column 3) shows the estimates to the first nondimensional buckling load for L^*=1 and I^*=0.25. Retaining two terms in the trial function, yields an error of 21% (compared with 31% for the previous solution), while retaining the first 20 terms brings the error down to slightly above 1%.

14.5 Buckling of a Heavy Stepped Column

It should be noted that one must be careful in deriving the governing equation for buckling of a heavy nonuniform column; indeed, several authors have made errors in this regard. For example, Eq. (14.29) in Elishakoff and Boutur [28] is clearly incorrect as can be seen by physical considerations. If we consider the loading in the lower column, there is the contribution from the weight of the entire upper column as well as from a portion of the lower column. Hence Eq. (14.29) must show dependence on the parameter p_1 in addition to p_2. Below we derive the buckling equation from fundamental principles.

Figure 14.3 shown an internal element AB of a column with variable bending stiffness $EI(x)$ and weight per unit length $p(x)$. We denote the internal bending moment and shear force at A by $M(x)$ and $S(x)$ respectively. The vertical force $Q(x)$ is due to the column weight from point A to the column tip at C. A similar situation holds at point B. The external load W represents the weight of the segment AB

$$Q(x) = \int_x^L p(X)\mathrm{d}X, \quad W = \int_x^{x+\Delta x} p(X)\mathrm{d}X \tag{14.40}$$

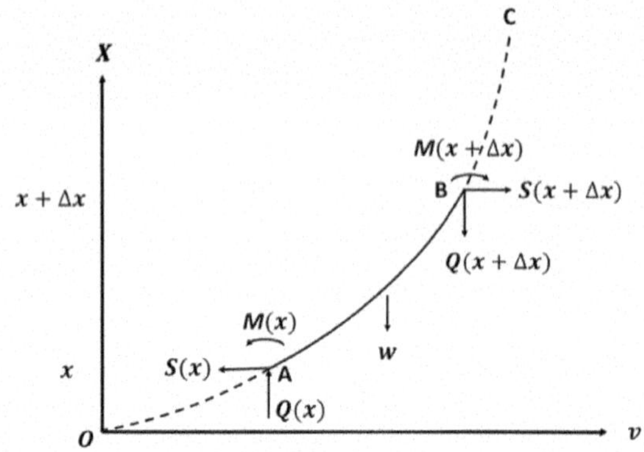

Fig. 14.3 Free body diagram of a heavy beam segment

Summing forces along the horizontal direction (v), we conclude that S is constant along the column. Moment balance about point A requires

$$M(x) - M(x+\Delta x) - S\Delta x - Q(x+\Delta x)\,[v(x+\Delta x) - v(x)] - \int_x^{x+\Delta x} p(X)\,[v(X) - v(x)]\,\mathrm{d}X \tag{14.41}$$

By the mean value theorem for integrals,

$$\int_x^{x+\Delta x} p(X)\,[v(X) - v(x)]\,\mathrm{d}X = p(\bar{X})\left[v(\bar{X}) - v(x)\right]\Delta x \quad \text{where } x < \bar{X} < x + \Delta x$$

Dividing Eq. (14.41) by Δx and taking the limit as $\Delta x \to 0$, we obtain

$$\frac{\mathrm{d}}{\mathrm{d}x}\left[EI\frac{\mathrm{d}^2 v}{\mathrm{d}x^2}\right] + S + Q(x)\frac{\mathrm{d}v}{\mathrm{d}x} = 0 \tag{14.42}$$

where we have employed the standard relation expressing the bending moment in terms of the bending stiffness and curvature.

14.5.1 Exact Solution

Figure 14.4 shows a two-segment heavy column where each segment is uniform. The base of the column at $x = 0$ is fixed and the tip at $x = L$ is free. The bottom segment is of length aL with bending stiffness EI_1 and weight per unit length p_1 while the top segment is of length $L(1-a)$ with bending stiffness EI_2 and weight per unit length p_2. Since the column is free at $x = L$, S vanishes identically. It follows from Eq. (14.40) that $Q(x) = p_2(L-x)$ in the top segment. Hence for the top segment, Eq. (14.42) assumes the form

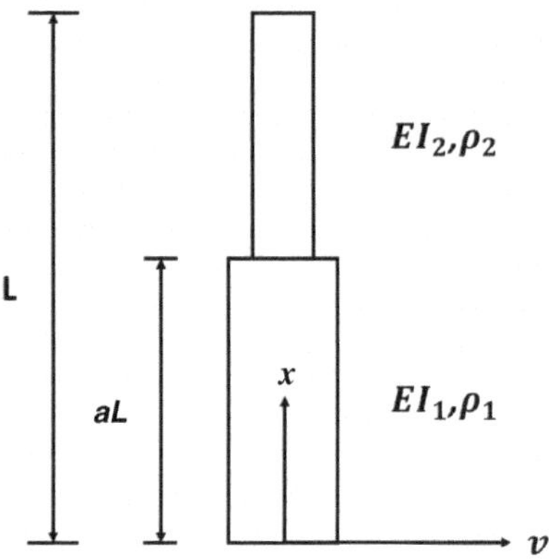

Fig. 14.4 A two-segment heavy stepped column

$$EI_2 \frac{d^3 v_2}{dx^3} + p_2 (L - x) \frac{dv_2}{dx} = 0 \tag{14.43}$$

Denoting the slope dv_2/dx by $w_2(x)$ and introducing the dimensionless axial coordinate $\xi = x/L$, the above equation assumes the form

$$\frac{d^2 w_2}{d\xi^2} + \lambda (1 - \xi) w_2 = 0, \quad a < \xi < 1 \tag{14.44}$$

where we have introduced the nondimensional buckling load

$$\lambda = \frac{p_2 L^3}{EI_2} \tag{14.45}$$

Making the change of independent variable $\eta = \lambda^{1/3} (\xi - 1)$, the differential equation (14.44) assumes the form

$$\frac{d^2 w_2}{d\eta^2} - \eta w_2 = 0 \tag{14.46}$$

which we recognize as Airy's equation (see Lebedev [33]). Thus, the general solution to Eq. (14.44) is given by

$$w_2 (\xi) = c_1 \mathrm{Ai} \left[\lambda^{1/3} (\xi - 1) \right] + c_2 \mathrm{Bi} \left[\lambda^{1/3} (\xi - 1) \right] \tag{14.47}$$

where Ai, Bi denote the Airy functions of the first and second kind and c_1, c_2 are arbitrary constants. Care must be taken in evaluating $Q(x)$ in the bottom segment since p has a discontinuity at $x = aL$. Splitting the integral into two parts, we find $Q(x) = p_2 L(1 - a) + p_1(aL - x)$ for $0 < x < aL$. Thus for the bottom segment, Eq. (14.42) can be written in the form

$$EI_1 \frac{d^3 v_1}{dx^3} + \left[p_2 L (1 - a) + p_1 (aL - x) \right] \frac{dv_1}{dx} = 0. \tag{14.48}$$

With $w_1(x) = \mathrm{d}v_1/\mathrm{d}x$ the above equation can be expressed in the form

$$\frac{\mathrm{d}^2 w_1}{\mathrm{d}\xi^2} + \frac{p_2 L^3}{EI_1}\left[1 - a + \frac{p_1}{p_2}(a-\xi)\right] w_1 = 0 \tag{14.49}$$

Assuming that both segments are made of the same material and are of circular cross section but have different cross-sectional area, we can find a relation between the inertia ratio and weight density ratio. Specifically, if $EI_1/EI_2 = c$, then $p_1/p_2 = \sqrt{c}$. This relation allows us to express Eq. (14.49) in the form

$$\frac{\mathrm{d}^2 w_1}{\mathrm{d}\xi^2} + \frac{\lambda}{c}\left[1 - a + \sqrt{c}\,(a-\xi)\right] w_1 = 0. \quad 0 < \xi < a \tag{14.50}$$

The general solution to this equation is given by

$$w_1(\xi) = \mathrm{c}_3 \mathrm{Ai}\left[q(\xi)\right] + \mathrm{c}_4 \mathrm{Bi}\left[q(\xi)\right] \tag{14.51}$$

where

$$q(\xi) = \frac{\lambda}{(c\lambda)^{2/3}}\left(a - a\sqrt{c} + \sqrt{c}\,\xi - 1\right) \tag{14.52}$$

At the interface of the two column sections, we enforce continuity of slope and bending moment

$$w_1(a) = w_2(a), \quad c\frac{\mathrm{d}w_1}{\mathrm{d}\xi}(a) = \frac{\mathrm{d}w_2}{\mathrm{d}\xi}(a) \tag{14.53}$$

Since the column is fixed at its base and free at its tip, then $w_1(0) = 0$ and $w_2'(1) = 0$ so Eqs. (14.51) &(14.47) assume the forms

$$w_1(\xi) = \mathrm{c}_3\left\{\mathrm{Ai}\left[q(\xi)\right] - \frac{\mathrm{Ai}\left[q(0)\right]}{\mathrm{Bi}\left[q(0)\right]}\mathrm{Bi}\left[q(\xi)\right]\right\} \tag{14.54}$$

$$w_2(\xi) = \mathrm{c}_1\left\{\mathrm{Ai}\left[\lambda^{1/3}(\xi-1)\right] + \frac{1}{\sqrt{3}}\mathrm{Bi}\left[\lambda^{1/3}(\xi-1)\right]\right\} \tag{14.55}$$

Applying the continuity conditions Eqs. (14.53), we arrive at the system of homogeneous equations $\mathbf{Mc} = \mathbf{0}$ where $\mathbf{c} = (\mathrm{c}_1, \mathrm{c}_3)^{\mathrm{T}}$ and the elements of the matrix $\mathbf{M}(\lambda, c; a)$ are given by

$$\begin{aligned}
m_{11} &= \mathrm{Ai}\left[(a-1)\lambda^{1/3}\right] + \frac{1}{\sqrt{3}}\mathrm{Bi}\left[(a-1)\lambda^{1/3}\right], \\
m_{12} &= \frac{\mathrm{Ai}\left[q(0)\right]}{\mathrm{Bi}\left[q(0)\right]}\mathrm{Bi}\left[q(a)\right] - \mathrm{Ai}\left[q(a)\right], \\
m_{21} &= -\lambda^{1/3}\left\{\mathrm{Ai}'\left[(a-1)\lambda^{1/3}\right] + \frac{1}{\sqrt{3}}\mathrm{Bi}'\left[(a-1)\lambda^{1/3}\right]\right\}, \\
m_{22} &= c^{5/6}\lambda^{1/3}\left\{\mathrm{Ai}'\left[q(a)\right] - \frac{\mathrm{Ai}\left[q(0)\right]}{\mathrm{Bi}\left[q(0)\right]}\mathrm{Bi}'\left[q(a)\right]\right\}
\end{aligned} \tag{14.56}$$

Given the inertia and length ratios of the two segments of the column, we can compute the dimensionless critical buckling load $\lambda$by setting the determinant of **M** to zero. Figure 14.5 shows the variation in the dimensionless buckling load λ as a function of a for several values of the inertia ratio c. In the limit as c approaches unity (uniform column), the plot approaches the straight line $\lambda = 7.3874$ as expected (see Timoshenko and Gere [30]).

14.5.2 Galerkin Solution

We now attempt to solve the problem by the Galerkin method with

$$w \cong \sum_{j=1}^{N} a_j \varphi_j(\xi)$$

where $\varphi_j(\xi) = \sin[(2j-1)\pi\xi/2]$, thus satisfying the two boundary conditions: $w(0) = 0$ and $\mathrm{d}w/\mathrm{d}\xi(1) = 0$. Utilizing Eqs. (14.44) and (14.50) and rendering the error residual orthogonal to $\{\varphi_i(\xi)\}$, we arrive at the eigenvalue problem

$$\mathbf{Ra} = \lambda(\mathbf{P}+\mathbf{Q})\mathbf{a} \tag{14.57}$$

where

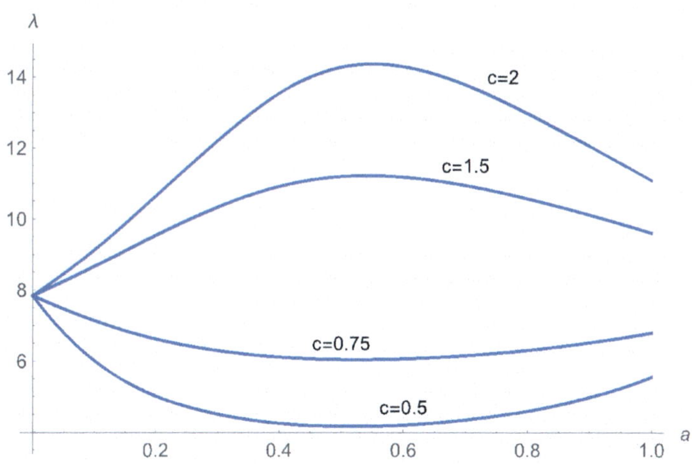

Fig. 14.5 Variation in buckling load for a heavy stepped column

$$r_{ij} = \int_0^1 \varphi_i'(\xi)\varphi_j'(\xi)\mathrm{d}\xi,$$
$$p_{ij} = \int_a^1 (1-\xi)\varphi_i(\xi)\varphi_j(\xi)\mathrm{d}\xi, \tag{14.58}$$
$$q_{ij} = \frac{1}{c}\int_0^a [1-a+\sqrt{c}(a-\xi)]\varphi_i(\xi)\varphi_j(\xi)\mathrm{d}\xi j$$

Once the inertia and length ratios of the two segments of the column are specified, we can solve the above eigenvalue problem for the critical buckling loads. Performing the integrations, the matrix elements are given by

$$\begin{aligned}
r_{ij} &= \frac{\pi^2}{8}(2i-1)^2\delta_{ij},\\
p_{ii} &= \frac{\mu_i[2\sin(\pi a(1-2i))+\mu_i]-2\cos(\pi a(1-2i))-2}{4\pi^2(2i-1)^2},\\
p_{ij} &= \frac{1}{2\pi^2(i-j)^2}\left[\cos(\pi a(i-j))+\frac{\gamma_{ij}+\mu_{ij}(i-j)}{(i+j-1)^2}\right] \quad (i\neq j),\\
q_{ii} &= \frac{\pi^2 a(2i-1)^2[a(\sqrt{c}-2)+2]-\sigma_i-\tau_i}{4c\pi^2(2i-1)^2},\\
q_{ij} &= \frac{1}{2\pi^2 c}\left[\pi(a-1)S_{ij}+\sqrt{c}\left(\frac{1}{(i-j)^2}-\frac{1}{(i+j-1)^2}\right)+C_{ij}\right] \quad (i\neq j)
\end{aligned} \tag{14.59}$$

where δ_{ij} is the Kronecker delta symbol and

$$\begin{aligned}
\mu_i &= \pi(a-1)(2i-1),\\
\gamma_{ij} &= (-1)^{i+j+1}[2i(i-1)+2j(j-1)+1],\\
\sigma_{ij} &= (i+j-1)\sin(\pi a(i-j))+(j-i)\sin(\pi a(i+j-1),\\
\mu_{ij} &= \pi(a-1)(i+j-1)\sigma_{ij}+(j-i)\cos(\pi a(i+j-1)),\\
\sigma_i &= 2\pi(a-1)(2i-1)\sin(\pi a(1-2i)),\\
\tau_i &= 2\sqrt{c}[1-\cos(\pi a(1-2i))],\\
S_{ij} &= \frac{\sin(\pi a(i+j-1))}{i+j-1}-\frac{\sin(\pi a(i-j))}{i-j},\\
C_{ij} &= \frac{\sqrt{c}\cos(\pi a(i+j-1))}{(i+j-1)^2}-\frac{\sqrt{c}\cos(\pi a(i-j))}{(i-j)^2}.
\end{aligned} \tag{14.60}$$

Numerical estimates of the first non-dimensional critical buckling load λ based on this classical application of Galerkins method is shown in Table 14.3 (columns labeled *Class*) for selected values of a and c where N denotes the number of terms retained in the trial function. It is clear that the solution does not converge to the known exact value. This behavior of the classical Galerkin method for columns with discontinuous material properties has been observed by previous investigators (see for example Elishakoff and Boutur [28], Elishakoff, Arvan and Marzani [3], Elishakoff, Amato, Arvan and Marzani [35]).

Table 14.3 Convergence of Bubnov-Galerkin method to first critical buckling load for a clamped-free heavy stepped column

	$c=2,\ a=0.25$		$c=2,\ a=0.75$		$c=0.5,\ a=0.25$		$c=0.5,\ a=0.75$	
N	*Class.*	*Mod.*	*Class.*	*Mod.*	*Class.*	*Mod.*	*Class.*	*Mod.*
10	8.18	11.53	11.859	13.399	7.17	4.76	4.860	4.447
20	8.18	11.48	11.859	13.398	7.17	4.73	4.860	4.447
30	8.18	11.46	11.859	13.398	7.17	4.73	4.860	4.447
40	8.18	11.45	11.859	13.398	7.17	4.72	4.860	4.446
60	8.18	11.44	11.859	13.397	7.17	4.72	4.860	4.446
exact	11.43	11.43	13.397	13.397	4.71	4.71	4.446	4.446

To improve the estimates of the critical buckling loads as provided by the Galerkin method, we modify the governing equations by the use of generalized functions as suggested in Bastatsky and Khvoles [36] (see also Vainberg and Roitfarb [37]). The governing differential equation is Eq. (14.42) with $S=0$ where $Q(x)$ is given by Eq. (14.40). The bending stiffness and loading per unit length are now expressed in the form

$$\begin{aligned} EI(x) &= EI_1+(EI_2-EI_1)H(x-aL), \\ p(x) &= p_1+(p_2-p_1)H(x-aL) \end{aligned} \tag{14.61}$$

where $H(x)$ is the Heaviside step function. Performing the integration in Eq. (14.42), we obtain

$$Q(x)=(p_2-p_1)(aL-x)H(x-aL)+aL(p_1-p_2)+p_2L-p_1x \tag{14.62}$$

Expanding the derivative appearing in the first term of Eq. (14.45) and setting $\mathrm{d}v/\mathrm{d}x=w$, we obtain

$$EI(x)\frac{\mathrm{d}^2w}{\mathrm{d}x^2}+(EI_2-EI_1)\,\delta(x-aL)\frac{\mathrm{d}w}{\mathrm{d}x}+Q(x)w=0 \tag{14.63}$$

where $\delta(x)$denotes the Dirac delta function. Introducing the dimensionless axial coordinate $\xi=x/L$, and recalling Eq. (14.45) and the definition of c following Eq. (14.49), we find after a bit of algebra, that Eq. (14.63) transforms into the dimensionless form

$$\begin{aligned} &[c+(1-c)H(\xi-a)]\frac{\mathrm{d}^2w}{\mathrm{d}\xi^2}+(1-c)\,\delta(\xi-a)\frac{\mathrm{d}w}{\mathrm{d}\xi}+ \\ &\lambda\left[(1-\sqrt{c})(a-\xi)H(\xi-a)+a\left(\sqrt{c}-1\right)+1-\sqrt{c}\,\xi\right]w=0 \end{aligned} \tag{14.64}$$

As before, we adopt the approximation

$$w = \sum_{j=1}^{N} a_j \varphi_j(\xi)$$

where

$$\varphi_j(\xi) = \sin[(2j-1)\pi\xi/2].$$

Rendering the error residual orthogonal to $\{\varphi_i(\xi)\}$, we arrive at the generalized eigenvalue problem

$$[\mathbf{P}+\mathbf{Q}]\,\mathbf{a}+\lambda\,[\mathbf{R}+\mathbf{S}]\,\mathbf{a}=\mathbf{0} \tag{14.65}$$

where

$$\begin{aligned}
p_{ij} &= \int_0^1 \varphi_i(\xi)\varphi_j''(\xi)\,[c+(1-c)H(\xi-a)]\,\mathrm{d}\xi,\\
q_{ij} &= (1-c)\int_0^1 \varphi_i(\xi)\varphi_j'(\xi)\delta(\xi-a)\,\mathrm{d}\xi,\\
r_{ij} &= (1-\sqrt{c})\int_0^1 \varphi_i(\xi)\varphi_j(\xi)\,(a-\xi)\,H(\xi-a)\,\mathrm{d}\xi,\\
s_{ij} &= \int_0^1 \varphi_i(\xi)\varphi_j(\xi)\left[a\left(\sqrt{c}-1\right)+1-\sqrt{c}\xi\right]\mathrm{d}\xi
\end{aligned} \tag{14.66}$$

Performing the integrations, the matrix elements are given by

$$\begin{aligned}
p_{ii} &= -\pi/8\,(2i-1)\,(\pi(2i-1)(a(c-1)+1)+(c-1)\sin(\pi a(1-2i))),\\
p_{ij} &= \frac{c-1}{8}\pi(2j-1)^2\left(\frac{\sin(\pi a(i+j-1))}{i+j-1}-\frac{\sin(\pi a(i-j))}{i-j}\right) \quad (i\neq j),\\
q_{ij} &= \frac{\pi}{4}(1-c)\,(2j-1)\,[\sin(\pi a(i-j))+\sin(\pi a(i+j-1))],\\
r_{ii} &= \frac{(\sqrt{c}-1)\left[\pi^2(a-1)^2(2i-1)^2+2\cos(\pi a(1-2i))+2\right]}{4\pi^2(2i-1)^2},\\
r_{ij} &= \frac{(\sqrt{c}-1)\left[-a_{ij}+(i-j)^2 b_{ij}+(-1)^{i+j}(i+j-1)^2\right]}{2\pi^2(i-j)^2(i+j-1)^2} \quad (i\neq j),
\end{aligned} \tag{14.67}$$

where

$$\begin{aligned}
a_{ij} &= (i+j-1)^2\cos(\pi a(i-j)),\\
b_{ij} &= (-1)^{i+j}+\cos(\pi a(i+j-1)),\\
s_{ii} &= \frac{1}{4}\left[2a(\sqrt{c}-1)-\sqrt{c}\left(\frac{4}{\pi^2(2i-1)^2}+1\right)+2\right],\\
s_{ij} &= \frac{\sqrt{c}\left[(2i-1)(2j-1)+(-1)^{i+j+1}(2i(i-1)+2j(j-1)+1)\right]}{2\pi^2(i-j)^2(i+j-1)^2} \quad (i\neq j)
\end{aligned}$$

Numerical results for this modified version of the Galerkin method are shown in Table 14.3 (column labeled *mod*). We see that excellent results are obtained in all cases even with relatively few terms in the trial function.

14.6 Conclusion

We provide exact and approximate solutions for heavy simply supported-sliding column as well as for the heavy stepped column. To improve the estimates of the critical buckling loads as provided by the Galerkin method, modification of the governing equations by the use of generalized functions as suggested in Bastatsky and Khvoles [36] is conducted with attendant excellent results in all cases even with relatively few terms in the trial function. In a general sense this paper falls into the category of high-precision methods in eigenvalue problems, as discussed by Akulenko and Nesterov [38].

Acknowledgements The Authors express deep gratitude to prof. Dr. Holm Altenbach for making available portions of Ref. [31], namely the excellent textbook by G. Bürgermeister and H. Steup.

References

1. L. Euler. Sur la force des callones, *Memories de L'Academie des Sciences et Belles-Letteres*, Berlin, Vol. 13, 252-282, 1759 (in French)
2. F. Engesser. Die Knickfestigkeit gerader Stäbe, *Centralblatt der Bauverwaltung*, Vol. 35, pp. 483-486, 1891.
3. N.J. Duncan. Galerkin's Method in Mechanics and Differential Equations, *Aeronautical Research Committee Reports and Memoranda*, No. 1798:484-515, 1937.
4. I. Elishakoff. Euler's Problem Reconsidered – 222 Years Later, *Meccanica*, 35:375-380, 2000.
5. I. Elishakoff. *Eigenvalues of Inhomogeneous Structures: Unusual Closed-form Solutions*, CRC, Boca Raton (FL), 2005.
6. I. Elishakoff, D. Pentaras and C. Gentilini. *Mechanics of Functionally Graded Material Structures*, World Scientific, Singapore, 2015.
7. I. Elishakoff, M. Eisenberger and A. Delmas. Buckling and Vibration of Functionally Graded Columns Sharing Duncan's Mode Shape, and New Cases, *Structures*, 5:170-174, 2016.
8. M. Ayadoğlu. Semi-Inverse Method for Vibration and Buckling of Axially Functionally Graded Beams, *Journal of Reinforced Plastics and Composites*, 27(7):683-689, 2008.
9. Q.S. Li. Exact Solutions for the Generalized Euler's Problem, *Journal of Applied Mechanics*, 76:041015, 2009.
10. K.Y. Maalawi. Optimization of Elastic Columns using Axial Grading Concept, *Engineering Structures*, 31(12):2922-2929, 2009.
11. K.V. Singh and G. Li. Buckling of Functionally Graded and Elastically Restrained Nonuniform Columns, *Composites: Part B Engineering*, 40:393-403, 2009.
12. C.B. Coskun. Determination of Critical Buckling Loads for Euler Columns of Variable Flexural Stiffness with Continuous Elastic Restraint using Homotopy Perturbation Method, *International Journal of Nonlinear Science and Numerical Simulation*, 10:191-197, 2009.
13. S.M. Darbandi, R.D. Firouz-Abadi and H. Haddadpour. Buckling of Variable Section Columns under Axial Loading, *Journal of Engineering Mechanics*, 136(4):472-476, 2010.
14. Y. Huang and X.F. Li. Buckling Analysis of Non-Uniform and Axially Graded Columns with Varying Flexural Rigidity, *Journal of Engineering Mechanics*, 137:73-81, 2011.
15. Y. Huang and Q.Z. Luo. A Simple Method to Determine the Critical Buckling Loads for Axially Inhomogeneous Beams with Elastic Restraint, *Computer Mathematics with Applications*, 61:2510-2517, 2011.
16. E. Babilio. Dynamics of an Axially Functionally Graded Beam under Axial Load, *The European Physical Journal, Special Topics*, 222(7):1519-1539, 2013.

17. W. Shan and Z. Chen. Mechanical Instability of Thin Elastic Rods, *Journal of Postdoctoral Research*, 1(2):1-8, 2013.
18. Yu. S. Krutyi. Construction of the Stability Problem Solution of a Column with Continuous Variation of Parameters (in Russ.), *Mathematical Methods in Physics and Mechanics*, 59(2):142-151, 2016.
19. N.I. Ioakimidis. Application of Quantifier Elimination to Inverse Buckling Problems, *Acta Mechanica*, 228:3709-3724, 2017.
20. E. Byskov. Applicability of an Asymptotic Expansion for Elastic Buckling Problems with Mode Interaction, *AIAA Journal*, 17(6):630-633, 1979.
21. R.L. Fox and M.P. Kapoor. Rates of Change of Eigenvalues and Eigenvectors, *AIAA Journal*, 6(12):2426-24298, 1968.
22. A.M. Vinogradov. Buckling of Viscoelastic Beam Columns, *AIAA Journal*, 25(3):479-483, 1987.
23. C.W. Bert, S.K. Jang and A.G. Striz. Two New Approximate Methods for Analyzing Free Vibration of Structural Components, *AIAA Journal*, 26(5):612-618, 1980.
24. E. Livne. Accurate Calculation of Control-Augmented Structural Eigenvalue Sensitivities using Reduced Order Models, *AIAA Journal*, 27(7):947-954, 1989.
25. R.A. Canfield. High-Quality Approximation of Eigenvalues in Structural Optimization, *AIAA Journal*, 28(6):1116-1122, 1990.
26. I. Elishakoff, N. Zaza, J. Curtin and J. Hashemi. Apparently First Closed-Form Solution for Vibration of Functionally Graded Rotating Beams, *AIAA Journal*, 52(11):2587-2593, 2014.
27. J. Storchand I. Elishakoff. Buckling of Axially Graded Columns: A Fifth Order Polynomial Mode Shape, *AIAA Journal*, 56(6):2509-2513, 2018.
28. I. Elishakoff and D. Boutur. Rigorous Implementation of the Galerkin Method for Uniform and Stepped Columns, *AIAA Journal*, 58(5):2260-2268, 2000.
29. M.M. Carroll. Foundations of Solid Mechanics, *Applied Mechanics Reviews*, 38(3):1301-12308, 1985.
30. S.P. Timoshenko and J.M. Gere. *Theory of Elastic Stability*, 2nd ed., McGraw-Hill book company, Inc., New York and London, 1961.
31. G. Bürgermeister and H. Steup. *Stabilitätstheorie*, Akademie Verlag, Berlin, 1957.
32. E.D. Rainville. *Special Functions*, The Macmillan Company, New York, 1960.
33. N.N. Lebedev. *Special Functions and Their Applications*, Dover Publications, New York, 1972.
34. I. Elishakoff, A.P. Arvan and A. Marzani. Rigorous versus naïve implementation of the Galerkin method for stepped beams, *Acta Mechanica*, 230:3861-3873, 2019.
35. I. Elishakoff, M. Amato, A.P. Arvan and A. Marzani. Rigorous implementation of the Galerkin method for stepped structures needs generalized functions, *Journal of Sound and Vibration*, 490, 115708, 2021.
36. B.N. Bastatsky and A.R. Khvoles. On Some Specifics of Applying Bubnov-Galerkin Method to Practical Analysis of Structures (in Russ.), *Stroitelnaya Mekhanika i Raschet Sooruzhenii (Structural Mechanics and Analysis of Constructions*, (2):55-59, 1972.
37. D.V. Vainberg and I.Z. Roitfarb. Analysis of Plates and Shells with Discontinuous Parameters (in Russ.), *Raschet Prostranstvennykh Konstrukzii*, Vol. 10, Stroiizdat Publishing House, Moscow, pp. 3-80, 1965.
38. L.D. Akulenko and S.V. Nesterov. *High-Precision Methods in Eigenvalue Problems and Their Applications*. Chapman & Hall/CRC Press, Boca Raton, 2005.

Chapter 15
Inter-laminar Delamination in Composite Laminates: Role of Buckling in its Growth

Jianxiang Wang and Bhushan L. Karihaloo

Abstract The Rayleigh-Ritz method is used to investigate the buckling of a thin delaminated sub-laminate from a thick substrate under axial compression representing an inter-laminar two-dimensional delamination near the surface of a composite laminate. An anti-interpenetration model is introduced to prevent potential non-physical interpenetration of material during buckling. The post-buckling response of a simpler one-dimensional strip delamination from a thick substrate is reported next without details. This response depends on the exerted compressive stress, the thickness of the sub-laminate and the inter-laminar critical energy release rate of the material.

Key words: Composite laminate, Delamination, Buckling

15.1 Introduction

The study of delamination buckling and propagation has attracted widespread attention for two reasons. Firstly, because composite laminates are most widely used structural elements in aeronautical engineering and wind-turbine blades where structural integrity is vital. Secondly, because layered structures are very prone to inter-laminar delamination under out-of-plane loading or under stress concentration caused by intra-laminar cracks [1] which makes them susceptible to transverse impact damage with a resulting drastic reduction of their in-plane compressive strength. As low-velocity impact usually causes inter-laminar delamination in com-

Jianxiang Wang
Department of Mechanics and Engineering Science, College of Engineering, Peking University, Beijing 100871, P.R. China,
e-mail: jxwang@pku.edu.cn

Bhushan L. Karihaloo
School of Engineering, Cardiff University, Cardiff CF24, 3AA, UK
e-mail: KarihalooB@cardiff.ac.uk

H. Altenbach et al. (eds.), *Advances in Solid and Fracture Mechanics*,
Advanced Structured Materials 180,
https://doi.org/10.1007/978-3-031-18393-5_15

posite laminates, a delamination model has been used to predict their post-impact compressive strength. The rationality of this and other models has been discussed by Wang [2].

The analysis of the buckling of a thin sub-laminate from a thick substrate has many useful applications such as in coatings (e.g. [3]), and in fibre-reinforced composite laminates when a manufacturing or a service-induced delamination is close to the surface.

Two-dimensional delamination buckling of a sub-laminate on a substrate or a thick plate has been studied both numerically (mostly by finite element method, e.g. [4, 5], among others) and analytically (usually the Rayleigh-Ritz method, e.g. [6, 7, 8, 9, 10], among others), although some other methods have also been used (e.g. [11, 12]).

The Rayleigh-Ritz method is inexpensive, simple to use, and accurate except for highly anisotropic delaminated sub-laminates. Therefore, it has been widely used to predict the buckling load/strain. Yin and Jane [9], and Jane and Yin [10] used a refined Rayleigh-Ritz method based on the von Kármán non-linear plate theory and high-order polynomial approximations of the displacement functions to study the buckling of a delaminated sub-laminate of circular/elliptic shapes with isotropic/anisotropic properties. They found that the low-order polynomial approximations of the displacement functions give estimates of the deflection and buckling strain that are close to those given by high-order approximations. Guided by this observation, in this paper we shall first use a low-order four-term polynomial approximation for the deflection to predict the buckling strain of a thin elliptic delaminated sub-laminate. It is found that for some aspect ratios of the ellipse, the sub-laminate penetrates into the substrate. An 'anti-interpenetration model' is introduced to prevent this non-physical interpenetration of material.

The post-buckling behaviour of laminates containing an interior delamination is then considered for which it is necessary to use a non-linear analysis (see, e.g. [13, 14]). For brevity, the main results of analysis are given without details for the pre- and post-buckling response of the simpler one-dimensional strip delamination in a laminate. The aim of this analysis is to check whether such a delamination threatens the operation of the laminate. No attempt is made to investigate other more complicated delamination-induced buckling responses (see, e.g. [15]). In a real composite laminate such as that used for wind-turbine blades, an inter-laminar strip delamination near the surface of the laminate may or may not buckle. That depends on many factors, such as the size of the delamination, its location from the surface of the laminate, the level of the local compressive stress field, and its eccentricity relative to the substrate. The strip delamination is unlikely to buckle if it is very small in size and situated well below the surface of the laminate. However, it may buckle because it is large in size, is located close to the surface, and is subjected to high in-plane compressive stress. The response will then depend on an additional fracture property of the composite material, viz. its inter-laminar critical energy release rate.

15.2 Mathematical Preliminaries

Consider the buckling of an elliptic delaminated sub-laminate on a thick substrate, as shown in Fig. 15.1. The delaminated sub-laminate of thickness h is relatively thin compared to the substrate, so that the small non-uniformity of the in-plane deformation in the substrate due to the buckling of the sub-laminate can be neglected.

To calculate the buckling strain and to examine the buckling mode for various shapes (aspect ratios) and anisotropic properties of the sub-laminate, the Rayleigh-Ritz method and the linear thin plate theory is used. For this, a kinematically admissible function for the deflection of the sub-laminate in the z-direction is chosen, following Jane and Yin [10]:

$$w(x,y) = \left(1 - \frac{x^2}{a^2} - \frac{y^2}{b^2}\right)\left(C_0 + C_1\frac{x^2}{a^2} + C_2\frac{xy}{ab} + C_3\frac{y^2}{b^2}\right) \tag{15.1}$$

where x and y are the coordinates in the local coordinate system xoy (Fig. 15.1). The extra term C_2 allows for the asymmetric buckling mode.

The total potential energy of the delaminated sub-laminate is

$$\Pi = U + V \tag{15.2}$$

where U is the strain energy of the sub-laminate and V the potential energy of the applied loads which can be expressed in the usual manner through the bending stiffness coefficients D_{ij} $(i,j = 1,2,6)$ of the sub-laminate, and in-plane stress resultants N_x, N_y and N_{xy}, respectively. If the delaminated sub-laminate is a general composite laminate, the stress resultants are related to the strain and curvature by the relation

$$\{N_{x,y,xy}\} = [A]\{\varepsilon_{x,y,xy}\} + [B]\{\kappa_{x,y,xy}\} \tag{15.3}$$

where

$$\begin{aligned}
\{N_{x,y,xy}\} &= \{N_x,\ N_y,\ N_{xy}\}^{\mathrm{T}},\\
\{\varepsilon_{x,y,xy}\} &= \{\varepsilon_x,\ \varepsilon_y,\ \varepsilon_{xy}\}^{\mathrm{T}},\\
\{\kappa_{x,y,xy}\} &= \{\kappa_x,\ \kappa_y, \kappa_{xy}\}^{\mathrm{T}}
\end{aligned}$$

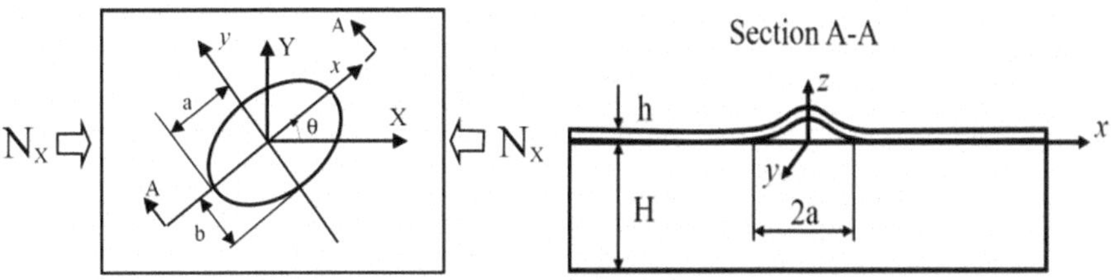

(a) (b)

Fig. 15.1 A thick substrate with a thin delaminated sub-laminate under uniaxial compression.

$[A]$ and $[B]$ are the in-plane stiffness matrix and the bending-stretching (stretching-bending) coupling matrix, respectively. Note that $[B]$ vanishes for a symmetric sub-laminate. Assume that the strain components in the global coordinate system are related to each other by $\varepsilon_Y = \rho_Y \varepsilon_X$ and $\varepsilon_{XY} = \rho_{XY} \varepsilon_Y$. The strain components in the local coordinate system can be easily related to those in the global coordinate system (see Fig. 15.1). The coefficients ρ_Y and ρ_{XY} depend on the global loading form and the stiffness of the substrate. We consider the case $\theta = 0$ and express the stress resultants via the global strains and the components of the in-plane stiffness matrix $[A]$. For expediency, we introduce two coefficients α_y and α_{xy} such that $N_y = \alpha_y N_x$, $N_{xy} = \alpha_{xy} N_x$. The strain energy of the sub-laminate U and the potential energy of the loads V can now be formally written as

$$U = \{C\}[K^U]\{C\}^{\mathrm{T}} \tag{15.4}$$

where $\{C\} = \{C_0, C_1, C_2, C_3\}$. $[K^U]$ is the stiffness matrix, and

$$V = -N_x\{C\}[K^V]\{C\}^{\mathrm{T}} \tag{15.5}$$

where $[K^V]$ is the geometric stiffness matrix. The buckling load N_x is calculated from the eigenvalue problem [16]

$$\mathrm{Det}([K^U)] - N_x[K^V]) = 0 \tag{15.6}$$

The corresponding buckling mode can be calculated after obtaining the buckling load N_x, which can be converted into the global strain ε_X.

15.3 Numerical Results

Typical results of the buckling analysis of the delaminated sub-laminate are given below. Although this problem has been studied before (e.g. [7, 17, 18], etc.), we highlight here the possible interpenetration of the two surfaces besides the delamination in some cases, and propose an anti-interpenetration model for the solution of this problem. We assume throughout that the delaminated sub-laminate is a uni- or multidirectional laminate composed of unidirectional fibre-reinforced laminae with the following properties [7]: $E_L = 131$ GPa, $E_T = 13$ GPa, $G_{LT} = 6.4$ GPa, $\mu_{LT} = 0.34$. Figure 15.2(a) shows the variation of the buckling strain $\varepsilon_{X\mathrm{cr}}$ of a quasi-isotropic sub-laminate with b for a constant $a = 25.4$ mm and a thickness $h = 0.5$ mm. The elastic constants of the quasi-isotropic sub-laminate were calculated from the above properties of the unidirectional lamina using the classical lamination theory. The value of $\varepsilon_{X\mathrm{cr}}$ shown in this figure is the absolute value of the global compressive strain in the X-direction. As $b/a \rightarrow \infty$, the buckling strain approaches that for a strip $\varepsilon_{X\mathrm{cr}} = 0.00032$.

We found that for some values of b, the fundamental buckling mode may involve positive and negative deflections. Figure 15.2b shows the fundamental buck-

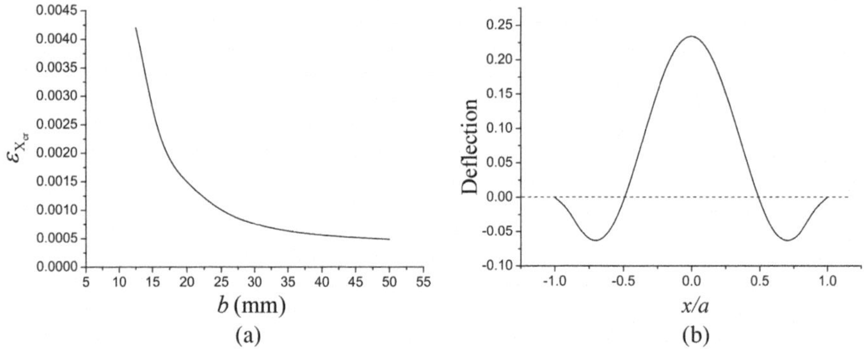

Fig. 15.2 (a) Variation of buckling strain with b of a quasi-isotropic elliptic delaminated sub-laminate under global uniaxial compression. (b) Deflection at $y = 0$ and $x \in [-a, a]$ of a quasi-isotropic elliptic delaminated sub-laminate under global uniaxial compression.

ling mode at $y = 0$ and $x \in [-a, a]$ for $a = 25.4$ mm and $b = 12.7$ mm. The values of the ordinate (deflection) in Fig. 15.2b are calculated based upon the normalized eigenvector $\{C\}^{\mathrm{T}}$. It is seen that close to $x = \pm a$, there is interpenetration of surfaces, which is physically impossible. A remedy to the non-physical interpenetration of surfaces is to use the 'anti-interpenetration model' proposed by Wang and Tong [19]. The difficulty in all contact-separation problems is that the contact/separation regions are unknown a priori. Therefore, although some schemes have been proposed to prevent the interpenetration (e.g. [18, 20], etc.), they usually result in an iterative procedure for determining the final boundary between the contact and separation regions. The model proposed by Wang and Tong [19] overcomes this difficulty by introducing a novel unified constraint function.

For two surfaces that can come in contact, assume that the contact force per unit area is

$$q(x, y) = \begin{cases} f_1 = 0, & \text{for } w > 0 \\ f_2[w(x, y)], & \text{for } w < 0 \end{cases} \tag{15.7}$$

The function $f_2(w)$ can be chosen as a linear spring function or a nonlinear Hertz-type contact function [22]

$$f_2[w(x, y)] = \begin{cases} -kw \\ -k_h \operatorname{sign}(w)(|w|)^{3/2} \end{cases} \tag{15.8}$$

where $\operatorname{sign}(w)$ is equal to $+1$ for $w > 0$, and -1 for $w < 0$. k and k_h are two positive numbers which can be obtained by experiment (cf. Wang and Tong [19]). The piecewise function $q(x, y)$ in Eq. (15.7) can be approximated to any desired accuracy by a weighted expression, first proposed by Wang and Karihaloo [21]

$$q^*(x, y) = \frac{1}{2}[1 + \tanh(A f_2)] f_2, \forall w \tag{15.9}$$

where, A is a judiciously chosen large positive number whose value depends on the desired accuracy, say, $A = 20000$ when f_2 is in the order of 1. It can be easily verified that the function q^* in Eq. (15.9) will approach zero for $w > 0$ and approach f_2 in Eq. (15.8) for $w < 0$. Obviously, when the anti-interpenetration model is imposed, the problem will become highly nonlinear but it can be easily solved. We have also calculated the buckling strains and the buckling modes for other lamination configurations of the delaminated sub-laminate, where the fibres in the sub-laminate are oriented at 0°, 45° and 90° with respect to the local x-axis in Fig. 15.1a. For these fibre orientations and the values of $a = 25.4$ mm, $b \in [12.5, 50]$ mm, no interpenetration was found. Therefore, it appears that the quasi-isotropic sub-laminate is the only one in which interpenetration is likely.

The growth of an inter-laminar delamination under an in-plane compressive stress may be dangerous for the safe operation of a composite laminate. This problem was investigated in detail by Karihaloo and Stang [14] who were interested in understanding whether a strip-like delamination left accidentally undetected during the manufacture of a long wind turbine blade (these blades are usually handmade and it is quite conceivable that a short length of tape protecting the fibres is accidentally not peeled off before resin penetration) or detected during its service poses a serious threat to the safe operation of the blade. They considered a strip delamination of length $2a$ across the width $2b$ of a long isotropic composite laminate (that is, the delamination in Fig. 15.1(b) is a rectangular area $2a \times 2b$, instead of an ellipse). They regarded the delamination as an interface crack between the thin (h) and thick (H) (Fig. 15.1b) parts of the laminate under plane strain conditions and therefore examined only a longitudinal section of unit width.

They studied the equilibrium of one-half of the homogeneous, isotropic and linear elastic laminate before the buckling of the strip delamination and after its buckling. They used the von Kármán non-linear plate theory for pre-buckling analysis and the bi-layer solution of [22] to characterize the interface plane-strain crack tip field after buckling. They indicated several ways of calculating the energy release rate, and used the simplest one which involved equating the difference of the strain energies per unit area in the edges far behind and far ahead of the crack front. They then invoked the criterion for an interface crack to grow, viz. when the energy release rate reaches a critical value for the composite material which in turn depends on the fracture mode mixity at the crack front. They produced guidelines and a series of nomograms to assess how dangerous a strip delamination is to the safe operation of a composite laminate. For this assessment, it is only necessary to know the location and size of the delamination and the intensity of the local compressive stress field.

15.4 Conclusions

A thin delaminated sub-laminate may buckle away from its thick base laminate under compression. Its role in the failure of the composite is discussed. However, the buckling modes may involve non-physical interpenetration of the two surfaces on the sides of the delamination. It is shown how to avoid the non-physical interpenetration in the theoretical analysis work via a novel "anti-interpenetration model". A strip delamination may or may not grow post-buckling. That will depend upon the applied compressive stress level and its eccentricity, the thickness of the delaminated sub-laminate and the inter-laminar critical energy release rate for the fracture mode mixity at the delamination front.

References

1. J. Wang, B. Karihaloo. Matrix crack-induced delamination in composite laminates under transverse loading. *Composite Structures*, 38(1):661-666, 1997.
2. J. Wang. Prediction of post-impact compressive strength of composite laminates us-ing an inhomogeneity model. *Journal of Composite Materials*, 33(24):2226-2247, 1999.
3. M.W. Moon, J.W. Chung, K.R. Lee, K.H. Oh, A.G. Evans. An experimental study of the influence of imperfections on the buckling of compressed thin films. *Acta Materialia*, 50(5):1219-1227, 2002.
4. J.D. Whitcomb. Finite element analysis of instability related delamination growth. *Journal of Composite Materials*, 15(5):403-426, 1981.
5. K.F. Nilsson, L.E. Asp, A. Sjögren. On transition of delamination growth behaviour for compression loaded composite panels. *International Journal of Solids and Structures*, 38(46-47):8407-8440, 2001.
6. H. Chai, C. Babcock. Two-dimensional modelling of compressive failure in delami-nated laminates. *Journal of Composite Materials*, 19(1):67-98, 1985.
7. K. Shivakumar, J. Whitcomb. Buckling of a sublaminate in a quasi-isotropic compo-site laminate. *Journal of Composite Materials*, 19(1):2-18, 1985.
8. B.D. Davidson. Delamination buckling: theory and experiment. *Journal of Composite Materials*, 25(10):1351-1378, 1991.
9. W.L. Yin, K.C. Jane. Refined buckling and postbuckling analysis of two-dimensional delaminations — I. Analysis and validation. *International Journal of Solids and Structures*, 29(5):591-610, 1992.
10. K.C. Jane, W.L. Yin. Refined buckling and postbuckling analysis of two-dimensional delaminations — II. Results for anisotropic laminates and conclusion. *International Journal of Solids and Structures*, 29(5):611-639, 1992.
11. X. Zhang, S. Yu. The growth simulation of circular buckling-driven delamination. *International Journal of Solids and Structures*, 36(12):1899-1821, 1999.
12. S. Moradi, F. Taheri. Postbuckling analysis of delaminated composite beams by differential quadrature method. *Composite Structures*, 46(1):33-39, 1999.
13. B. Storåkers, B. Andersson. Nonlinear plate theory applied to delamination in composites. *Journal of the Mechanics and Physics of Solids*, 36(6):689-718, 1988.
14. B.L. Karihaloo, H. Stang. Buckling-driven delamination growth in composite laminates: guidelines for assessing the threat posed by interlaminar matrix delamination. *Composites Part B: Engineering*, 39(2):386-395, 2008.
15. P.M. Rao, T. Wenge, D. Shu. Buckling analysis of tri-layer beams with enveloped delaminations. *Composites Part B: Engineering*, 36(1):33-39, 2005.
16. C.L. Dym. *Stability Theory and its Applications to Structural Mechanics*. Leyden, Noordhoff International Publishing, 1974.
17. L. Gui, Z. Li. Delamination buckling of stitched laminates. *Composites Science and Technology*, 61(5):629-636, 2001.
18. H. Sekine, N. Hu, M.A. Kouchakzadeh. Buckling analysis of elliptically delaminated composite laminates with consideration of partial closure of delamination. *Journal of Composite Materials* 34(7):551-574, 2000.

19. J. Wang, L. Tong. A study of the vibration of delaminated beams using a nonlinear anti-interpenetration constraint model. *Composite Structures*, 57(1):483-488, 2002.
20. H. Chen, M. Hong, Y. Liu. Dynamic behavior of delaminated plates considering progressive failure process. *Composite Structures*, 66(1):459-466, 2004.
21. J. Wang, B.L. Karihaloo. Optimum in situ strength design of composite laminates. Part II: Optimum design. *Journal of Composite Materials*, 30(12):1338-1358, 1996.
22. J.W. Hutchinson, Z.G. Suo. Mixed mode cracking in layered materials, in: *Advances in Applied Mechanics*, vol. 29 (editors: J.W. Hutchinson & T.Y. Wu). pp. 63-191, Elsevier, 1991.

Chapter 16
Suppression of Oscillations of a Loaded Flexible Robotic "Arm" as a Generalized Chebyshev Problem

Mikhail P. Yushkov and Sergei O. Bondarenko

Abstract We consider the problem of suppression of oscillations of a loaded flexible robotic "arm" that carries a load in the horizontal plane. It is required to find an optimal control force applied to the massive load of the "arm" that moves a mechanical system, within a given time period, from the initial state of rest to the new state of rest. A flexible robotic arm is considered, in an approximate model, as a set of three sequentially linked rods connected with each other and with the base by three spiral springs. First, the problem is solved via the Pontryagin maximum principle with minimization of the functional of the squared control force. Next, we pose the generalized Chebyshev problem based on the generalized Gauss principle. Calculations by these two methods are compared. The second method is shown as being superior to the first one.

Key words: Control, Suppression of oscillations, Generalized Gauss principle, Pontryagin maximum principle, Generalized Chebyshev problem

16.1 Statement of the Problem and Motion Equations of the Mechanical System

Let us consider suppression of oscillation of a flexible robotic "arm" that transfers a load in the horizontal plane Oxy. The problem is to find an optimal control force F applied to the massive base of the "arm" along the Ox-axis and that moves a mechanical system within a given time period T from the initial state of rest to the new required state of rest. Here, a flexible robotic "arm" in the approximative model is considered as a tuple of three sequentially linked rods with masses m_i and lengths $L_i\,(i = \overline{1,3})$, which are connected with each other and with the base by three spiral

Mikhail P. Yushkov · Sergei O. Bondarenko
St. Petersburg State University, 7/9 University Embankment, St. Petersburg, Russian Federation,
e-mail: yushkovmp@mail.ru, sergey.bondarenko.051@gmail.com

H. Altenbach et al. (eds.), *Advances in Solid and Fracture Mechanics*,
Advanced Structured Materials 180,
https://doi.org/10.1007/978-3-031-18393-5_16

springs of stiffness $\varkappa_i$ (see Fig. 16.1). We assume that the mass of the base, which has to be moved by a distance S, is m_0, and the transferred mass is a material point of mass m_4. The Oz-axis points up.

To compose the Lagrange equations of the second kind it is convenient to introduce the coordinates of centers of mass of the rods and the load. Denote them by x_i and y_i, $(i = \overline{1,4})$. Now, for small oscillations of the system, the kinetic and potential energies are as follows

$$\mathrm{T} = \frac{m_0\dot{x}^2}{2} + \sum_{i=1}^{4}\left(\frac{1}{2}m_i\left(\dot{x}_i^2+\dot{y}_i^2\right)\right) + \sum_{i=1}^{3}\frac{1}{2}J_{zi}\dot{\varphi}_i^2,$$

$$\Pi = \frac{1}{2}\left(\varkappa_1\varphi_1^2+\varkappa_2(\varphi_2-\varphi_1)^2+\varkappa_3(\varphi_3-\varphi_2)^2\right)$$

with

$$J_{zi} = \frac{L_i^2 m_i}{12}, \quad i = \overline{1,3}.$$

Small oscillations of the system are described by the differential equations

$$M\ddot{x} - L_1\left(\frac{m_1}{2}+m_2+m_3+m_4\right)\ddot{\varphi}_1 - L_2\left(\frac{m_2}{2}+m_3+m_4\right)\ddot{\varphi}_2$$
$$- L_3\left(\frac{m_3}{2}+m_4\right)\ddot{\varphi}_3 = F;$$

$$D_{j0}\ddot{x} + \sum_{i=1}^{3}\left(D_{ji}\ddot{\varphi}_i+\widetilde{D}_{ji}\varphi_i\right) = 0, \quad j = \overline{1,3}. \tag{16.1}$$

Here, $D_{j0}, D_{ji}, \widetilde{D}_{ji}$ are the known functions of geometrical and mass characteristics of the system. The system of differential equations (16.1) should be solved under the boundary conditions

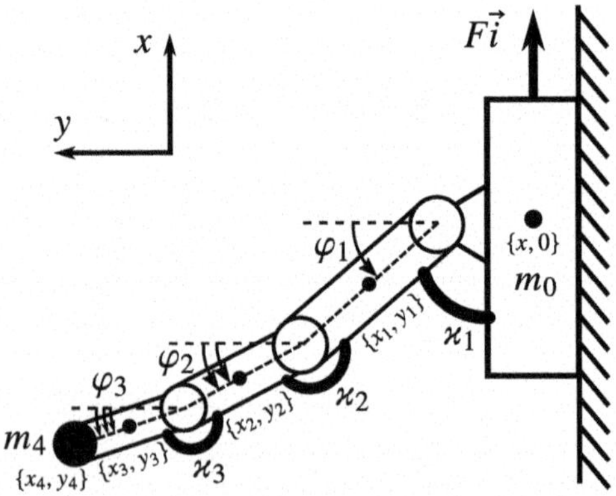

Fig. 16.1 Approximate model of a loaded flexible robotic "arm" in the Oxy plane

$$x(0) = 0; \dot{x}(0) = 0; \varphi_i(0) = 0; \dot{\varphi}_i(0) = 0;$$
$$x(T) = S; \dot{x}(T) = 0; \varphi_i(T) = 0; \dot{\varphi}_i(T) = 0, \quad i = \overline{1,3}. \tag{16.2}$$

In what follows, the control force F will be sought via the Pontryagin maximum principle [1]. Note that the system of four equations (16.1) contains five unknowns, and hence it can be augmented with the requirement that some functional be minimal. According to [2], as such functional in our problem one can take the definite integral of the minus squared required control force

$$I[F] = -\int_0^T F^2\,\mathrm{d}t. \tag{16.3}$$

For convenience of further analysis, we rewrite the last three equations of system (16.1) so that only the rods rotation angles would be preserved. To this end, we find the expression for $\ddot{x}$ from the first equation of this system and substitute it into the next three equations. Another thing is worth pointing out: the first equation itself expresses the theorem on the center mass motion of the entire system in the projection to the Ox-axis. So it is appropriate to denote that projection by x_c. Hence, finally, system (16.1) assumes the form

$$M\ddot{x}_c = F; \quad \sum_{i=1}^{3} \left(B_{ji}\ddot{\varphi}_i + \widetilde{B}_{ji}\varphi_i\right) = \beta_j F, \quad j = \overline{1,3}. \tag{16.4}$$

Here, the coefficients $B_{ij}, \widetilde{B}_{ij}, \beta_i$ depend, in the known way, on geometrical and mass properties of the system.

Let us change to the principal coordinates in the system of differential equations (16.4). Note that x_c can be taken as one of the principal coordinates, because it is not involved in the three remaining equations. Next, in these three equations we change from the φ_i-coordinates to the principal coordinates θ_i ($i = \overline{1,3}$). To this end, we find three nonzero frequencies Ω_i, and from them we define three principal forms of oscillations, which define the relation between the principal coordinates and the original angular ones. In addition, we change by scaling from θ_i to the ξ_i-coordinates so that these equations would contain equal inhomogeneities in their right-hand sides. As a result, system (16.4) can be written as four independent equations

$$\frac{\ddot{x}_c}{L_1} = \frac{F}{ML_1}, \quad \ddot{\xi}_i + \Omega_i^2 \xi_i = \frac{F}{ML_1}, \quad i = \overline{1,3}.$$

To write these equations in the dimensionless form, we make the following changes:

$$\xi_0 = \frac{x_c}{L_1}, \quad \tau = \Omega_1 t, \quad \omega_i = \frac{\Omega_i}{\Omega_1}, \quad u = \frac{F}{ML_1\Omega_1^2}, \quad \mathbb{T} = \Omega_1 T.$$

Now we have the system of dimensionless differential equations in the principal coordinates (the primes denotes derivatives with respect to the dimensionless time τ):

$$\xi_0'' = u, \quad \xi_i'' + \omega_i^2 \xi_i = u, \qquad i = \overline{1,3}. \tag{16.5}$$

The boundary conditions (16.2) now assume the form

$$\begin{aligned} &\xi_i(0) = 0; \quad \xi_i'(0) = 0, \quad i = \overline{0,3}; \\ &\xi_0(\mathbb{T}) = a; \quad \xi_0'(\mathbb{T}) = 0; \quad \xi_i(\mathbb{T}) = 0; \quad \xi_i'(\mathbb{T}) = 0, \quad i = \overline{1,3}. \end{aligned} \tag{16.6}$$

16.2 Application of the Pontryagin Maximum Principle. Relation to the Nonholonomic Problem

In the case of dimensionless variables, the functional (16.3) takes the form

$$\widetilde{I}[u] = -\int_0^{\mathbb{T}} u^2 \, d\tau. \tag{16.7}$$

Solving the above boundary-value problem of control theory (16.5), (16.6) by the classical approach with the help of the Pontryagin maximum principle for functional (16.7), we get the required dimensionless optimal control in the form [1]

$$u(\tau) = C_1 + C_2\tau + \sum_{i=1}^{3} \big(C_{2i+1} \sin(\omega_i \tau) + C_{2i+2} \cos(\omega_i \tau)\big). \tag{16.8}$$

This expression is substituted in the right-hand sides of Eqs. (16.5). Taking into account the four conditions from (16.6), one can write the solutions of these equations via the Duhamel integrals

$$\begin{aligned} \xi_0(\tau) &= \int_0^{\tau} u(\tau_0)(\tau - \tau_0)\, d\tau_0; \\ \xi_i(\tau) &= \frac{1}{\omega_i} \int_0^{\tau} u(\tau_0) \sin[\omega_i(\tau - \tau_0)]\, d\tau_0, \quad i = 1,2,3. \end{aligned} \tag{16.9}$$

Substituting functions (16.9) in the second group of boundary conditions (16.6), we find the eight arbitrary constants in the expression for the dimensionless control (16.8). Now we completely know the control force (16.8) and the motion of the system (16.9). The results of calculations via these formulas are shown below in figures by solid lines.

Let us now dwell on the **interesting phenomenon**, which is characteristic of the motion of a mechanical system obtained via the Pontryagin maximum principle. Note that the resulting control (16.8) is a solution of a tenth-order differential equation, which in the dimensional form can be written as [5]

$$\frac{\mathrm{d}^2}{\mathrm{d}t^2}\left(\frac{\mathrm{d}^2}{\mathrm{d}t^2}+\Omega_1^2\right)\left(\frac{\mathrm{d}^2}{\mathrm{d}t^2}+\Omega_2^2\right)\left(\frac{\mathrm{d}^2}{\mathrm{d}t^2}+\Omega_3^2\right)F=0. \tag{16.10}$$

Inserting in (16.10) the expression for F taken from the first equation of system (16.1), we get the following differential equation with known constant coefficients (which are not presented due to their bulkiness):

$$\sum_{i=2}^{5}\left(a_{2i,x}\frac{\mathrm{d}^{2i}x}{\mathrm{d}t^{2i}}+a_{2i,\varphi_1}\frac{\mathrm{d}^{2i}\varphi_1}{\mathrm{d}t^{2i}}+a_{2i,\varphi_2}\frac{\mathrm{d}^{2i}\varphi_2}{\mathrm{d}t^{2i}}+a_{2i,\varphi_3}\frac{\mathrm{d}^{2i}\varphi_3}{\mathrm{d}t^{2i}}\right)=0. \tag{16.11}$$

But the differential equation (16.11) thus obtained can be considered as a **tenth-order (!)** linear nonholonomic constraint, which is continuously satisfied in the motion of a mechanical system obtained by application of the Pontryagin maximum principle. This suggests the following generalized Chebyshev problem [3, 4] in place of the original statement of the problem: *find the controlled motion of a guide with three rods bearing a given load, provided that the motion program is specified as some additional differential equation with constant tenth-order coefficients.* In this statement of the problem, the differential equation in question is considered as a tenth-order linear nonholonomic constraint, whose reaction is the required control force.

16.3 Application of the Generalized Gauss Principle for Solving the Above Generalized Chebyshev Problem

Let us now examine the above generalized Chebyshev problem, which will be solved as the corresponding problem of nonholonomic mechanics under a tenth-order nonholonomic constraint. For theoretical background of solutions of similar problems, see, for example, [3, 4, 5, 6, 7], in which the generalized Gauss principle plays a central role. This principle was proposed already in 1983 in [8]. When dealing with a problem involving a tenth-order linear nonholonomic constraint, one should apply the eights-order generalized Gauss principle, according to which the requirement

$$\delta^{(10)}\left(\mathbf{R}_{(8)}\right)^2=0 \tag{16.12}$$

should be satisfied. In other words, one varies the tenths derivatives in the expression for the squared eights derivatives of the reaction $\mathbf{R}$ of the applied high-order constraint. As noted above, in the generalized Chebyshev problem, the constraint reaction is considered as the sought-for control force $u(\tau)$, and hence from (16.12) we find that

$$\frac{d^8 u}{d\tau^8} = 0 \tag{16.13}$$

resulting in

$$u(\tau) = \sum_{i=1}^{8} C_i \tau^{i-1}. \tag{16.14}$$

As in the case of the Pontryagin maximum principle, the arbitrary constants in control (16.14) can be found from the second group of boundary-value conditions (16.6), after which we evaluate the motion of the system via the Duhamel integrals (16.9). The resulting solutions are shown in figures by dashed lines.

Note that in the case of a short-time motion with $\mathbb{T} = \frac{\pi}{4}$, all the graphs (not provided in the paper) differ insignificantly and tend to each other as the dimensionless motion time decreases. In other words, for short-time motion, the results of the new method are close to those obtained by the classical method of control theory.

However, results become to differ significantly if one considers more extensive motions (see Figs. 16.2 and 16.3). This can be explained by the fact that, under the first method, the resulting control (16.8) contains harmonics with eigenfrequencies of the system, and consequently, the resulting control force tends to bring the system

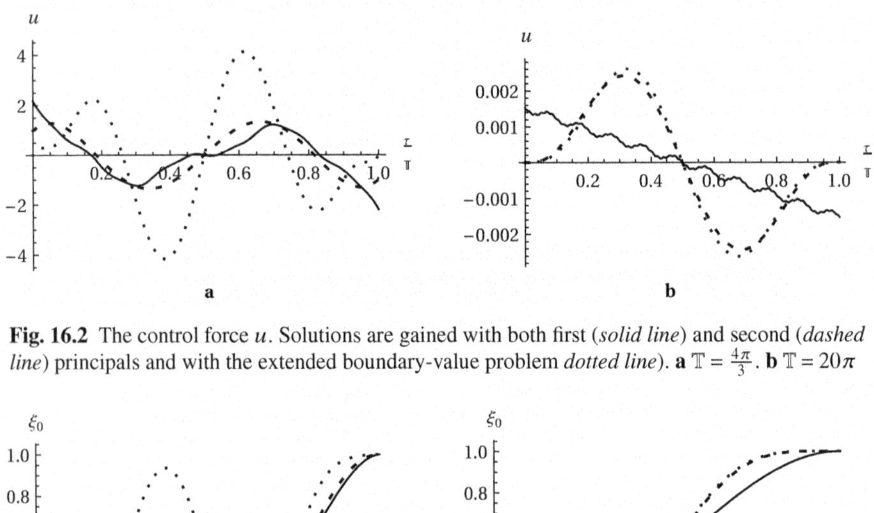

Fig. 16.2 The control force u. Solutions are gained with both first (*solid line*) and second (*dashed line*) principals and with the extended boundary-value problem *dotted line*). **a** $\mathbb{T} = \frac{4\pi}{3}$. **b** $\mathbb{T} = 20\pi$

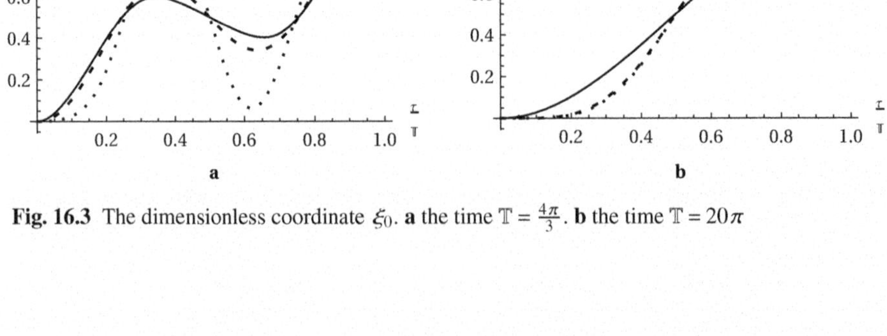

Fig. 16.3 The dimensionless coordinate ξ_0. **a** the time $\mathbb{T} = \frac{4\pi}{3}$. **b** the time $\mathbb{T} = 20\pi$

into resonance. In contrast, in the second method, the control is found in the form of polynomial (16.14), which ensures a relatively smooth motion of the system.

16.4 Statement and Solution of the Extended (Generalized) Boundary-value Problem

Note that the application of the Pontryagin maximum principle always involves jumps in the control force at the beginning and end of the motion of the system (see Fig. 16.2). At the same time, the solution of the generalized Chebyshev problem has jumps on the left panel of Fig. 16.2 (the case of "short" motion), while on the right panel (the case of "long" motion) they disappear. This leads us to the problem of removal of jumps in the solution of the generalized Chebyshev problem also in the case of a "short-time" motion. To this end, we augment the available dimensional boundary conditions (16.2) with the conditions that the acceleration of the base should be zero at the beginning and end of motion:

$$\ddot{x}(0) = 0, \qquad \ddot{x}(T) = 0.$$

We rewrite these conditions in the dimensionless form:

$$\xi_0''(0) = 0, \qquad \xi_0''(\mathbb{T}) = 0. \tag{16.15}$$

Let us now solve the boundary-value problem of control theory (16.5), (16.6), (16.15). We call this problem the first-order extended (generalized) boundary-value problem. Note that this problem is not amenable to the Pontryagin maximum principle, because the resulting control would have an insufficient number of arbitrary constants for satisfaction of all applicable boundary-value conditions (16.6), (16.15). At the same time, the solution can be found by the generalized Gauss principle by using the two orders higher principle. The resulting solutions are shown in figures by dotted lines. The left panel in Fig. 16.2 shows that, for "short-time" motion, it indeed proved possible to eliminate the jumps of the control force. Note that by posing higher-orders generalized boundary-value problems (it is required that higher-order derivatives would also vanish) one may achieve an additional smoothness of the control force at the beginning and end of motion.

16.5 Calculation Results

The dimensionless parameters of the system are related to the dimensional ones as follows:

$$\mu_i = \frac{m_i}{M}, \quad i = \overline{0,4},$$
$$\alpha_i = \frac{L_i}{L_1}, \qquad \gamma_i = \frac{\varkappa_i}{L_1^2 M \Omega_1^2}, \quad i = \overline{1,3}.$$

For calculations, we used the following values of the dimensionless parameters:

$$\mu_0 = \frac{5}{13}, \mu_1 = \frac{2}{13}, \mu_2 = \frac{3}{26}, \mu_3 = \frac{1}{26}, \mu_4 = \frac{4}{13},$$
$$\alpha_2 = 0.7, \alpha_3 = 0.5, \gamma_1 = 1.69, \gamma_2 = 0.63, \gamma_3 = 0.25.$$

The time $\mathbb{T}$ during which the base travels the distance $a = 1$ will be increased for comparison of the resulting solutions. The calculations results depicted in Figs. 16.2 – 16.6 were discussed above.

16.6 Conclusions

Two completely different methods are used for solution of one practical problem of finding an optimal control force that transfers a mechanical system, within a given

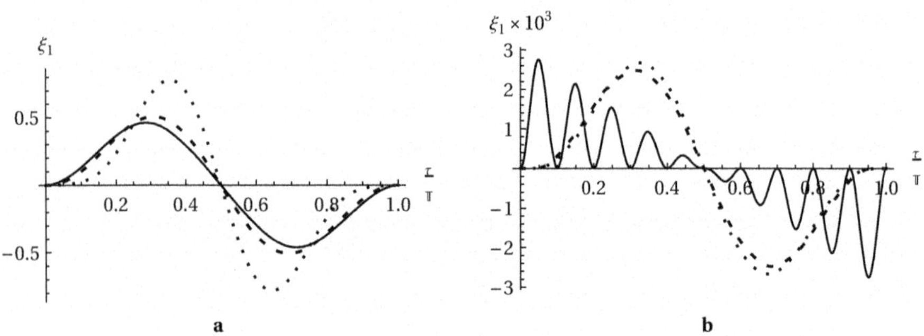

Fig. 16.4 The dimensionless coordinate ξ_1. **a** the time $\mathbb{T} = \frac{4\pi}{3}$. **b** the time $\mathbb{T} = 20\pi$

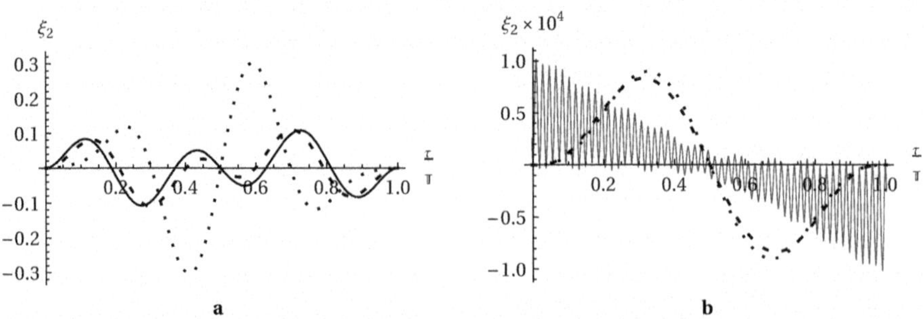

Fig. 16.5 The dimensionless coordinate ξ_2. **a** the time $\mathbb{T} = \frac{4\pi}{3}$. **b** the time $\mathbb{T} = 20\pi$

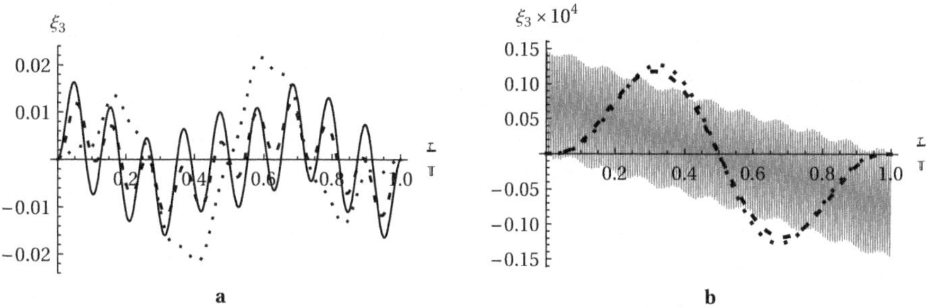

Fig. 16.6 The dimensionless coordinate ξ_3. **a** the time $\mathbb{T} = \frac{4\pi}{3}$. **b** the time $\mathbb{T} = 20\pi$

time period, from a given phase state to a given finial phase state. The first approach is a classical method of control theory depending on the Pontryagin maximum principle for minimization of some functional. It was found that, during the entire motion time, a high-order linear nonholonomic constraint applies. This enabled us to formulate the original problem of control theory as a generalized Chebyshev problem, in which this constraint is considered as a motion program of the system, and its reaction is the sought-for control force. This problem is solved via the generalized Gauss principle. Numerous calculations via these two methods have shown that the results are practically equal for short-time motions, but they were found to be drastically different for long-time motions. This is explained by the fact that, under the first method, the evaluated control force contains harmonics with eigenfrequencies of system (which, for long-time motion, this tends to bring the system into resonance), while in the second method the control is constructed in the form of a polynomial of time, which ensures a relatively smooth motion of the system. Moreover, by formalizing and solving the generalized boundary-value problem of control theory, it is possible to achieve that the system moves arbitrarily smoothly at the beginning and end of the motion. It is worth also pointing out that the control force via the first method always has jumps at these time instants.

References

1. L.S. Pontryagin, V.G. Boltyanskii, R.V. Gamkrelidze, E.F. Mishchenko. *The Mathematical Theory of Optimal Processes*. Gordon and Breach, New York, 1987.
2. F.L. Chernous'ko, L.D. Akulenko, B.N. Sokolov. *Control of Oscillations (in Russ.)*. Moscow: Nauka, 1980.
3. M.P. Yushkov. Formulation and solution of a generalized Chebyshev problem: First Part. *Vestnik St. Petersburg University, Mathematics*, 52(4):436–451, 2019.
4. M.P. Yushkov. Formulation and solution of a generalized Chebyshev problem: Second Part. *Vestnik St. Petersburg University, Mathematics*, 53(4):459–472, 2020.
5. S. Zegzhda, M. Yushkov, Sh. Soltakhanov, N. Naumova, T. Shugaylo. A novel approach to suppression of oscillations. *ZAMM (Zeitschrift für angew. Math. und Mech.)*. 98(5):781–788, 2018.

6. Sh. Kh. Soltakhanov, M.P. Yushkov, S.A. Zegzhda. *Mechanics of Non-Holonomic Systems. A New Class of Control Systems*. Berlin, Heidelberg: Springer-Verlag, 2009.
7. S.A Zegzhda, M.P. Yushkov, Sh. Kh. Soltakhanov, E.A. Shatrov. *Nonholonomic Mechanics and Control Theory (in Russ.)*. Moscow: Fizmatlit, 2018.
8. N.N. Polyakhov, S.A. Zegzhda, M.P. Yushkov. Generalization of Gauss's principle to nonholonomic systems of higher orders (in Russ.). *Dokl. Akad. Nauk SSSR*, 269(6):1328–1330, 1983.

Chapter 17
Theory of Cosserat-type Elastic Shells with Distributed Dislocations and Disclinations

Leonid M. Zubov and Mikhail I. Karyakin

Abstract A mathematical model is constructed for Cosserat elastic shells containing continuously distributed dislocations and disclinations. Displacements, rotations, and strains are assumed to be small, i.e., the geometrically linear 6-parameter theory of shells is used. The system of equations describing the static deformations of the shells is derived by passing to the limit from a discrete set of isolated dislocations and disclinations to their continuous distribution. Deformational boundary conditions are derived, and a boundary value problem is formulated for the equilibrium of an elastic shell with given dislocation and disclination densities. With the help of stress functions, a variational statement of the boundary value problem of the statics of shells with distributed defects is given. A static-geometry analogy of the Cosserat shells theory considering distributed dislocations and disclinations is established. The dual boundary value problems of shell statics are formulated. The dual problems are mathematically equivalent, but completely different in their physical formulation. The problem of the stress state of a spherical shell with uniformly distributed dislocations and disclinations is solved.

Key words: Dislocation and disclination densities, Deformational boundary conditions, Stress functions, Static-geometry analogy, Dual boundary value problems

17.1 Introduction

Since the first exfoliation of single-layer graphene, two-dimensional (2D) materials have attracted worldwide attention due to their unique structures and remarkable applications. The large family of discovered two-dimensional materials covers an extremely wide range of properties, allowing the exploration of new physi-

Leonid M. Zubov · Mikhail I. Karyakin
Southern Federal University, Russian Federation,
e-mail: lmzubov@sfedu.ru, karyakin@sfedu.ru

H. Altenbach et al. (eds.), *Advances in Solid and Fracture Mechanics*,
Advanced Structured Materials 180,
https://doi.org/10.1007/978-3-031-18393-5_17

cal effects and the fabrication of new devices with novel capabilities. In particular, 2D atomically-thick materials, graphene and graphene-like layered transition metal dichalcogenides, have showed vast potential as novel energy materials due to their unique physicochemical properties [24].

Defects in the crystal structure of bulk materials have been studied extensively and in detail since the middle of the last century, due to their decisive role in the formation of the most important physical and mechanical properties of these materials. Structural defects in 2D materials have attracted much attention since the era of graphene has arisen. During the synthesis of 2D materials, various types of defects are inevitably generated. A wide variety of dislocations was observed in two-dimensional crystals of different nature [4, 6]. Experimental studies show that these defects significantly affect physical, chemical, and mechanical properties of 2D materials [26]. Dislocation-type defects play an essential role in the mechanical behavior of surface crystals, nanotubes, nanofilms, and other two-dimensional physical systems [14].

The design based on topological defects in 2D materials has become a hot topic in the field of mechanics and material sciences [5, 20]. Defect engineering, among others, is known as a promising technique to manipulate and tune the properties of available and yet undiscovered 2D materials [3].

2D nanomaterials are an emerging class of biomaterials with remarkable potential for biomedical applications [19]. Lattice defects affect the theranostic performance of 2D nanomaterials significantly by altering their electronic properties and chemical binding. Recent investigations have shown that defect-rich 2D nanomaterials are capable of enhancing treatment techniques and improving diagnostics [22].

Recent interest in defects in thin sheets comes from the modeling of biological growth processes [7]. A detailed theoretical perspective of the interplay between symmetry, topological defects, and mechanical properties of virus shells (capsids) is provided in [27]. Various aspects of the biological relevance of geometric defects in the virus shell are described in [25]. Icosahedral alphaviruses and flaviviruses, among others, have capsids with geometric defects that may facilitate assembly, dissociation, or accessibility of cellular proteins to virion components. The studies of viral capsids within the framework of the nonlinear theory of elastic shells [15, 23, 21] demonstrated good fits to experimentally determined virus shapes.

In this paper, we apply the geometrically linear theory of elastic shells of the Cosserat type. Based on the concept of an isolated defect (Volterra dislocation) in a multiply connected two-dimensional elastic body, we construct a model of Cosserat elastic shell with distributed dislocations and disclinations by passing to the limit from a discrete set of isolated defects to their continuous distribution. Previously, the theory of continuously distributed dislocations and disclinations was developed in[30] within the framework of the classical Kirchhoff-Love shell model.

17.2 Initial Relations of the Geometrically Linear Theory of the Cosserat-type Shells

The system of equations describing the static deformations of an elastic shell of the Cosserat type [28, 1, 10, 2] consists of equilibrium equations for forces and moments

$$\operatorname{div}\mathbf{T}+\boldsymbol{f}=0, \tag{17.1}$$

$$\operatorname{div}\mathbf{M}+\mathbf{T}_{\times}+\boldsymbol{l}=0, \tag{17.2}$$

$$\operatorname{div}\mathbf{F}\triangleq\boldsymbol{r}^{\gamma}\cdot\frac{\partial\mathbf{F}}{\partial y^{\gamma}},\ \boldsymbol{r}^{\gamma}\cdot\boldsymbol{r}_{\varkappa}=\delta^{\gamma}_{\varkappa},\ \boldsymbol{r}^{\gamma}\cdot\boldsymbol{n}=0,$$

$$\boldsymbol{r}_{\varkappa}=\frac{\partial\boldsymbol{r}}{\partial y^{\varkappa}},\ \gamma,\varkappa=1,2,$$

constitutive relations (equations of state)

$$\mathbf{T}=\frac{\partial W}{\partial\mathbf{E}},\ \mathbf{M}=\frac{\partial W}{\partial\mathbf{K}},\ W=W(\mathbf{E},\mathbf{K}), \tag{17.3}$$

and geometric relationships

$$\mathbf{E}=\operatorname{grad}\boldsymbol{u}+\mathbf{g}\times\boldsymbol{\omega},\ \mathbf{K}=\operatorname{grad}\boldsymbol{\omega}, \tag{17.4}$$

$$\operatorname{grad}\mathbf{F}\triangleq\boldsymbol{r}^{\gamma}\otimes\frac{\partial\mathbf{F}}{\partial y^{\gamma}},\ \mathbf{g}=\mathbf{I}-\boldsymbol{n}\otimes\boldsymbol{n}.$$

In (17.1)–(17.4) $\mathbf{T}$ is the stress tensor, $\mathbf{M}$ is the couple stress tensor, $\boldsymbol{f}$ is the vector intensity of the external forces distributed over shell surface σ, $\boldsymbol{l}$ is the vector intensity of the external couples distributed over shell surface σ, $\mathbf{E}$ is the linear strain tensor of the shell, $\mathbf{K}$ is the tensor of curvature changes, W is specific (per surface σ area unit) strain energy, y^1, y^2 are Gaussian coordinates on the surface σ, $\mathbf{I}$ is three-dimensional unit tensor, $\boldsymbol{n}$ is the unit normal to σ, $\mathbf{g}$ is the first fundamental (metric) tensor of the surface, $\boldsymbol{r}(y^1,y^2)$ is the radius vector of a surface point, $\boldsymbol{r}_{\varkappa}$ and $\boldsymbol{r}^{\gamma}$ are the main and reciprocal vector bases on σ, $\mathbf{F}(y^1,y^2)$ is a differentiable field of arbitrary rank, grad and div are gradient and divergence operations on the surface, $\delta^{\gamma}_{\varkappa}$ is the Kronecker symbol, $\boldsymbol{u}(y^1,y^2)$ is the displacement vector field of the shell σ, $\boldsymbol{\omega}(y^1,y^2)$ is the linear vector of shell particles rotations. The vector field $\boldsymbol{\omega}$ is kinematically independent of the displacement field $\boldsymbol{u}$. The symbol $\mathbf{T}_{\times}$ means the vector invariant of the second rank tensor $\mathbf{T}$. In particular, for a dyad of vectors, we have $(\boldsymbol{a}\otimes\boldsymbol{b})_{\times}=\boldsymbol{a}\times\boldsymbol{b}$. The tensors $\mathbf{E}$ and $\mathbf{K}$ vanish during rigid motion of the shell, when $\boldsymbol{u}=\boldsymbol{u}_0+\boldsymbol{\omega}_0\times\boldsymbol{r}$, $\boldsymbol{\omega}=\boldsymbol{\omega}_0$, $\boldsymbol{u}_0$ and $\boldsymbol{\omega}_0$ are constant vectors.

The tensors of strains, curvature changes, stress, and couple stress satisfy the relations

$$\boldsymbol{n}\cdot\mathbf{E}=\boldsymbol{n}\cdot\mathbf{K}=\boldsymbol{n}\cdot\mathbf{T}=\boldsymbol{n}\cdot\mathbf{M}=0.$$

The vector $\mathbf{E}\cdot\boldsymbol{n}$ is a quantitative characteristic of the transverse shear in an elastic shell, and the vector $\mathbf{T}\cdot\boldsymbol{n}$ characterizes shear forces.

The most common types of boundary conditions for a shell are kinematic boundary conditions, when displacements and rotations are specified on some part of the boundary contour

$$\boldsymbol{u} = \boldsymbol{u}^*(s), \ \boldsymbol{\omega} = \boldsymbol{\omega}^*(s), \tag{17.5}$$

and dynamic boundary conditions, when force load $\boldsymbol{\tau}$ and couple load $\boldsymbol{\mu}$ distributed along the curve are stated on a part of the boundary

$$\boldsymbol{m} \cdot \mathbf{T} = \boldsymbol{\tau}(s), \ \boldsymbol{m} \cdot \mathbf{M} = \boldsymbol{\mu}(s). \tag{17.6}$$

Here $\boldsymbol{m}$ is the unit normal to the boundary contour satisfying the condition $\boldsymbol{m} \cdot \boldsymbol{n} = 0$ and directed to the side external to the surface σ, $\boldsymbol{u}^*$, $\boldsymbol{\omega}^*$, $\boldsymbol{\tau}$ and $\boldsymbol{\mu}$ are given functions of the current arc length s.

Although the fields of displacements, rotations, and strains are assumed to be small, the dependencies of the stress and couple stress tensors **T** and **M** on the strain tensors **E** and **K** can be nonlinear, i.e., the physical nonlinearity of the shell material is allowed.

17.3 Continuously Distributed Dislocations and Disclinations in an Elastic Shell

To introduce the concept of dislocations in an elastic shell, consider the problem of determining the displacement field $\boldsymbol{u}$ from given single-valued differentiable fields of the strain tensor **E** and the rotation vector $\boldsymbol{\omega}$. Based on (17.4) we have

$$\operatorname{grad} \boldsymbol{u} = \mathbf{E} - \mathbf{g} \times \boldsymbol{\omega}. \tag{17.7}$$

The solution of the equation (17.7) is written as a curvilinear integral

$$\boldsymbol{u}(\boldsymbol{r}) = \int_{\boldsymbol{r}_0}^{\boldsymbol{r}} d\boldsymbol{r} \cdot (\mathbf{E} - \mathbf{g} \times \boldsymbol{\omega}) + \boldsymbol{u}(\boldsymbol{r}_0). \tag{17.8}$$

If the condition

$$\operatorname{div}(\mathbf{d} \cdot \mathbf{E} - \mathbf{d} \times \boldsymbol{\omega}) = 0, \ \mathbf{d} = -\mathbf{I} \times \boldsymbol{n} \tag{17.9}$$

holds, then the integral (17.8) in the case of a simply connected domain does not depend on the choice of the curve connecting the points $\boldsymbol{r}$ and $\boldsymbol{r}_0$ on the surface σ. This means that for a given initial value $\boldsymbol{u}(\boldsymbol{r}_0)$, the expression (17.8) defines a single-valued function $\boldsymbol{u}(\boldsymbol{r})$. The tensor **d** in (17.9) is called the surface discriminant tensor [29, 31].

Let us now consider some section σ_0 of the shell surface and assume that the domain σ_0 is multiply connected and homeomorphic to a circle with N circular holes, and the functions $\mathbf{E}(\boldsymbol{r})$ and $\boldsymbol{\omega}(\boldsymbol{r})$ are single-valued in a multiply connected domain. In this case, the property of displacement single-valuedness is, generally, lost. The multivaluedness can be eliminated by turning the domain σ_0 into a simply connected one by drawing the necessary number of cuts (partitions). In this case, the values of the function $\boldsymbol{u}(\boldsymbol{r})$ will differ on different sides of the cut. It follows from (17.8) that the displacement jump at the intersection of each cut is described by the formula

$$\boldsymbol{u}_+ - \boldsymbol{u}_- = \boldsymbol{b}_k, \tag{17.10}$$

where $\boldsymbol{b}_k$ $(k = 1, 2, \ldots, N)$ are vectors constant for each cut, called Burgers vectors. These vectors do not depend on the choice of the cut system and, according to (17.8), are expressed in terms of the fields $\mathbf{E}(\boldsymbol{r})$ and $\omega(\boldsymbol{r})$ by contour integrals

$$\boldsymbol{b}_k = \oint_{\Gamma_k} d\boldsymbol{r} \cdot (\mathbf{E} - \mathbf{g} \times \omega), \tag{17.11}$$

where Γ_k is any simple closed contour enveloping only one k-th hole. Nonzero values of $\boldsymbol{b}_k$ mean that the shell contains isolated dislocations.

If the number of dislocations in a limited part of the shell is very large, it is expedient to pass to a continuous distribution of defects. The total Burgers vector of a discrete set of N isolated dislocations contained in the subdomain σ_0, according to (17.11), is given by

$$\boldsymbol{B} = \sum_{k=1}^{N} \boldsymbol{b}_k = \sum_{k=1}^{N} \oint_{\Gamma_k} d\boldsymbol{r} \cdot (\mathbf{E} - \mathbf{g} \times \omega). \tag{17.12}$$

Due to the well-known properties of curvilinear integrals and the uniqueness of the integrand, the sum of the integrals in (17.12) can be replaced by a single integral over the closed contour Γ_0, which encloses all holes in the region σ_0

$$\boldsymbol{B} = \oint_{\Gamma_0} d\boldsymbol{r} \cdot (\mathbf{E} - \mathbf{g} \times \omega). \tag{17.13}$$

To go from a discrete set of dislocations to their continuous distribution, we tend the hole diameters to zero and transform the contour integral by the divergence theorem [29, 31] into a surface integral over the region σ_0 bounded by the contour Γ_0

$$\boldsymbol{B} = \iint_{\sigma_0} \alpha \, d\sigma, \quad \alpha = \operatorname{div}(\mathbf{d} \cdot \mathbf{E}) + (\mathbf{d} \cdot \mathbf{K})_\times. \tag{17.14}$$

Here we use the easily verified identity

$$\operatorname{div}(\mathbf{d} \times \omega) = -(\mathbf{d} \cdot \operatorname{grad} \omega)_\times,$$

and the geometric relation (17.4).

Since $\boldsymbol{B}$ is the total Burgers vector of all dislocations contained in an arbitrary region σ_0, the vector field $\boldsymbol{\alpha}$ should be called the dislocation density.

Further, we assume that a part of the shell containing continuously distributed dislocations occupies a multiply connected region and set the problem of determining the field of rotations $\boldsymbol{\omega}$ in this multiply connected region from the given single-valued and differentiable fields of the strain tensor $\mathbf{E}$ and the tensor of curvature changes $\mathbf{K}$. We now abandon the requirement that rotations be single-valued. Integrating the relationship $(17.4)_2$, we get

$$\boldsymbol{\omega}(\boldsymbol{r}) = \int_{\boldsymbol{r}_0}^{\boldsymbol{r}} d\boldsymbol{r} \cdot \mathbf{K} + \boldsymbol{\omega}(\boldsymbol{r}_0). \tag{17.15}$$

In a multiply connected domain, the rotation field given by (17.15) is, generally, non-unique. When traversing a closed contour enveloping one of the holes, the vector $\boldsymbol{\omega}$ receives an increment called the Frank vector

$$\boldsymbol{q}_s = \oint_{\Gamma_s} d\boldsymbol{r} \cdot \mathbf{K}, \; s = 1, 2, \ldots, M. \tag{17.16}$$

Non-zero constants $\boldsymbol{q}_s$ mean the existence of isolated disclinations in a multiply connected shell with continuously distributed dislocations. The total Frank vector for a discrete set of disclinations is expressed as

$$\boldsymbol{Q} = \sum_{s=1}^{M} \boldsymbol{q}_s = \sum_{s=1}^{M} \oint_{\Gamma_s} d\boldsymbol{r} \cdot \mathbf{K} = \oint_{\Gamma_0} d\boldsymbol{r} \cdot \mathbf{K}. \tag{17.17}$$

Arguing similarly to the previous one, we pass from a discrete set of disclinations to their continuous distribution. We get

$$\boldsymbol{Q} = \iint_{\sigma_0} \operatorname{div}(\mathbf{d} \cdot \mathbf{K})\, d\sigma. \tag{17.18}$$

The integrand in (17.18) should be called the disclination density

$$\boldsymbol{\beta} = \operatorname{div}(\mathbf{d} \cdot \mathbf{K}).$$

In what follows, the dislocation and disclination densities, $\boldsymbol{\alpha}$ and $\boldsymbol{\beta}$, will be considered as given functions of the Gaussian coordinates y^1, y^2, similarly to the external loads $\boldsymbol{f}$ and $\boldsymbol{l}$ distributed over the shell surface.

In the presence of distributed defects in the form of dislocations and disclinations, the fields of displacements $\boldsymbol{u}$ and rotations $\boldsymbol{\omega}$ do not exist, so the geometric relations (17.4) do not make sense. Instead, one should use the equations following from the definition of the densities $\boldsymbol{\alpha}$ and $\boldsymbol{\beta}$, called the incompatibility equations

$$\operatorname{div}(\mathbf{d} \cdot \mathbf{E}) + (\mathbf{d} \cdot \mathbf{K})_{\times} = \boldsymbol{\alpha}, \tag{17.19}$$

$$\mathrm{div}(\mathbf{d}\cdot\mathbf{K}) = \boldsymbol{\beta}. \tag{17.20}$$

For $\alpha = \boldsymbol{\beta} = 0$, the equations (17.19), (17.20) take the form

$$\mathrm{div}(\mathbf{d}\cdot\mathbf{E}) + (\mathbf{d}\cdot\mathbf{K})_\times = 0, \ \mathrm{div}(\mathbf{d}\cdot\mathbf{K}) = 0 \tag{17.21}$$

and are called the compatibility equations. They can be easily obtained by eliminating the vectors $\boldsymbol{u}$ and $\boldsymbol{\omega}$ from the relations (17.4). When the expressions (17.4) are substituted into the compatibility equations (17.21), the latter turn into identities.

The dislocation and disclination densities, as well as the surface load intensities, can be generalized functions. This allows us to consider the case of concentrated defects. The density $\boldsymbol{\alpha}$ corresponding to a dislocation concentrated at the point $\boldsymbol{r}_*$ has the form $\boldsymbol{\alpha} = \boldsymbol{b}\delta(\boldsymbol{r}-\boldsymbol{r}_*)$, where $\delta(\boldsymbol{r})$ is the Dirac delta function, and the density $\boldsymbol{\beta}$ for a localized disclination is described by the formula $\boldsymbol{\beta} = \boldsymbol{q}\delta(\boldsymbol{r}-\boldsymbol{r}_*)$. Here $\boldsymbol{b}$ and $\boldsymbol{q}$ are the Burgers and Frank vectors of concentrated defects. The axis of a dislocation or disclination concentrated at a given point on the shell surface should be thought of as a straight line orthogonal to the surface at that point. Therefore, following the terminology of the theory of linear defects in a three-dimensional medium [8, 9], a concentrated dislocation whose Burgers vector is directed along the normal to the surface will be called a screw dislocation, and a dislocation with a Burgers vector lying in the tangent plane will be called an edge dislocation. A disclination with the Frank vector directed along the normal to the surface is called a wedge disclination, a disclination whose Frank vector is orthogonal to the normal $\boldsymbol{n}$ is called a torsional disclination.

This terminology extends to continuously distributed dislocations and disclinations. So, if the dislocation density has the form $\boldsymbol{\alpha} = a(y^1, y^2)\boldsymbol{n}$, then we can talk about screw dislocations distributed in the shell with a scalar density $a(y^1, y^2)$.

Note that the shell models of the Cosserat and Kirchhoff-Love types differ, in particular, in that the latter cannot consider continuously distributed screw dislocations [30]. In addition, the theory of Kirchhoff-Love shells does not allow the application of distributed over the surface couple load, the vector of which is directed along the normal to the surface.

17.4 Boundary-value Problem of Equilibrium of a Shell with Distributed Dislocations and Disclinations

The kinematic boundary conditions (17.5) can be replaced by constraints on the contour values of the strain tensors $\mathbf{E}$ and $\mathbf{K}$. If the equation of a curve that is part of the shell boundary is given as $\boldsymbol{r} = \boldsymbol{r}(s)$, where s is the current arc length, then the unit vector of the tangent to this curve is $\boldsymbol{t} = \partial\boldsymbol{r}/\partial s$. Therefore, based on (17.4), (17.5) we have

$$\boldsymbol{t}\cdot\mathbf{E} = \boldsymbol{v}(s), \ \boldsymbol{t}\cdot\mathbf{K} = \boldsymbol{w}(s), \tag{17.22}$$

$$\boldsymbol{v}(s) = \frac{d\boldsymbol{u}^*}{ds} + \boldsymbol{t}\times\boldsymbol{\omega}^*, \ \boldsymbol{w}(s) = \frac{d\boldsymbol{\omega}^*}{ds}. \tag{17.23}$$

Boundary conditions (17.22) are called deformational conditions [10]. In the general case, they are not equivalent to kinematic ones. Indeed, let us find the functions $\boldsymbol{u}^*(s)$ and $\boldsymbol{\omega}^*(s)$ from the differential equations (17.23), assuming the functions $\boldsymbol{v}(s)$ and $\boldsymbol{w}(s)$ are known. Integrating the (17.23) equations, we obtain

$$\boldsymbol{\omega}^*(s) = \boldsymbol{\omega}_0^* + \int_{s_0}^{s} \boldsymbol{w}(s')ds',$$

$$\boldsymbol{u}^*(s) = \boldsymbol{u}_0^* + \boldsymbol{\omega}_0^* \times (\boldsymbol{r} - \boldsymbol{r}_0) + \int_{s_0}^{s} \boldsymbol{v}(s')ds' + \int_{s_0}^{s} \boldsymbol{w}(s') \times (\boldsymbol{r} - \boldsymbol{r}')ds', \qquad (17.24)$$

$$\boldsymbol{r} = \boldsymbol{r}(s),\ \boldsymbol{r}_0 = \boldsymbol{r}(s_0),\ \boldsymbol{u}_0^* = \boldsymbol{u}^*(s_0),\ \boldsymbol{\omega}_0^* = \boldsymbol{\omega}^*(s_0).$$

Here $\boldsymbol{u}_0^*$ and $\boldsymbol{\omega}_0^*$ are arbitrary vector constants, and the integration variable is marked with a prime.

The expression (17.24) shows that the kinematic boundary conditions are restored from the deformational ones within the accuracy up to the motion of an absolutely rigid body. If the deformational conditions are set on a connected (solid) segment of the shell boundary, then we can put $\boldsymbol{u}_0^* = \boldsymbol{\omega}_0^* = 0$, because the rigid motion of the entire shell does not affect its stressed state. Thus, if the part of the shell boundary on which the deformation conditions are satisfied is connected, then they are equivalent to the kinematic boundary conditions (17.5).

If there are distributed defects in the shell, the kinematic boundary conditions do not make sense. Their role shifts to deformational conditions. The functions $\boldsymbol{v}(s)$ and $\boldsymbol{w}(s)$ can be specified arbitrarily, and the part of the boundary where these functions are specified does not necessarily consist of a single connected segment.

The complete system of equations describing the state of an elastic shell with distributed defects consists of equilibrium equations for forces and moments (17.1), (17.2), the incompatibility equations (17.19), (17.20) and constitutive relations (17.3). If the tensors $\mathbf{T}$ and $\mathbf{M}$ are expressed in terms of $\mathbf{E}$ and $\mathbf{K}$ using (17.3), then the equilibrium equations (17.1), (17.2) will be written in terms of strain tensors. Since, in the general case, each of the tensors $\mathbf{E}$ and $\mathbf{K}$ has six independent components, the equilibrium equations, together with the incompatibility equations, form a system of twelve equations with twelve unknown functions $\boldsymbol{r}_\gamma \cdot \mathbf{E} \cdot \boldsymbol{r}_\varkappa$, $\boldsymbol{r}_\gamma \cdot \mathbf{E} \cdot \boldsymbol{n}$, $\boldsymbol{r}_\gamma \cdot \mathbf{K} \cdot \boldsymbol{r}_\varkappa$, $\boldsymbol{r}_\gamma \cdot \mathbf{K} \cdot \boldsymbol{n}$ $(\gamma, \varkappa = 1, 2)$.

Assume that the shell surface σ is simply connected and homeomorphic to a circle, and the boundary of this surface χ consists of two non-intersecting connected parts: $\chi = \chi_1 \cup \chi_2$. One of the possible formulations of the boundary value problem of the equilibrium of a shell with continuously distributed dislocations and disclinations is that the unknown functions $\mathbf{E}$ and $\mathbf{K}$ must satisfy the equilibrium equations, incompatibility equations, deformational boundary conditions (17.22) on χ_1 and dynamic boundary conditions (17.6) on χ_2 written in terms of tensors $\mathbf{E}$ and $\mathbf{K}$.

It is possible to formulate the boundary value problem with another choice of unknown functions. Let $\mathbf{T}'$ and $\mathbf{M}'$ be some particular solution of the equilibrium equations (17.1), (17.2). Then the tensors $\mathbf{T}^{\wedge} = \mathbf{T} - \mathbf{T}'$ and $\mathbf{M}^{\wedge} = \mathbf{M} - \mathbf{M}'$ satisfy homogeneous equilibrium equations

$$\operatorname{div}\mathbf{T}^{\wedge} = 0, \ \operatorname{div}\mathbf{M}^{\wedge} + \mathbf{T}^{\wedge}_{\times} = 0. \tag{17.25}$$

There is an analogy between the equilibrium equations (17.25) and the strain compatibility equations (17.21). The systems (17.21) and (17.25) pass one into the other during the following mutual substitutions

$$\mathbf{T}^{\wedge} \rightleftarrows \mathbf{d} \cdot \mathbf{K}, \ \mathbf{M}^{\wedge} \rightleftarrows \mathbf{d} \cdot \mathbf{E}.$$

From this, by analogy with (17.4), we can conclude that the equations (17.25) are identically satisfied by substitution [28]

$$\mathbf{T}^{\wedge} = \mathbf{d} \cdot \operatorname{grad} \boldsymbol{\eta}, \ \mathbf{M}^{\wedge} = \mathbf{d} \cdot \operatorname{grad} \boldsymbol{\lambda} + \mathbf{d} \times \boldsymbol{\eta}, \tag{17.26}$$

where $\boldsymbol{\eta}$ and $\boldsymbol{\lambda}$ are arbitrary twice continuously differentiable functions of coordinates on the surface σ.

It is known in the theory of elasticity [18], the functions by which the general solution of the equilibrium equations is expressed are usually called stress functions. An example is the Airy stress function in a plane problem of elasticity theory, the Prandtl stress function in the theory of torsion of prismatic bodies, the Finzi's general tensor of stress functions [18]. Therefore, the functions $\boldsymbol{\eta}$ and $\boldsymbol{\lambda}$ will be referred to as stress functions. Note that stress functions were first introduced in shell theory by Lurie and Goldenweiser [16, 11] within the Kirchhoff-Love shell model.

The strain tensors $\mathbf{E}$ and $\mathbf{K}$ can be expressed using the constitutive relations (17.3) and the formulas (17.26) in terms of stress functions $\boldsymbol{\eta}$ and $\boldsymbol{\lambda}$. Consequently, the incompatibility equations (17.19), (17.20) will be written regarding the stress functions and form a system of six scalar equations with six unknown functions – components of the vectors $\boldsymbol{\eta}$ and $\boldsymbol{\lambda}$. The deformational boundary conditions (17.22) are also formulated in terms of the functions $\boldsymbol{\eta}$ and $\boldsymbol{\lambda}$. Let's write the dynamic boundary conditions (17.6) in the modified form,

$$\boldsymbol{m} \cdot \mathbf{T}^{\wedge} = \boldsymbol{\tau}^{\wedge}(s), \ \boldsymbol{m} \cdot \mathbf{M}^{\wedge} = \boldsymbol{\mu}^{\wedge}(s), \tag{17.27}$$

where the functions $\boldsymbol{\tau}^{\wedge}$ and $\boldsymbol{\mu}^{\wedge}$ are expressed in terms of the known vectors $\boldsymbol{\tau}$ and $\boldsymbol{\mu}$ and the tensors $\mathbf{T}'$ and $\mathbf{M}'$. Based on (17.26) and the equality $\mathbf{d} = \boldsymbol{m} \otimes \boldsymbol{t} - \boldsymbol{t} \otimes \boldsymbol{m}$, the boundary conditions (17.27) become

$$\frac{d\boldsymbol{\eta}}{ds} = \boldsymbol{\tau}^{\wedge}(s), \ \frac{d\boldsymbol{\lambda}}{ds} + \boldsymbol{t} \times \boldsymbol{\eta} = \boldsymbol{\mu}^{\wedge}(s). \tag{17.28}$$

These boundary conditions can be converted to a form that does not contain derivatives of stress functions. For this, it is necessary to integrate the system of differential equations (17.28). The result will be following

$$\boldsymbol{\eta}(s) = \boldsymbol{\eta}_0 + \int_{s_0}^{s} \boldsymbol{\tau}(s')ds',$$

$$\begin{aligned}\boldsymbol{\lambda}(s) &= \boldsymbol{\lambda}_0 + \boldsymbol{\eta}_0 \times (\boldsymbol{r} - \boldsymbol{r}_0) + \int_{s_0}^{s} \boldsymbol{\mu}^{\wedge}(s')ds' \\ &+ \int_{s_0}^{s} \boldsymbol{\tau}^{\wedge}(s') \times (\boldsymbol{r} - \boldsymbol{r}')ds', \end{aligned} \tag{17.29}$$

$$\boldsymbol{r} = \boldsymbol{r}(s), \ \boldsymbol{r}_0 = \boldsymbol{r}(s_0).$$

Here $\boldsymbol{\eta}_0$, $\boldsymbol{\lambda}_0$ are arbitrary vector constants. With the help of (17.26), it is easy to verify that adding the terms $\boldsymbol{\eta}_0$ and $\boldsymbol{\lambda}_0 + \boldsymbol{\eta}_0 \times \boldsymbol{r}$ to the stress functions $\boldsymbol{\eta}$ and $\boldsymbol{\lambda}$, respectively, does not affect the stress tensor fields $\mathbf{T}^{\wedge}$ and $\mathbf{M}^{\wedge}$. Therefore, in the case of a connected section of the boundary χ_2, we can put $\boldsymbol{\eta}_0 = \boldsymbol{\lambda}_0 = 0$. Thus, due to (17.29), the dynamic boundary conditions are reduced to specifying the values of the stress functions $\boldsymbol{\eta}$ and $\boldsymbol{\lambda}$ on a part of the shell boundary.

17.5 Variational Formulation of the Equilibrium Problem for an Elastic Shell with Distributed Dislocations and Disclinations

Let us consider a shell whose surface σ is simply connected and homeomorphic to a circle. The shell boundary χ consists of two non-intersecting parts: $\chi = \chi_1 \cup \chi_2$. On the curve χ_1, the deformational boundary conditions (17.22) are satisfied. The dynamic boundary conditions on χ_2, as shown above, are reduced to setting the boundary values of the stress functions

$$\boldsymbol{\eta}\big|_{\chi_2} = \boldsymbol{\eta}^*(s), \ \boldsymbol{\lambda}\big|_{\chi_2} = \boldsymbol{\lambda}^*(s) \tag{17.30}$$

where $\boldsymbol{\eta}^*$ and $\boldsymbol{\lambda}^*$ are known functions. Next, suppose that the constitutive relations of the shell, i.e., the dependencies $\mathbf{T} = \mathbf{T}(\mathbf{E}, \mathbf{K})$, $\mathbf{M} = \mathbf{M}(\mathbf{E}, \mathbf{K})$, admit a unique inversion regarding the tensors $\mathbf{E}$ and $\mathbf{K}$, i.e., there are single-valued dependencies

$$\mathbf{E} = \mathbf{E}(\mathbf{T}, \mathbf{M}), \ \mathbf{K} = \mathbf{K}(\mathbf{T}, \mathbf{M}).$$

Let us introduce the specific complementary energy of the shell $V(\mathbf{T}, \mathbf{M})$ as a function related to the specific strain energy by the Legendre transformation,

$$V = \mathbf{T} \odot \mathbf{E} + \mathbf{M} \odot \mathbf{K} - W. \tag{17.31}$$

Here the symbol $\mathbf{A} \odot \mathbf{B} = \mathrm{tr}\,(\mathbf{A} \cdot \mathbf{B}^{\mathrm{T}})$ means the full multiplication of second rank tensors [31]. By the property of the Legendre transformation, we have

$$\mathbf{E} = \frac{\partial V}{\partial \mathbf{T}}, \ \mathbf{K} = \frac{\partial V}{\partial \mathbf{M}}. \tag{17.32}$$

Let us express stresses and couple stresses in terms of the stress functions by the formulas

$$\mathbf{T} = \mathbf{d} \cdot \operatorname{grad} \boldsymbol{\eta} + \mathbf{T}', \ \mathbf{M} = \mathbf{d} \cdot (\operatorname{grad} \boldsymbol{\lambda} + \mathbf{g} \times \boldsymbol{\eta}) + \mathbf{M}' \tag{17.33}$$

and consider the functional

$$\Pi[\boldsymbol{\eta}, \boldsymbol{\lambda}] = \iint\limits_{\sigma} V d\sigma - \iint\limits_{\sigma} (\boldsymbol{\alpha} \cdot \boldsymbol{\eta} + \boldsymbol{\beta} \cdot \boldsymbol{\lambda}) d\sigma + \int\limits_{\chi_1} (\boldsymbol{v} \cdot \boldsymbol{\eta} + \boldsymbol{w} \cdot \boldsymbol{\lambda}) ds, \tag{17.34}$$

which is defined on the set of twice differentiable stress functions satisfying the dynamic boundary conditions (17.30) on χ_2. Let us show that the requirement of stationarity of the functional Π is equivalent to the incompatibility equations (17.19), (17.20) and deformational boundary conditions (17.22).

Let us calculate the variation of the functional (17.34), considering that the functions $\boldsymbol{\alpha}$, $\boldsymbol{\beta}$, $\boldsymbol{v}$, $\boldsymbol{w}$, $\mathbf{T}'$ and $\mathbf{M}'$ are given and hence do not vary:

$$\delta\Pi = \iint\limits_{\sigma} \delta V d\sigma - \iint\limits_{\sigma} (\boldsymbol{\alpha} \cdot \delta\boldsymbol{\eta} + \boldsymbol{\beta} \cdot \delta\boldsymbol{\lambda}) d\sigma + \int\limits_{\chi_1} (\boldsymbol{v} \cdot \delta\boldsymbol{\eta} + \boldsymbol{w} \cdot \delta\boldsymbol{\lambda}) ds. \tag{17.35}$$

Based on (17.32), (17.33) we have

$$\delta V = \mathbf{E} \odot \delta\mathbf{T} + \mathbf{K} \odot \delta\mathbf{M} = \mathbf{E} \odot (\mathbf{d} \cdot \operatorname{grad} \delta\boldsymbol{\eta}) + \mathbf{K} \odot (\mathbf{d} \cdot \operatorname{grad} \delta\boldsymbol{\lambda}) + \mathbf{K} \odot (\mathbf{d} \cdot \delta\boldsymbol{\eta}). \tag{17.36}$$

Applying the standard technique of the calculus of variations, we transform the expression (17.36) to the following form

$$\begin{aligned} \delta V = {} & [\operatorname{div} (\mathbf{d} \cdot \mathbf{E})] \cdot \delta\boldsymbol{\eta} - \operatorname{div} (\mathbf{d} \cdot \mathbf{E} \cdot \delta\boldsymbol{\eta}) + \\ & + [\operatorname{div} (\mathbf{d} \cdot \mathbf{K})] \cdot \delta\boldsymbol{\lambda} - \operatorname{div} (\mathbf{d} \cdot \mathbf{K} \cdot \delta\boldsymbol{\lambda}) + (\mathbf{d} \cdot \mathbf{K})_\times \cdot \delta\boldsymbol{\eta}. \end{aligned} \tag{17.37}$$

Substituting (17.37) into (17.35) and using the surface divergence theorem, we get

$$\begin{aligned} \delta\Pi = {} & \iint\limits_{\sigma} [\operatorname{div} (\mathbf{d} \cdot \mathbf{E}) + (\mathbf{d} \cdot \mathbf{K})_\times - \boldsymbol{\alpha}] \cdot \delta\boldsymbol{\eta} \, d\sigma + \\ & + \iint\limits_{\sigma} [\operatorname{div} (\mathbf{d} \cdot \mathbf{K}) - \boldsymbol{\beta}] \cdot \delta\boldsymbol{\lambda} \, d\sigma + \\ & + \int\limits_{\chi_1} (\boldsymbol{v} - \boldsymbol{t} \cdot \mathbf{E}) \cdot \delta\boldsymbol{\eta} \, ds + \int\limits_{\chi_1} (\boldsymbol{w} - \boldsymbol{t} \cdot \mathbf{K}) \cdot \delta\boldsymbol{\lambda} \, ds. \end{aligned} \tag{17.38}$$

Since the functions $\delta\boldsymbol{\eta}$ and $\delta\boldsymbol{\lambda}$ are arbitrary on χ_1 and on σ, the requirement $\delta\Pi = 0$ implies incompatibility equations and deformational boundary conditions. Conversely, if these equations and boundary conditions are satisfied, then $\delta\Pi = 0$.

17.6 Static-geometry Analogy

In the theory of elastic shells of the Kirchhoff-Love type, the static-geometric analogy is well known [12, 17]. It consists in the identity of the equilibrium equations for forces and moments and the compatibility equations for metric strains and strains, related to the curvature changes. The extension of the static-geometric analogy to Kirchhoff-Love shells with distributed dislocations and disclinations is given in [30]. In this section, we will study the static-geometric analogy for Cosserat shells, taking into account distributed dislocations and disclinations.

There is an analogy between the equilibrium equations (17.1), (17.2) and the incompatibility equations (17.19), (17.20). Namely, these systems of equations go over to one another under the following mutual replacements:

$$\mathbf{T} \rightleftarrows \mathbf{d}\cdot\mathbf{K},\ \mathbf{M} \rightleftarrows \mathbf{d}\cdot\mathbf{E},\ \boldsymbol{f} \rightleftarrows -\boldsymbol{\beta},\ \boldsymbol{l} \rightleftarrows -\boldsymbol{\alpha}. \tag{17.39}$$

In the absence of external loads distributed over σ, i.e., for $\boldsymbol{f} = \boldsymbol{l} = 0$, the stress and couple stress tensors are expressed in terms of the stress functions by the formulas

$$\mathbf{T} = \mathbf{d}\cdot\operatorname{grad}\boldsymbol{\eta},\ \mathbf{M} = \mathbf{d}\cdot\operatorname{grad}\boldsymbol{\lambda} + \mathbf{d}\times\boldsymbol{\eta}, \tag{17.40}$$

and in the absence of distributed dislocations and disclinations, i.e., for $\boldsymbol{\alpha} = \boldsymbol{\beta} = 0$, the strain tensors are expressed in terms of displacement and rotation fields as follows:

$$\mathbf{d}\cdot\mathbf{K} = \mathbf{d}\cdot\operatorname{grad}\boldsymbol{\omega},\ \mathbf{d}\cdot\mathbf{E} = \mathbf{d}\cdot\operatorname{grad}\boldsymbol{u} + \mathbf{d}\times\boldsymbol{\omega}. \tag{17.41}$$

The expressions (17.40) and (17.41) are similar, with the correspondence

$$\boldsymbol{\eta} \rightleftarrows \boldsymbol{\omega},\ \boldsymbol{\lambda} \rightleftarrows \boldsymbol{u}. \tag{17.42}$$

There is also an analogy between different boundary conditions. Indeed, the kinematic boundary conditions

$$\boldsymbol{u}\big|_{\chi_1} = \boldsymbol{u}^*(s),\ \boldsymbol{\omega}\big|_{\chi_1} = \boldsymbol{\omega}^*(s),$$

are identical to the dynamic boundary conditions formulated in terms of stress functions

$$\boldsymbol{\lambda}\big|_{\chi_2} = \boldsymbol{\lambda}^*(s),\ \boldsymbol{\eta}\big|_{\chi_2} = \boldsymbol{\eta}^*(s),$$

Deformational boundary conditions

$$\mathbf{m}\cdot\mathbf{d}\cdot\mathbf{E} = \boldsymbol{v}(s),\ \mathbf{m}\cdot\mathbf{d}\cdot\mathbf{K} = \boldsymbol{w}(s)$$

are similar to dynamic boundary conditions

$$\mathbf{m}\cdot\mathbf{M} = \boldsymbol{\mu}(s),\ \mathbf{m}\cdot\mathbf{T} = \boldsymbol{\tau}(s),$$

with the correspondence $\boldsymbol{v} \rightleftarrows \boldsymbol{\mu},\ \boldsymbol{w} \rightleftarrows \boldsymbol{\tau}$.

Let us assume that dislocations, disclinations, surface force and couple loads are distributed in a simply connected shell, and the deformational boundary conditions $\boldsymbol{t}\cdot\mathbf{E}=\boldsymbol{v}(s)$, $\boldsymbol{t}\cdot\mathbf{K}=\boldsymbol{w}(s)$ are satisfied on the entire closed curve χ bounding the surface σ. It turns out that functions $\boldsymbol{\alpha}(y^1,y^2)$, $\boldsymbol{\beta}(y^1,y^2)$, $\boldsymbol{v}(s)$, and $\boldsymbol{w}(s)$ cannot be set completely arbitrarily, but must satisfy certain relations, the derivation of which follows.

Integrating the incompatibility equations (17.19), (17.20) over the area of the shell and using the divergence theorem, we obtain the equalities

$$\oint\limits_{\chi} d\boldsymbol{r}\cdot\mathbf{E}+\iint\limits_{\sigma}(\mathbf{d}\cdot\mathbf{K})_{\times}d\sigma=\iint\limits_{\sigma}\boldsymbol{\alpha}\,d\sigma, \tag{17.43}$$

$$\oint\limits_{\chi} d\boldsymbol{r}\cdot\mathbf{K}=\iint\limits_{\sigma}\boldsymbol{\beta}\,d\sigma. \tag{17.44}$$

The second integral on the left-hand side of (17.43) can be expressed in terms of the given known functions $\boldsymbol{\alpha}$, $\boldsymbol{\beta}$, $\boldsymbol{v}$, $\boldsymbol{w}$. To do this, consider the relation following from (17.20)

$$\iint\limits_{\sigma}[\operatorname{div}(\mathbf{d}\cdot\mathbf{K})]\times\boldsymbol{r}\,d\sigma=\iint\limits_{\sigma}\boldsymbol{\beta}\times\boldsymbol{r}\,d\sigma. \tag{17.45}$$

We transform the integrand in (17.45) as follows

$$[\operatorname{div}(\mathbf{d}\cdot\mathbf{K})]\times\boldsymbol{r}=\operatorname{div}(\mathbf{d}\cdot\mathbf{K}\times\boldsymbol{r})-\boldsymbol{r}^{\gamma}\cdot\mathbf{d}\cdot\mathbf{K}\times\boldsymbol{r}_{\gamma}. \tag{17.46}$$

Using a directly verifiable identity

$$-\boldsymbol{r}^{\gamma}\cdot\mathbf{d}\cdot\mathbf{K}\times\boldsymbol{r}_{\gamma}=(\mathbf{d}\cdot\mathbf{K})_{\times}$$

and substituting (17.46) into (17.45), taking (17.22) into account, we get

$$\iint\limits_{\sigma}(\mathbf{d}\cdot\mathbf{K})_{\times}\,d\sigma=\iint\limits_{\sigma}\boldsymbol{\beta}\times\boldsymbol{r}\,d\sigma-\oint\limits_{\chi}\boldsymbol{w}\times\boldsymbol{r}\,ds. \tag{17.47}$$

Due to (17.22), (17.47), the equalities (17.43), (17.44) become

$$\iint\limits_{\sigma}(\boldsymbol{\alpha}+\boldsymbol{r}\times\boldsymbol{\beta})d\sigma=\oint\limits_{\chi}(\boldsymbol{v}+\boldsymbol{r}\times\boldsymbol{w})ds, \tag{17.48}$$

$$\iint\limits_{\sigma}\boldsymbol{\beta}\,d\sigma=\oint\limits_{\chi}\boldsymbol{w}\,ds. \tag{17.49}$$

The relations (17.48), (17.49), which are the necessary conditions for the solvability of the boundary value problem of shell equilibrium, can be called the defects balance equations. If the dislocation and disclination densities are equal to zero, then the conditions of the defects balance reduce to the equalities

$$\oint_{\chi} \boldsymbol{w}\, ds = 0, \quad \oint_{\chi} (\boldsymbol{v} + \boldsymbol{r} \times \boldsymbol{w}) ds = 0.$$

This result can be obtained in a different way. In the absence of dislocations and disclinations, there exist the fields $\boldsymbol{u}(\boldsymbol{r})$ and $\boldsymbol{\omega}(\boldsymbol{r})$ and the functions $\boldsymbol{v}$ and $\boldsymbol{w}$ are expressed through the boundary values of displacements and rotations according to the formulas (17.23). Therefore, we have

$$\oint_{\chi} \boldsymbol{w}(s)\, ds = \oint_{\chi} \frac{d\boldsymbol{\omega}^*}{ds} ds = 0,$$

$$\oint_{\chi} [\boldsymbol{v}(s) + \boldsymbol{r}(s) \times \boldsymbol{w}(s)]\, ds = \oint_{\chi} \frac{d\boldsymbol{u}^*}{ds} ds + \oint_{\chi} \left(\boldsymbol{t} \times \boldsymbol{\omega}^* + \boldsymbol{r} \times \frac{d\boldsymbol{\omega}^*}{ds} \right) ds =$$

$$= \oint_{\chi} \left(\frac{d\boldsymbol{r}}{ds} \times \boldsymbol{\omega}^* + \boldsymbol{r} \times \frac{d\boldsymbol{\omega}^*}{ds} \right) ds = \oint_{\chi} \frac{d}{ds} (\boldsymbol{r} \times \boldsymbol{\omega}^*)\, ds = 0.$$

If the shell surface is closed, i.e., has no boundary, then the defects balance equations are reduced to the equalities

$$\iint_{\sigma} \boldsymbol{\beta}\, d\sigma = 0, \quad \iint_{\sigma} (\boldsymbol{\alpha} + \boldsymbol{r} \times \boldsymbol{\beta})\, d\sigma = 0. \tag{17.50}$$

Next, consider another problem for the same shell, which differs from the previous one by boundary conditions. Now the dynamical boundary conditions (17.6) are satisfied on the entire boundary χ. We integrate the equilibrium equations (17.1), (17.2) over the surface area σ and, after performing simple transformations, we arrive at the necessary conditions for the solvability of this problem

$$\iint_{\sigma} \boldsymbol{f}\, d\sigma + \oint_{\chi} \boldsymbol{\tau}\, ds = 0. \tag{17.51}$$

$$\iint_{\sigma} (\boldsymbol{l} + \boldsymbol{r} \times \boldsymbol{f})\, d\sigma + \oint_{\chi} (\boldsymbol{\mu} + \boldsymbol{r} \times \boldsymbol{\tau})\, ds = 0. \tag{17.52}$$

The relations (17.51) and (17.52) have a clear physical meaning. They imply the requirement of self-balance of external loads, i.e., the equality to zero, correspondingly, of the resultant vector and the resultant moment of all loads. There is an analogy between the defects balance equations (17.48), (17.49) and the self-balance conditions (17.51), (17.52). The former go into the latter and vice versa with the following mutual substitutions:

$$\boldsymbol{f} \rightleftarrows -\boldsymbol{\beta},\ \boldsymbol{l} \rightleftarrows -\boldsymbol{\alpha},\ \boldsymbol{\tau} \rightleftarrows \boldsymbol{w},\ \boldsymbol{\mu} \rightleftarrows \boldsymbol{v}.$$

Thus, the static-geometry analogy in the theory of Cosserat-type shells with defects extends not only to differential equations in the domain σ, but also to boundary conditions, and to necessary conditions for the solvability of boundary value problems of equilibrium.

The established static-geometric analogy entails the existence of dual boundary value problems in the theory of shells. The dual problems are mathematically equivalent, but different in their physical formulation. As before, suppose that the boundary of a simply connected shell consists of two connected non-intersecting curves: $\chi = \chi_1 \cup \chi_2$. Let us introduce the notation $\mathbf{E}^\circ = \mathbf{d}\cdot\mathbf{E}$, $\mathbf{K}^\circ = \mathbf{d}\cdot\mathbf{K}$ and consider two boundary-value problems of the shell equilibrium.

In the first problem, there are no distributed dislocations and disclinations and the surface loads $\boldsymbol{f}$ and $\boldsymbol{l}$ are given. The kinematic boundary conditions are satisfied on the curve χ_1, while the dynamic boundary conditions are satisfied on χ_2 with the given contour loads $\boldsymbol{\tau}(s)$ and $\boldsymbol{\mu}(s)$. The equations of this problem consist of equilibrium equations for forces and moments, constitutive relations expressing stresses and couple stresses in terms of strains, and geometric relations expressing strain tensors in terms of displacements and rotations.

In the second problem, there are no external loads distributed over the shell surface, but the densities of distributed dislocations and disclinations are specified. The values of the stress function are given on the curve χ_1, i.e., the dynamic boundary conditions are satisfied. Deformational boundary conditions are set on the part of the boundary χ_2. The equations of the second problem comprise incompatibility equations, constitutive relations expressing strain tensors in terms of stress and couple stress tensors, and representations of the latter in terms of stress functions.

We will assume that all equations and boundary conditions are reduced to a dimensionless form, i.e., they represent relationships between dimensionless quantities. Let us write out the complete systems of equations and boundary conditions for these two problems in the form of a table with two columns. The left column corresponds to the first task, the right – to the second task.

$\operatorname{div}\mathbf{T} + \boldsymbol{f} = 0,$ $\operatorname{div}\mathbf{M} + \mathbf{T}_\times + \boldsymbol{l} = 0;$	$\operatorname{div}\mathbf{K}^\circ - \boldsymbol{\beta} = 0,$ $\operatorname{div}\mathbf{E}^\circ + \mathbf{K}^\circ_\times - \boldsymbol{\alpha} = 0;$	(17.53)
$\mathbf{T} = \mathbf{G}(\mathbf{K}^\circ, \mathbf{E}^\circ),$ $\mathbf{M} = \mathbf{H}(\mathbf{K}^\circ, \mathbf{E}^\circ);$	$\mathbf{K}^\circ = \mathbf{G}_1(\mathbf{T}, \mathbf{M}),$ $\mathbf{E}^\circ = \mathbf{H}_1(\mathbf{T}, \mathbf{M});$	(17.54)
$\mathbf{K}^\circ = \mathbf{d}\cdot\operatorname{grad}\boldsymbol{\omega},$ $\mathbf{E}^\circ = \mathbf{d}\cdot(\operatorname{grad}\boldsymbol{u} + \mathbf{g}\times\boldsymbol{\omega});$	$\mathbf{T} = \mathbf{d}\cdot\operatorname{grad}\boldsymbol{\eta},$ $\mathbf{M} = \mathbf{d}\cdot(\operatorname{grad}\boldsymbol{\lambda} + \mathbf{g}\times\boldsymbol{\eta});$	(17.55)
$\boldsymbol{u} = \boldsymbol{u}^*,\ \boldsymbol{\omega} = \boldsymbol{\omega}^*$ on χ_1, $\boldsymbol{m}\cdot\mathbf{T} = \boldsymbol{\tau},\ \boldsymbol{m}\cdot\mathbf{M} = \boldsymbol{\mu}$ on χ_2;	$\boldsymbol{\lambda} = \boldsymbol{\lambda}^*,\ \boldsymbol{\eta} = \boldsymbol{\eta}^*$ on χ_1, $\boldsymbol{m}\cdot\mathbf{K}^\circ = \boldsymbol{w},\ \boldsymbol{m}\cdot\mathbf{E}^\circ = \boldsymbol{v}$ on χ_2.	(17.56)

Two elastic shells with the same surface σ are called conjugate if the functions **G** and **H** in the constitutive relations (17.54) of the first shell coincide respectively with the functions $\mathbf{G}_1$ and $\mathbf{H}_1$ of the second shell. Conjugate shells generally have different physical properties, i.e., they are made from different materials.

The relation (17.53)–(17.56) implies the following duality theorem for the theory of Cosserat-type elastic shells:

Theorem 17.1. *The boundary value problem of the equilibrium of a shell with given densities of dislocations and disclinations in the absence of surface loads with stress functions specified on the part of the boundary χ_1 and with deformational boundary conditions on the part of the boundary χ_2 is mathematically equivalent to the boundary value problem of the equilibrium of a conjugate shell with given surface loads in the absence of dislocations and disclinations, with kinematic boundary conditions on χ_1 and with contour loads specified on χ_2.*

17.7 Spherical Shell with Uniformly Distributed Dislocations and Disclinations

In a closed spherical shell of radius r_0, we take the geographic coordinates $0 \leqslant \varphi \leqslant 2\pi$ (longitude) and $-\pi/2 \leqslant \theta \leqslant \pi/2$ (latitude) as Gaussian coordinates. The position of the sphere point in space is determined by the formulas

$$x_1 = r_0 \cos\varphi \cos\theta, \; x_2 = r_0 \sin\varphi \cos\theta, \; x_3 = r_0 \sin\theta,$$

where x_1, x_2, x_3 are Cartesian coordinates. Let us introduce an orthonormal vector basis $\boldsymbol{e}_r, \boldsymbol{e}_\varphi, \boldsymbol{e}_\theta$, related to the vectors of the Cartesian coordinates $\boldsymbol{i}_k$ by the relations

$$\begin{aligned} \boldsymbol{e}_r &= (\boldsymbol{i}_1 \cos\varphi + \boldsymbol{i}_2 \sin\varphi)\cos\theta + \boldsymbol{i}_3 \sin\theta, \\ \boldsymbol{e}_\varphi &= -\boldsymbol{i}_1 \sin\varphi + \boldsymbol{i}_2 \cos\varphi, \\ \boldsymbol{e}_\theta &= -(\boldsymbol{i}_1 \cos\varphi + \boldsymbol{i}_2 \sin\varphi)\sin\theta + \boldsymbol{i}_3 \cos\theta. \end{aligned} \tag{17.57}$$

The vector $\boldsymbol{e}_r$ is the normal to the surface of the shell, and the vectors $\boldsymbol{e}_\varphi$ and $\boldsymbol{e}_\theta$ are directed along the tangents to the coordinate lines. The metric **g** and discriminant **d** tensors on the sphere have the form

$$\begin{aligned} \mathbf{g} &= \mathbf{I} - \boldsymbol{e}_r \otimes \boldsymbol{e}_r = \boldsymbol{e}_\varphi \otimes \boldsymbol{e}_\varphi + \boldsymbol{e}_\theta \otimes \boldsymbol{e}_\theta, \\ \mathbf{d} &= -\mathbf{I} \times \boldsymbol{e}_r = \boldsymbol{e}_\varphi \otimes \boldsymbol{e}_\theta - \boldsymbol{e}_\theta \otimes \boldsymbol{e}_\varphi, \end{aligned} \tag{17.58}$$

and the gradient operation is given by the relations

$$\mathrm{grad} = \frac{1}{r_0 \cos\theta}\boldsymbol{e}_\varphi \frac{\partial}{\partial\varphi} + \frac{1}{r_0}\boldsymbol{e}_\theta \frac{\partial}{\partial\theta}. \tag{17.59}$$

The dislocation and disclination densities, as well as external loads, will be taken as spherically symmetric vector fields

$$\boldsymbol{\alpha} = \alpha_0 \boldsymbol{e}_r,\ \boldsymbol{\beta} = \beta_0 \boldsymbol{e}_r,\ \boldsymbol{f} = f_0 \boldsymbol{e}_r,\ \boldsymbol{l} = l_0 \boldsymbol{e}_r, \tag{17.60}$$

where α_0, β_0, f_0, l_0 are constants. These formulas mean that screw dislocations and wedge disclinations are uniformly distributed in the shell, and the load is reduced to the hydrostatic pressure f_0 and the uniform drilling moment l_0.

We will search for strain tensors in the form of spherically symmetric tensor fields [13]

$$\mathbf{E} = E_1 \mathbf{g} + E_2 \mathbf{d},\ \mathbf{K} = K_1 \mathbf{g} + K_2 \mathbf{d}. \tag{17.61}$$

Here E_γ and K_γ ($\gamma = 1,2$) are constants. In what follows, we will use the linear constitutive relations of the isotropic elastic Cosserat shell [10]

$$\begin{aligned} \mathbf{T} &= a_1(\operatorname{tr}\mathbf{E})\mathbf{g} + a_2(\mathbf{E}\cdot\mathbf{g})^{\mathrm{T}} + a_3\mathbf{E}\cdot\mathbf{g} + a_4\mathbf{E}\cdot\boldsymbol{n}\otimes\boldsymbol{n}, \\ \mathbf{M} &= b_1(\operatorname{tr}\mathbf{K})\mathbf{g} + b_2(\mathbf{K}\cdot\mathbf{g})^{\mathrm{T}} + b_3\mathbf{K}\cdot\mathbf{g} + b_4\mathbf{K}\cdot\boldsymbol{n}\otimes\boldsymbol{n}, \end{aligned} \tag{17.62}$$

where a_s, b_s ($s = 1,2,3,4$) are the material constants.

The equations of state (17.62) correspond to a quadratic form of the specific strain energy

$$\begin{aligned} 2W &= a_1\operatorname{tr}^2\mathbf{E} + a_2\operatorname{tr}(\mathbf{E}\cdot\mathbf{g})^2 + a_3\operatorname{tr}\left(\mathbf{E}\cdot\mathbf{g}\cdot\mathbf{E}^{\mathrm{T}}\right) + a_4\boldsymbol{n}\cdot\mathbf{E}^{\mathrm{T}}\cdot\mathbf{E}\cdot\boldsymbol{n} \\ &+ b_1\operatorname{tr}^2\mathbf{K} + b_2\operatorname{tr}(\mathbf{K}\cdot\mathbf{g})^2 + b_3\operatorname{tr}\left(\mathbf{K}\cdot\mathbf{g}\cdot\mathbf{K}^{\mathrm{T}}\right) + b_4\boldsymbol{n}\cdot\mathbf{K}^{\mathrm{T}}\cdot\mathbf{K}\cdot\boldsymbol{n}. \end{aligned}$$

In the spherically symmetric problem considered here, the tensors $\mathbf{E}$ and $\mathbf{K}$ according to (17.61) have the property $\mathbf{E}\cdot\boldsymbol{n} = \mathbf{K}\cdot\boldsymbol{n} = 0$, so the constants a_4, b_4 do not participate in the constitutive relations. Based on (17.61), (17.62) we get (T_γ and M_γ are constants):

$$\mathbf{T} = T_1\mathbf{g} + T_2\mathbf{d},\ \mathbf{M} = M_1\mathbf{g} + M_2\mathbf{d}. \tag{17.63}$$

Using (17.61)–(17.63) we obtain the constitutive relations in the scalar form

$$\begin{aligned} &T_1 = A_1E_1,\ T_2 = A_2E_2,\ M_1 = B_1K_1,\ M_2 = B_2K_2, \\ &A_1 = 2a_1 + a_2 + a_3,\ A_2 = a_3 - a_2,\ B_1 = 2b_1 + b_2 + b_3,\ B_2 = b_3 - b_2. \end{aligned} \tag{17.64}$$

Using formulas for the basis vectors differentiating

$$\begin{aligned} &\frac{\partial \boldsymbol{e}_\varphi}{\partial\varphi} = \sin\theta\,\boldsymbol{e}_\theta - \cos\theta\,\boldsymbol{e}_r, && \frac{\partial \boldsymbol{e}_\varphi}{\partial\theta} = 0, && \frac{\partial \boldsymbol{e}_\theta}{\partial\varphi} = -\sin\theta\,\boldsymbol{e}_\varphi, \\ &\frac{\partial \boldsymbol{e}_\theta}{\partial\theta} = -\boldsymbol{e}_r, && \frac{\partial \boldsymbol{e}_r}{\partial\varphi} = \cos\theta\,\boldsymbol{e}_\varphi, && \frac{\partial \boldsymbol{e}_r}{\partial\theta} = \boldsymbol{e}_\theta, \end{aligned}$$

we obtain the relations

$$\begin{aligned} &\operatorname{div}\mathbf{g} = -2r_0^{-1}\boldsymbol{e}_r, \quad \operatorname{div}\mathbf{d} = 0, \\ &\operatorname{div}(\mathbf{d}\cdot\mathbf{E}) + (\mathbf{d}\cdot\mathbf{K})_\times = 2\left(r_0^{-1}E_2 + K_1\right)\boldsymbol{e}_r, \quad \operatorname{div}(\mathbf{d}\cdot\mathbf{K}) = 2r_0^{-1}K_2\boldsymbol{e}_r. \end{aligned} \tag{17.65}$$

Based on (17.60), (17.61), (17.65), the incompatibility equations (17.19), (17.20) are reduced to two scalar equations

$$2\left(r_0^{-1}E_2+K_1\right)=\alpha_0,\ 2r_0^{-1}K_2=\beta_0. \tag{17.66}$$

The equilibrium equations (17.1), (17.2), considering (17.60), are reduced to two relations

$$T_1=\frac{1}{2}r_0f_0,\ M_1=r_0\left(T_2+\frac{1}{2}l_0\right). \tag{17.67}$$

Let us expand the stress and couple stress tensors in terms of the basis of spherical coordinates

$$\mathbf{T}=t_{\varphi\varphi}\boldsymbol{e}_\varphi\otimes\boldsymbol{e}_\varphi+t_{\varphi\theta}\boldsymbol{e}_\varphi\otimes\boldsymbol{e}_\theta+t_{\theta\varphi}\boldsymbol{e}_\theta\otimes\boldsymbol{e}_\varphi+t_{\theta\theta}\boldsymbol{e}_\theta\otimes\boldsymbol{e}_\theta,$$
$$\mathbf{M}=m_{\varphi\varphi}\boldsymbol{e}_\varphi\otimes\boldsymbol{e}_\varphi+m_{\varphi\theta}\boldsymbol{e}_\varphi\otimes\boldsymbol{e}_\theta+m_{\theta\varphi}\boldsymbol{e}_\theta\otimes\boldsymbol{e}_\varphi+m_{\theta\theta}\boldsymbol{e}_\theta\otimes\boldsymbol{e}_\theta.$$

The components $m_{\varphi\varphi}$ and $m_{\theta\theta}$ are torques, and $m_{\varphi\theta}$ and $m_{\theta\varphi}$ are bending moments. From the system of equations (17.64), (17.66), (17.67), considering (17.58), (17.63) we find the components of stress tensors

$$t_{\varphi\varphi}=t_{\theta\theta}=\frac{1}{2}r_0f_0,\qquad t_{\varphi\theta}=-t_{\theta\varphi}=\frac{r_0A_2\left(B_1\alpha_0-r_0l_0\right)}{2\left(r_0^2A_2+B_1\right)},$$
$$m_{\varphi\varphi}=m_{\theta\theta}=\frac{r_0B_1\left(r_0A_2\alpha_0+l_0\right)}{2\left(r_0^2A_2+B_1\right)},\qquad m_{\varphi\theta}=-m_{\theta\varphi}=\frac{1}{2}B_2r_0\beta_0. \tag{17.68}$$

Note some properties of the spherically symmetric stress state of the elastic shell, following from the formulas (17.68).

1. If no external loads present, only $t_{\varphi\varphi}$ and $t_{\theta\theta}$ stresses are absent. The remaining components of the stress and couple stress tensors are nonzero, i.e., they are generated by distributed defects.
2. Distributed disclinations with density β_0 generate only bending moments $m_{\varphi\theta}$ and $m_{\theta\varphi}$.
3. When $\beta_0=0$ and $\alpha_0=-l_0/(r_0A_2)$, a momentless stress state of the shell occurs, in which

$$\mathbf{M}=0,\ \mathbf{T}=\frac{1}{2}r_0f_0\left(\boldsymbol{e}_\varphi\otimes\boldsymbol{e}_\varphi+\boldsymbol{e}_\theta\otimes\boldsymbol{e}_\theta\right)-\frac{1}{2}l_0\left(\boldsymbol{e}_\varphi\otimes\boldsymbol{e}_\theta-\boldsymbol{e}_\theta\otimes\boldsymbol{e}_\varphi\right).$$

   At zero hydrostatic pressure, the stress tensor is antisymmetric.
4. In the absence of hydrostatic load ($f_0=0$) and $\alpha_0=r_0l_0/B_1$, a pure moment state of the shell takes place

$$\mathbf{T}=0,\ \mathbf{M}=\frac{1}{2}r_0l_0\left(\boldsymbol{e}_\varphi\otimes\boldsymbol{e}_\varphi+\boldsymbol{e}_\theta\otimes\boldsymbol{e}_\theta\right)+\frac{1}{2}B_2r_0\beta_0\left(\boldsymbol{e}_\varphi\otimes\boldsymbol{e}_\theta-\boldsymbol{e}_\theta\otimes\boldsymbol{e}_\varphi\right).$$

   If there are no dislocations and disclinations in the shell, then a pure moment state is impossible. Indeed, $(17.68)_1$ and $(17.68)_2$ implies that if $\alpha_0=0$ then $t_{\varphi\theta}=0$ only if $l_0=0$. But then, due to $(17.68)_3$ and $(17.68)_4$ and the condition $\beta_0=0$, it follows that $\mathbf{M}=0$.

5. Let us consider the case when distributed defects are reduced to dislocations, i.e., $\beta_0 = 0$. Then according to (17.65), (17.66) $K_2 = 0$, and the second compatibility equation in (17.21) holds. This means there is a rotation vector ω such that $\mathbf{K} = \operatorname{grad}\omega$. This vector is easy to find and is equal to

$$\omega = \Omega \boldsymbol{e}_r, \ \Omega = r_0 K_1 = \frac{r_0 M_1}{B_1} = \frac{r_0^2 (r_0 A_2 \alpha_0 + l_0)}{2\left(r_0^2 A_2 + B_1\right)}. \tag{17.69}$$

Thus, distributed screw dislocations cause the elementary areas of the sphere to rotate around the radial axes $\boldsymbol{e}_r$ by an angle Ω.

17.8 Conclusion

This article develops the theory of continuously distributed dislocations and disclinations in elastic shells, described by the Cosserat model with kinematically independent fields of displacements and rotations. Various formulations of the boundary value problem of the equilibrium of an elastic shell with given dislocation and disclination densities are specified. With the use of stress functions and the concept of complementary energy, a variational statement of the static problem is given, considering distributed dislocations and disclinations. A static-geometry analogy between the incompatibility equations and the equilibrium equations for forces and moments is established. The dislocation density in this analogy corresponds to the intensity of the couple load distributed over the surface of the shell, and the disclination density corresponds to the intensity of the distributed force load. An analogy is also found between the deformational and dynamic boundary conditions and the necessary conditions for the solvability of two shell theory boundary value problems. Statements of mathematically equivalent, but distinctive in physical formulation, boundary value problems of the Cosserat-type shells statics are given. The general theory is illustrated by determining the stress state of a spherical shell with a spherically symmetric distribution of dislocations and disclinations.

References

1. H. Altenbach and P.A. Zhilin. General theory of elastic simple shells. *Uspekhi Mechaniki (Advances in Mechanics)*, 11(4):107—148, 1988.
2. J. Altenbach, H. Altenbach, and V.A. Eremeyev. On generalized Cosserat-type theories of plates and shells: a short review and bibliography. *Archive of Applied Mechanics*, 80(1):73–92, 2010.
3. P. Ares and K.S. Novoselov. Recent advances in graphene and other 2d materials. *Nano Materials Science*, 4(1):3–9, 2022.
4. A. Azizi, X. Zou, P. Ercius, Z. Zhang, A.L. Elías, N. Perea-López, G. Stone, M. Terrones, B.I. Yakobson, and N. Alem. Dislocation motion and grain boundary migration in two-dimensional

tungsten disulphide. *Nature Communications*, 5(1):4867, 2014.
5. F. Banhart, J. Kotakoski, and A.V. Krasheninnikov. Structural defects in graphene. *ACS Nano*, 5(1):26–41, 2011.
6. B. Butz, C. Dolle, F. Niekiel, K. Weber, D. Waldmann, H.B. Weber, B. Meyer, and E. Spiecker. Dislocations in bilayer graphene. *Nature*, 505(7484):533–537, 2014.
7. J. Dervaux, P. Ciarletta, and M. Ben Amar. Morphogenesis of thin hyperelastic plates: A constitutive theory of biological growth in the Föppl–von Kármán limit. *Journal of the Mechanics and Physics of Solids*, 57(3):458–471, 2009.
8. R. deWit. Theory of disclinations: II. Continuous and discrete disclinations in anisotropic elasticity. *J Res Natl Bur Stand A Phys Chem*, 77A(1):49–100, 1973.
9. R. deWit. Theory of disclinations: III. Continuous and discrete disclinations in isotropic elasticity. *J Res Natl Bur Stand A Phys Chem*, 77A(3):359–368, 1973.
10. V.A. Eremeyev and L.M. Zubov. *Mechanics of Elastic Shells (in Russ.)*. Nauka, Moscow, 2008.
11. A.L. Goldenveizer. The equations of the theory of thin shells. *Prik. Mat. Mech. (PMM)*, 4(2):35–42, 1940.
12. A.L. Goldenveizer. *Theory of Elastic Thin Shells*. ASME, Pergamon Press, New York, 1961.
13. E.V. Goloveshkina and L.M. Zubov. Spherically symmetric tensor fields and their application in nonlinear theory of dislocations. *Symmetry*, 13(5):830, 2021.
14. M.Y. Gutkin and I.A. Ovid'ko. *Plastic Deformation in Nanocrystalline Materials*. Springer, Berlin, Heidelberg, 2004.
15. J. Lidmar, L. Mirny, and D.R. Nelson. Virus shapes and buckling transitions in spherical shells. *Phys. Rev. E*, 68:051910, 2003.
16. A.I. Lurie. General theory of elastic thin shells. *Prik. Mat. Mech. (PMM)*, 4(2):7–34, 1940.
17. A.I. Lurie. On the static-geometric analogy of the theory of shells. In *Muschilishvili Anniversary Volume*, pages 233–240. SIAM, Philadelphia, 1961.
18. A.I. Lurie. *Theory of Elasticity*. Springer, Berlin, 2005.
19. A. Murali, G. Lokhande, K.A. Deo, A. Brokesh, and A.K. Gaharwar. Emerging 2d nanomaterials for biomedical applications. *Materials Today*, 50:276–302, 2021.
20. B. Ni, T. Zhang, J. Li, X. Li, and H. Gao. Topological design of graphene. In T. Stauber, editor, *Handbook of Graphene*, volume 2, pages 1–44. Scrivener, New York, 2019.
21. L.E. Perotti, A. Aggarwal, J. Rudnick, R. Bruinsma, and W.S. Klug. Elasticity theory of the maturation of viral capsids. *Journal of the Mechanics and Physics of Solids*, 77:86–108, 2015.
22. Z. Qi, L. Li, and Z. Xu. Engineering lattice defects in 2d nanomaterials for enhancing biomedical performances. *Particuology*, 64:121–133, 2022.
23. A. Šiber. Buckling transition in icosahedral shells subjected to volume conservation constraint and pressure: Relations to virus maturation. *Phys. Rev. E*, 73:061915, 2006.
24. H. Wang, H. Feng, and J. Li. Graphene and graphene-like layered transition metal dichalcogenides in energy conversion and storage. *Small*, 10(11):2165–2181, 2014.
25. J.C. Wang, S. Mukhopadhyay, and A. Zlotnick. Geometric defects and icosahedral viruses. *Viruses*, 10(1):25, 2018.
26. Z. Xiong, L. Zhong, H. Wang, and X. Li. Structural defects, mechanical behaviors, and properties of two-dimensional materials. *Materials*, 14(5):1192, 2021.
27. R. Zandi, B. Dragnea, A. Travesset, and R. Podgornik. On virus growth and form. *Physics Reports*, 847:1–102, 2020.
28. P.A. Zhilin. Mechanics of deformable directed surfaces. *International Journal of Solids and Structures*, 12(9):635–648, 1976.
29. L.M. Zubov. *Methods of Nonlinear Theory of Elasticity in the Theory of Shells (in Russ.)*. Rostov Univ. Press., Rostov-on-Don, 1982.
30. L.M. Zubov. The linear theory of dislocations and disclinations in elastic shells. *Journal of Applied Mathematics and Mechanics*, 74(6):663–672, 2010.
31. L.M. Zubov and M.I. Karyakin. *Tensor Calculus (in Russ.)*. Vuzovskaya Kniga, Moscow, 2006.

The manufacturer's authorised representative in the EU is Springer Nature Customer Service Centre GmbH, Europaplatz 3, 69115 Heidelberg, Germany. If you have any concerns regarding our products, please contact ProductSafety@springernature.com

Printed and bound by CPI Group (UK) Ltd, Croydon, CR0 4YY
30/09/2026
02225304-0010